Climate Risks Management
Sustainable Pulse Production

The Editors

Dr A K Srivastava post graduated M.Sc.(Tech) and Ph.D. in meteorology from Banaras Hindu University, Varanasi (India). He has 24 years (Research, Teaching and Extension) experience and awarded Merit scholarship from UGC in M.Sc.(Tech.), Senior Research Associate for Ph.D. and Pool Scientist for Post doctoral work from CSIR, New Delhi. He has worked at IGKVV, IARI and currently working in JNKVV, Jabalpur. He has worked on various aspects of Agrometeorology and highlighted new insight in the area of crop modeling and pest monitoring. He is associated to various programmes related to weather forecasting, climate change/variability, and weather based insurance linked to agriculture.

He has been editor of International and National Journals and also got the best paper presentation award from Agro-physics Society of India, New Delhi and by Ministry of Earth Sciences, New Delhi in different years. He has published more than 50 research papers and articles in International and National Journals and attended many International and National seminar, symposiums and review meetings. He has five books chapters and one book in his credit. He has life membership of many scientific societies in India. He is Fellow of Society of Life Sciences and presently handling two externally funded projects.

He has been teaching to UG, PG and Ph.D programme and taught courses of Fundamental of Climatology and Meteorology, Agro-meteorology and climate change, Agricultural statistics, Elementary mathematics, Natural disaster management and Crop modeling. He has undergone 15 training courses from ICAR institute and Agricultural Universities. He has worked on Weather based crop management, Crop modeling, Climate change and its impact on crop production. He is also involved in writing blog on Agro-meteorological aspects and future weather events.

Dr Yogranjan has been awarded graduate and postgraduate degrees with Certificate of Honours from Jawaharlal Nehru Agricultural University, Jabalpur, M.P. and honoured with Ph.D. in Agricultural Biotechnology from the MG Rural University, Chitrakoot, Satna, M.P. He is having a brilliant academic record. He received ICAR scholarships during UG and PG and has been awarded the CSIR-Junior Research Fellow (JRF) and National Eligibility Test (NET) in the professional subject of Life Sciences. He has more than 10 years of extensive experiences of teaching and research. Currently, he is working as Assistant Professor in JNKVV, College of Agriculture, Tikamgarh (Madhya Pradesh). He also teaches different courses *viz.*, Biotechnology, Plant breeding, Bio-chemistry, Microbiology and Intellectual property right to the graduate and post graduate students and currently serves as the co-leader of the research team accountable for formulating region specific agro-climate wise research oriented programme and identification of potential threats of the zone. He has published more than 20 research papers in the reputed National and International journals. He has 4 book chapters in his credit. He has visited 4 countries and attended international trainings and workshops. He has also contributed in seminars and conferences in National and international levels. He has life membership of many national and international societies and scientific forums. His current research interests include gene technology and translational genomics for crop improvement, particularly in improving resistance against abiotic stress.

Climate Risks Management
Sustainable Pulse Production

— *Editors* —

A K Srivastava

Yogranjan

2018

Daya Publishing House®
A Division of

Astral International Pvt. Ltd.
New Delhi – 110 002

ISBN: 9789388173575 (Int. Edn.)

Publisher's Note:

Published by   :   **Daya Publishing House®**
                            A Division of
                            Astral International Pvt. Ltd.
                            – ISO 9001:2015 Certified Company –
                            4736/23, Ansari Road, Darya Ganj
                            New Delhi-110 002
                            Ph. 011-43549197, 23278134
                            E-mail: info@astralint.com
                            Website: www.astralint.com

Digitally Printed at   :   **Replika Press Pvt. Ltd.**

राजमाता विजयाराजे सिंधिया कृषि विश्वविद्यालय
रेस कोर्स रोड, ग्वालियर (म.प्र.) – 474002
Rajmata Vijayaraje Scindia Krishi Vishwa Vidyalaya,
Raja Pancham Singh Marg, Gwalior (M.P.) – 474 002
(An ISO Certified 9001:2008)
Tel: 0751 – 2970502, Fax: 0751 – 2970504, E-mail: vcrvsaugwa@mp.gov.in

प्रो. एस. के. राव
कुलपति
Prof. S. K. Rao

Foreword

India has diverse Agro-climatic conditions comprising 127 Agro-climatic zone with the unique socio-economic constraints and still there is a lag in technological adoption by the farming community. Global warming has significantly affected the farming and economy, thus farmers come across with increase climate risk/ variability. Owing to technological reach, farmers are able to manage these risks, but an adequate way of analyzing and reducing the risk of climate change is the need of hour for sustainable pulse production. Pulses in India are mostly grown in rain-fed areas and this increases the risk of sustainable pulse production. Pulse production in India is very much dependent on weather variability along with other factors like area under cultivation and potential yields. There are issues at each stage, starting from availability of quality seed, production risks, marketing, processing and risk transfer/ insurance.

The cereal based crop cultivation has altered the dietary consumption consumes and thus both human as well soil suffers from nutritional content. So to balance this nutritional deficit; nutritious crop like pulses could be grown as well as consume more. Pulses occupy a unique place in India's nutritional food security and also India is the largest producer, consumer and importer of pulses. Though the growth rate of pulses is 2.61 per cent from last two-three decades; per day availability of pulse has decreased from 70.3 gram from 1956 to 41.9 gram in 2013. To achieve sustainability in pulse production in the country amidst climate risks, focused technological advancement on adequate planning, extensive research, proper management and efficient extension models for dissemination of pulse-based technologies, value addition and risk transfer mechanisms are immediately and urgently required.

Responding to climate change will also require the sustained participation of all stakeholders, including government and the private sector. Addressing climate change is now a top priority of the government. To enhance our understanding of the risks and opportunities and find innovative and effective ways to address the climate change challenges in a manner that would result in a climate resilient pulse production will be more remunerative for growers. I firmly believe that this book will meet these requirements.

I would like to sincerely thank the authors and experts who contributed chapters to the book. I have no doubt that the book will serve as a reference book, and serve as a useful wealth of knowledge to various stakeholders. I wish to complement the editors of this book for bringing out this valuable document on time.

(S. K. Rao)

Preface

The evidence of adversities from climate change is beyond doubt. All human activities starting from food production, their processing and many other activities are changing the earth's climate. Increased green house gases (GHG) emissions from these actions are causing temperatures to increase, in turn, resulting in extreme and harmful conditions such as heat and cold waves, cyclones, droughts, floods, and rises in the sea level. These conditions put populations at greater risk of food supply and nutritional insecurity. No doubt, climate change and global warming have widened climate risks not only for overall food production encompassing pulses, cereals, vegetables and spices, but also, in parallel, influenced other natural resources too. In the days to come, the severity of climate change is predicted to be more challenging in context of their extent and spectrum, and impacting critically the productivity and stability of production of pulse crops.

Now it's the need of the hour to overview on some feasible technological options more than formal efforts for enhanced pulse production. Driven by advances in crop breeding, most notably the development and commercialization of hybrid and upcoming transgenic crop varieties, plus the rapid mechanization, global food production grew at an unprecedented rate. Although, these research endeavors are showing potential for adaptation to climate change through genetic improvement to incorporate such traits as tolerance to abiotic stresses, multidisciplinary but integrated approach need to be immediately started. The most suited farming practices for harvesting utmost yield, government policies for maximum adoption, extension approaches for wider reach and post harvest processing to ensure nutritional security are among the prime components to be made essential adjuncts of the said feasible option.

Limited studies have been reported on climatic risks and their impact on pulse production and are widely scattered in research papers. This book is complied upon the recent advances and includes materials on multidisciplinary aspects such that it could be utilized readily by research workers across the disciplines. This book contains thirteen chapters covering the significant pulse production scenario, climatic risks and its impact on pulsed productivity, the nutritional gains from the pulses, major genomic and genetic interventions made, extension approaches depicting policy frame works and a economically viable marketing strategies needed for sustainable pulse production. The chapters have been classified under five sections and, each chapter is fully self-contained and quite explicit. This book provides the latest perspectives on the multidisciplinary approaches that define the outlook of global agriculture. It argues that even in the panic of climate challenges, a balanced management package could be able to achieve sustainability in pulse production. Graphics and case studies from a wide variety of sources enrich the flow of the narratives in the chapters. The book is written with the general readers in mind, and offers insights and opinions from respected experts in the field. By integrating the diverse stream of agricultural science (including agro-meteorology, biochemistry, physiology, breeding and genomics, economics and extension) this book embraces a chronological perspective as well as a look to the future.

We heartedly take this opportunity to thanks to all contributors for furnishing their worthy contributions as chapters. We must not miss this chance to express our special thanks for cooperation rendered by our colleagues, who directly or indirectly have rendered their help to bring out this book. A very special thanks to Dr. L.M.Bal, Scientist, College of Agriculture, Tikamgarh,M.P., Mr. Manish Mishra, Assistant Director-Agriculture for their immense help, during the editing and compilation of this book. At last but not least, we bow at the feet of "**Almighty God**" with whose omnipresent blessings, the assignment of editing a book could be completed. Lastly, our sincere thanks to publisher Astral International (P) Limited, New Delhi who provided an opportunity to publish this book.

A K Srivastava

Yogranjan

Contents

Section C: Nutritional Heritage of Pulses

Section D: Technological Intervention and Approaches for Sustainable Pulse Production

Section E: Extension Approaches, Government Planning and Policies for Enhancing Pulse Production

Section A

Pulses: Current Scenario–
Global, National and Regional Status,
Niche Area of Research

2018, *Climate Risks Management: Sustainable Pulse Production*
Editors: A K Srivastava and Yogranjan
Published by: **ASTRAL INTERNATIONAL PVT. LTD., NEW DELHI**   *Pages 1–26*

Chapter 1

Pulse Production and Technological Advancement for Meeting the Growing Need: Assessment and Analysis

A K Srivastava, Suvashree R. Prusty and Yogranjan

College of Agriculture, Jawaharlal Nehru Agricultural University,
Tikamgarh – 472 001, M.P.

ABSTRACT

Climate change is envisaged to alter the growing conditions of pulse production quite variably in different regions of India. It is the earnest need of the hour for agriculture to meet the dire demand of pulses in the days to come. A macro view of the global as well as national pulse area, production scenario and their temporal variations are put forward. Area and yield variability were analyzed to see the impact of technological advancement on major pulse productivity. Attempts were made to describe the available technological options for minimizing the constraints of improved production. The export, import and supply demand issues were also discussed. Many of the technological solutions must be coupled with wise management strategies and adequate investments. An early action is utmost important, but a better mechanisms is also needed for long-term action and adoption. Altogether, there is the need to have a integrated multidisciplinary research efforts to find solution to the problems posed. The government should be consistent in its agriculture policies such as provision of credit and insurance facilities along with marketing privileges.

Keywords: Yield, Area production, Cultivars, Major pulses, Export, Import.

Legumes represent a vast family of plants including more than 600 genera and more than 13,000 species. Pulses are having high protein, fiber, low fat and complex carbohydrates making it a unique plant food. Pulses such as Pigeonpea, Chickpea, Green gram (Mungbean), Black gram (Urdbean), Field pea, Lentil *etc.*

are important source of many nutrients. Pulses also provide substantial amounts of micronutrients (vitamins and minerals) such as Vitamin E, Vitamin B6, folic acid, iron, potassium, magnesium, calcium, phosphorus, sulfur and zinc. Chickpea and pigeonpea are great sources of iron, manganese and zinc and can play a key role in countering iron deficiency *viz;* anemia – a serious health issue that ranges from 50-70 per cent in women and children, with pregnant women being the most susceptible.

Pulses can be grown on a range of soil and climatic conditions and have important roles in crop rotation, mixed and inter-cropping, maintaining soil fertility through nitrogen fixation, release of soil-bound phosphorus, and thus contribute significantly to sustainability of the farming systems. Pulses have several unique attributes other than high nutritional value, such as low input cost and long preservation period. India is the world's largest pulse producer country, accounting for almost 25 per cent of the global share, and is the largest consumer too. About 90 per cent of the global pigeonpea, 75 per cent of chickpea and 37 per cent of lentil area falls in India. Pulses continue to be an important component of the *rainfed* agriculture. Pulses are also known to reduce several non-communicable diseases such as colon cancer and cardiovascular diseases (Jukanti *et al.,* 2012).

In general, pulses are mostly grown in two seasons: (i) Rainy season or *kharif* (June-September), and (ii) Winter/dry season or *rabi* (October-April). Chickpea, lentil, and dry peas are grown in the *rabi* season, while pigeonpea, urdbean, mungbean, and cowpea are grown during the *kharif* season. Among various pulse crops, chickpea dominates with over 40 per cent share of total pulse production followed by pigeonpea (18-20 per cent), mungbean (11 per cent), urdbean (10-12 per cent), lentil (8-9 per cent) and other legumes (20 per cent) (IIPR Vision 2030; Gowda *et al.,* 2013).

These pulse crops have potential to withstand weather variability, require less water and enrich the soil and thus considered as climate smart. Pulses like chickpea, pigeon pea and lentil will contribute to reduce poverty and hunger, improve soil as well human health and may have the sustainable yield under changing climate scenarios. Endowed with the unique ability of biological nitrogen fixation, carbon sequestration, soil amelioration, low water requirement and capacity to withstand harsh climate, pulses have remained an integral component of sustainable crop production system since time immemorial, especially in the dry areas.

Diversified Uses

Demand of protein from majority of Indian population is supplemented through pulse. Pulses are also an excellent feed and fodder for livestock. Chickpea green leaves are used as a leafy vegetable. The green immature seed is used as a snack or vegetable. The split dry seed and its flour are used in a variety of food preparations. Pigeonpea also lends itself to various uses. The leaves and forage are high in protein and are largely used as fodder. The stalks are used for fencing, thatching and preparation of baskets. They make for excellent firewood as the calorific value of stalks is about half of that of the same weight of coal. It is also used as a shade crop in many countries.

Total pulses and pulse products were 19 to 20 Mt in 2013-14, out of which 35-40 per cent of total pulse produce was consumed at urban *i.e.* 6-7 Mt. The urban people mostly consume pulses in their diet. Pigeonpea in their food basket accounts for nearly 34 per cent (Figure 1.1). Moong, black gram, lentil, gram are other major pulses consumed mostly by urban consumers.

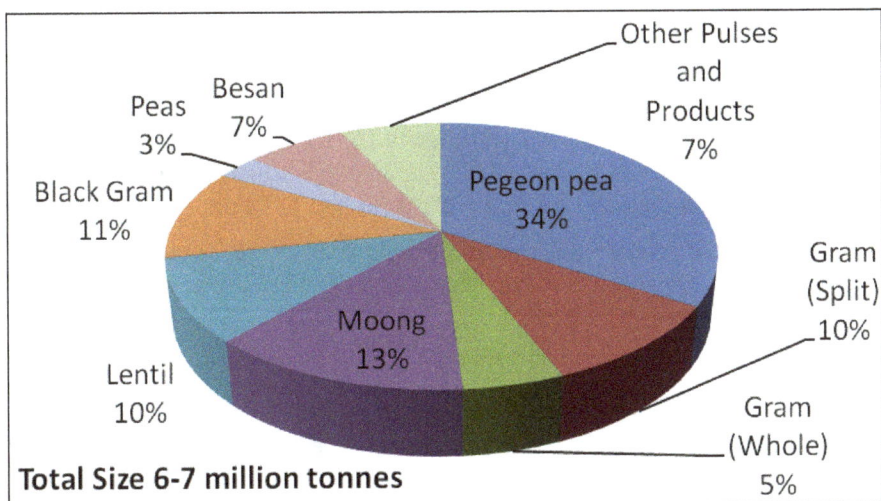

Figure 1.1: Urban Consumption Basket of Pulses in India.
(*Source*: NSSO, 2013).

Pulse Production Status

India is the largest producer of pulses in the world, with 24 per cent share in the global production. The important pulse crops are chickpea (48 per cent), pigeonpea (15 per cent), mungbean (7 per cent), urdbean (7 per cent), lentil (5 per cent) and fieldpea (5 per cent). The major pulse-producing states are Madhya Pradesh, Maharashtra, Rajasthan, Uttar Pradesh, Karnataka and Andhra Pradesh, which together account for about 80 per cent of the total production.

Globally, India has a pleasant position in pulse production and consumption. The contrast between pulses and cereals is striking. In the case of rice and wheat, of world's total production, Indian represents only about 20 and 13 per cent, respectively. India exports more staple crops in comparison to pulse crop and in case of import, the reverse is true. Moreover, India both exports and imports the pulses. Pulse production in India amplified from 8.41 Mt in 1950-51 to 19.78 Mt in 2013-14, showing a net increase of approximately 11 Mt. India is the largest pulses growing nation in the world with around 152.63 lakh ha of area and 20 to 24 per cent of global pulse production. However, the production of pulses in 2015- 16 was decreased to 17.33 Mt.

Global Scenario

Over the past two decades, the overall pulse production has increased at a rate higher than the growth rate in population both in developing and developed countries, and there are significant improvements in production due to the increased consumption trends globally. Production of pulses increased from 56 Mt in 2001 to 73 Mt in 2013, representing a growth of 2.18 per cent, yearly whereas, area and yield increased at the rate of 1.34 and 0.85 per cent, respectively (FAO,2012).

Pigeonpea and chickpea are major pulse crops contributing together about 25 per cent share in world's total pulses production and acreage. The Chickpea productivity has increased about 50 per cent during last five decades, whereas, pigeon pea productivity remained almost sluggish at global level (Table 1.1). Pigeonpea productivity decreased from 816 kg/ha to 762 kg/ha from year 1961 to 2013 but chickpea yield increased from 649 kg/ha to 978 kg/ha while the total pulse productivity increased from 673 kg/ha to 904 kg/ha. Chickpea is cultivated in 52 countries, contributes 16.59 per cent in area and 17.94 per cent in production.

Table 1.1: Area, Production and Productivity of Pulses in World
(A: Million ha, P: Million tonnes and Y: kg/ha)

Year	Pigeonpea			Chickpea			Total Pulses		
	A	P	Y	A	P	Y	A	P	Y
1961	2.73	2.23	816	11.84	7.68	649	64.01	40.78	637
1971	3.02	2.10	695	10.22	6.61	647	63.51	42.67	672
1981	3.20	2.17	678	8.94	5.77	645	62.56	41.63	665
1991	4.16	2.77	666	11.44	8.12	710	70.03	55.26	789
2001	4.58	2.97	648	9.46	6.97	734	66.57	55.94	840
2002	4.44	3.97	719	10.40	8.29	797	71.59	58.14	812
2003	4.55	3.10	681	9.64	7.11	738	72.11	59.31	822
2004	4.72	3.29	697	10.47	8.38	800	70.71	59.77	845
2005	4.77	3.34	700	10.20	8.44	827	71.55	61.20	855
2006	4.85	3.89	802	10.51	8.46	805	73.37	60.87	830
2007	4.82	3.54	734	11.27	9.75	865	75.36	61.69	812
2008	5.04	3.34	663	11.67	8.60	737	73.39	62.59	853
2009	4.64	3.60	776	11.51	10.44	907	69.77	64.01	917
2010	4.88	3.92	803	11.99	11.06	922	78.92	70.43	892
2011	5.95	4.52	760	13.27	11.75	885	79.80	68.93	864
2012	5.60	4.32	771	12.35	11.61	940	80.25	72.74	909
2013	6.22	4.74	762	13.40	13.10	978	80.75	73.01	904

Source: faostat.fao.org.

India is the largest producer of pulses contributing about 24.5 per cent of total production. Other major pulse growing countries are Myanmar contributing (7.3 per cent), Canada (7.25 per cent), China (6.4 per cent), Brazil (5.1 per cent), Nigeria (3.9 per cent), USA (3.2 per cent), Australia (3.1 per cent) and Ethiopia (2.9 per cent) of total global pulse production. India was the largest producer of pigeon pea, chickpea and total pulses in the world contributing about 67, 73 and 26 per cent respectively in the year 2013 (Table 1.2) followed by Myanmar. Uganda occupied fifth position among all the top five pulse producing countries. Canada, Myanmar, China and Niger were also important contributors in total pulses production. Malawi achieved the highest productivity of pigeon pea *i.e.* about 1326 kg/ha, followed by Myanmar (1230 kg/ha); whereas India remained far behind with 813 kg/ha. In case of chickpea, Myanmar and Australia achieved higher productivity over 1400 kg/ha whereas India remained with 960 kg/ha. The Canada has achieved the highest total pulses productivity of 2520 kg/ha, followed by China (1679 kg/ha) and Myanmar (1398 kg/ha), but India has achieved only 764 kg/ha even in good monsoon year of 2013. This indicates that, India needs to give more but focused attention to improve the pulse productivity by evolving better varieties and improving agronomical practices. Low pulse yield in India compared to other countries is attributed to poor spread of improved varieties and technologies, abrupt climatic changes, vulnerability to pests and diseases, and generally declining growth rate of total factor productivity.

Table 1.2: Top Five Pigeonpea, Chickpea and Total Pulses Producing Countries in the World
(A: Million ha, P: Million tonnes and Y: kg/ha)

Country	2001			2013		
	A	P	Y	A	P	Y
Pigeonpea						
1. India	3.33	2.26	679	3.91	3.17	813
2.Myanmar	0.35	0.32	879	0.65	0.80	1230
3.Malawi	0.14	0.11	780	0.22	0.29	1326
4.Tanzania	0.13	0.09	650	0.29	0.25	861
5.Uganda	0.08	0.08	1000	0.11	0.09	895
World total	5.58	2.97	648	6.22	4.74	762
Chickpea						
1. India	6.42	5.47	853	9.73	9.53	960
2.Myanmar	0.91	0.40	438	0.99	0.75	757
3.Malawi	0.20	0.26	1323	0.57	0.81	1417
4.Tanzania	0.65	0.54	829	0.42	0.51	1194
5.Uganda	0.16	0.12	711	0.34	0.49	1462
World total	9.47	6.97	734	13.40	13.10	978
Total pulses						
1. India	34.49	17.52	501	25.21	19.25	764
2.Myanmar	2.59	3.36	1299	2.42	6.11	2520
3.Malawi	2.55	2.01	788	3.88	5.44	1398
4.Tanzania	3.79	5.13	1353	2.90	4.87	1679
5.Uganda	3.79	2.26	597	3.94	3.01	765
World total	66.57	55.94	840	80.75	73.01	904

Source: Faostat.fao.org.

Pulse Production and Area: Indian Context

India is the largest producer (25 per cent of global production), consumer (27 per cent of world consumption) and importer (14 per cent) of pulses in the world. Pulses account for around 20 per cent of the area under food grains and contribute around 7-8 per cent of the total food grains production in the country. Pulses are grown in two seasons, *kharif* and *rabi* seasons. Only *rabi* pulses contribute more than 60 per cent of the total production and *kharif* pulse about 40 per cent (Table 1.3). The area under pulses has increased from 191 lakh ha in 1950-51 to 152.63 lakh ha in 2016-17, indicating a decrease of 22 per cent, whereas, the production of pulses during the same period has increased from 84.10 lakh tonnes to 168.1 lakh tones *i.e.* increasing 50 per cent and productivity from 441 kg/ha to 785 kg/ha *i.e.* increasing 78 per cent (Figure 1.2).

Table 1.3: Area, Production and Productivity of Pulses

	Season	2013-14	2014-15	2015-16	2016-17 (Targeted)
Area (Million/Ha)	Kharif	10.33	9.76	11.20	11
	Rabi	14.89	13.34	13.70	15
	Total	25.20	23.10	14.90	26
Productivity (kg/Ha)	Kharif	580	576	462	600
	Rabi	891	77	850	900
	Total	764	744	676	773
Production (Million Tonnes)	Kharif	5.99	5.62	5.16	6.6
	Rabi	13.26	11.57	11.65	13.5
	Total	19.25	17.19	16.81	20.1

Source: Directorate of Economics and Statistics, GOI, 2017.

In India, the area under chickpea has increased from 7.57 Mha in 1950–51 to 9.93 Mha in 2013–14. Similarly, the area of pigeon pea, increased slightly from 2.18 Mha to 3.9 Mha and yield from 788 kg/ha in 1950–51 to 813 kg/ha in 2013–14 (Figure 1.3).

Among all the states, Madhya Pradesh is the leader pulse producing state. The other top five pulse producing states are Rajasthan, Maharashtra, Karnataka, Uttar Pradesh, and Andhra Pradesh. Madhya Pradesh is producing about 27 per cent followed by Rajasthan (11 per cent) and Maharashtra (10 per cent) of total pulse production in India.

The production of pigeonpea in India amplified significantly in first decade of 21st century (Table 1.4). The state of Maharashtra occupied first position in pigeonpea production, followed by Karnataka and Madhya Pradesh, whereas least production was seen in the state of Gujarat. In 1970s, 1980s and 1990s, Uttar Pradesh was the leading state in pigeonpea production in India, but its production was drastically reduced in 2000s and onwards. In Andhra Pradesh, the production of pigeonpea increased from 1970 to 2010s, then decreased in the year 2011s, then remained stagnant in 2012s and 2013s. In Madhya Pradesh, the production is almost stagnant

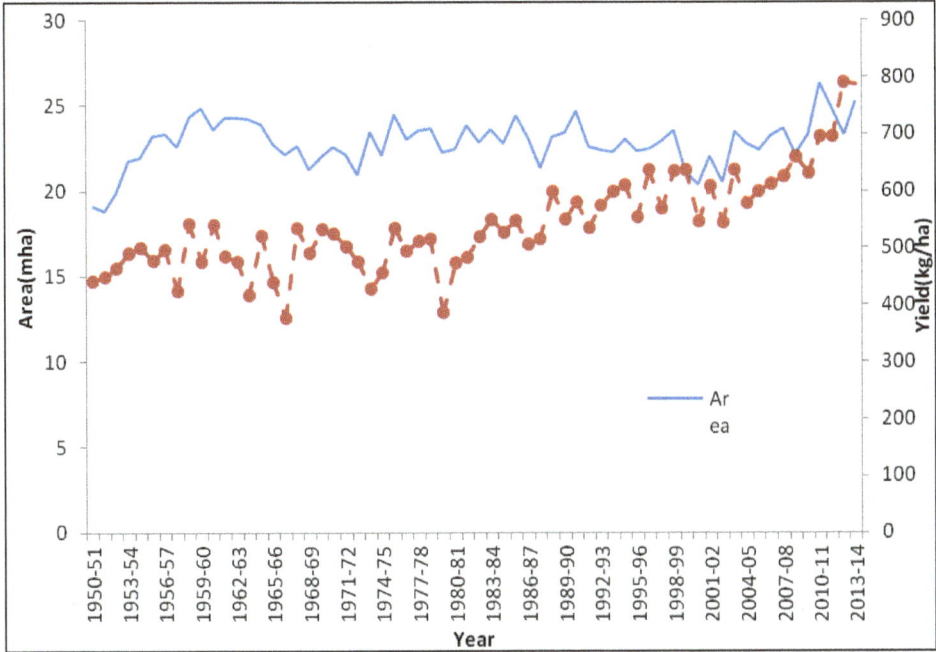

Figure 1.2: All India Variability in Pulse Yield and Area.

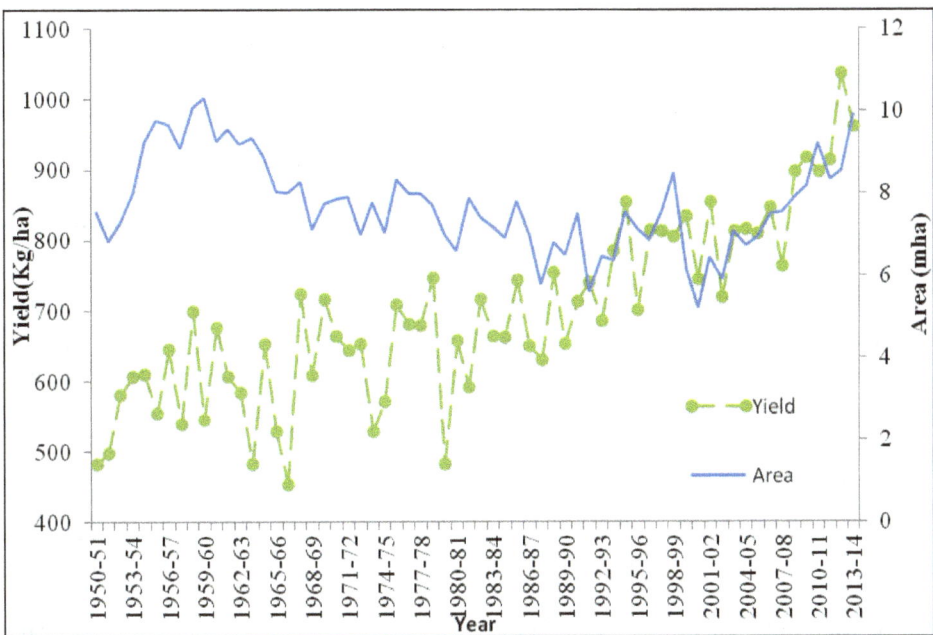

Figure 1.3: All India Variability in Chickpea Yield and Area.

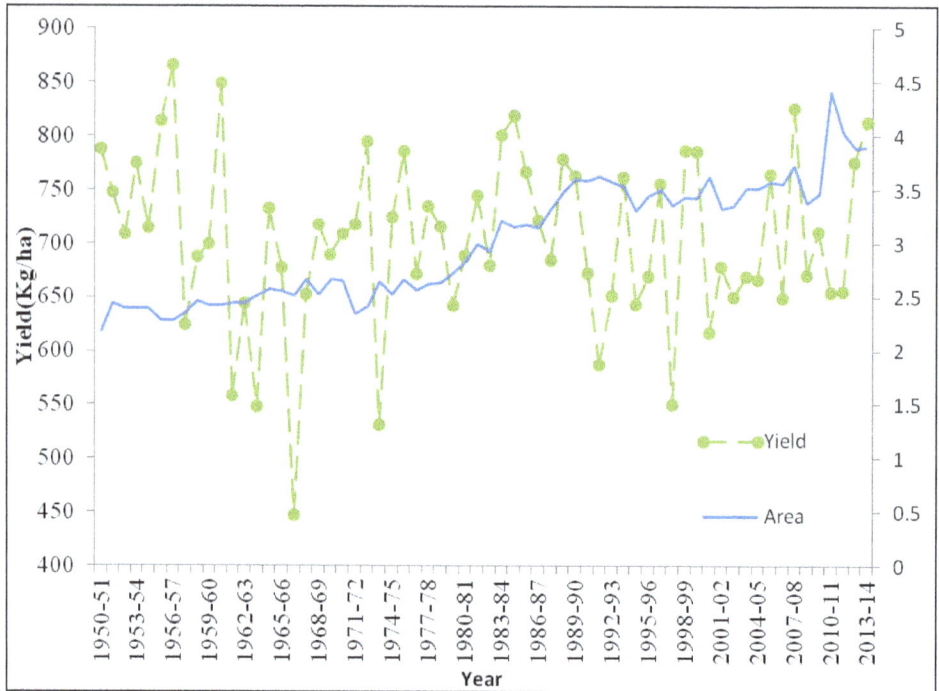

Figure 1.4: All India Variability in Pigeonpea Yield and Area.

from 1970 to 1990, the decrease up to 2010 and then increase up to 2013. In Gujarat, pigeonpea production was meager till 1970s and increased considerably in 1980s and 1990s then it became stagnant in 2000s and onwards.

Table 1.4: Major State-wise Production of Pigeonpea in India
(Lakh tonnes)

States	1970-71	1980-81	1990-91	2000-01	2010-11	2011-12	2012-13	2013-14
Gujarat	0.45	2.33	3.65	1.24	2.73	2.58	2.70	2.09
Andhra Pradesh	0.62	0.44	0.74	2.19	2.65	1.46	2.51	2.43
Karnataka	1.53	1.25	1.75	2.64	5.29	3.54	3.66	5.88
Madhya Pradesh	4.09	3.04	4.37	2.10	1.65	3.34	3.51	3.32
Maharashtra	3.05	3.60	4.21	6.60	9.76	8.71	9.66	10.34
Uttar Pradesh	6.78	7.57	5.78	5.10	3.09	3.34	3.25	2.71
All India	**18.83**	**19.57**	**24.17**	**22.46**	**28.61**	**26.54**	30.23	**31.34**

Source: Directorate of Economics and Statistics, GOI, 2017.

A total of 95.26 lakh tonnes of chickpea is produced by the major eight states in India. The state Madhya Pradesh is the highest producer of gram as arhar, contributing around 35 per cent in the country's total production, followed by Rajasthan and Maharashtra and least production is in the state Haryana *i.e.* all are contributing rest 65 per cent of total gram (Table 1.5). In last three decades, from 1970 to 2013, the production of chickpea in Uttar Pradesh and Haryana decreased whereas, it showed an increasing trend in Gujarat, Andhra Pradesh, Karnataka, Madhya Pradesh and Maharashtra.

Table 1.5: Major State-wise Production of Chickpea in India
(Lakh tonnes)

States	1970-71	1980-81	1990-91	2000-01	2010-11	2011-12	2012-13	2013-14
Gujarat	0.44	0.47	1.16	0.09	2.00	2.73	1.68	3.09
Andhra Pradesh	0.25	0.14	0.57	2.29	7.20	5.20	7.62	8.43
Haryana	7.74	4.66	4.69	0.80	1.10	0.72	0.53	0.75
Karnataka	0.62	0.58	0.68	2.39	6.31	4.68	6.23	7.16
Madhya Pradesh	8.56	10.63	18.92	24.08	26.87	32.90	38.12	32.99
Maharashtra	0.99	1.50	3.58	3.51	13.00	8.15	8.54	16.22
Rajasthan	11.95	8.54	10.11	3.97	16.01	10.61	12.77	16.40
Uttar Pradesh	15.44	12.88	11.22	7.03	5.30	6.84	6.76	4.75
All India	51.99	43.28	53.56	38.55	82.22	77.02	88.32	95.26

Source: Directorate of Economics and Statistics, GOI, 2017.

Decadal Variability of Area and Yield of Major Pulses

The decadal change in area and yield of major pulse growing states of India was carried out and depicted in Figures 1.5–1.8. There was a positive decadal growth in area and yield of chickpea in Madhya Pradesh (Figure 1.5). Lentil area has also sown positive growth in Uttar Pradesh as well as the productivity during past forty five years (Figure 1.6). The area of urdbean and mungbean have shown a negative growth over decades in the state Maharashtra and Madhya Pradesh respectively (Figures 1.7 and 1.8), though their yields have sown positive growth over decades. These trends clearly reflect the adoption of technological intervention in the cultivation of urdbean and mungbean the state Maharashtra and Madhya Pradesh.

In India, Madhya Pradesh, Maharashtra, Rajasthan, Uttar Pradesh and Karnataka, Gujarat and Andhra Pradesh are major pulse producing states. The state wise pulse productions across decades are shown in Table 1.6. The state Gujarat is the least pulse producing among all major pulse producing states but production showed an increasing trend from 2.01 lakh tonne in 1970 to 7.29 lakh tonne in 2013. Total pulse production has increased in India during last three decades. The total pulses production has increased in all major producing states Except, Uttar Pradesh in last three decades.

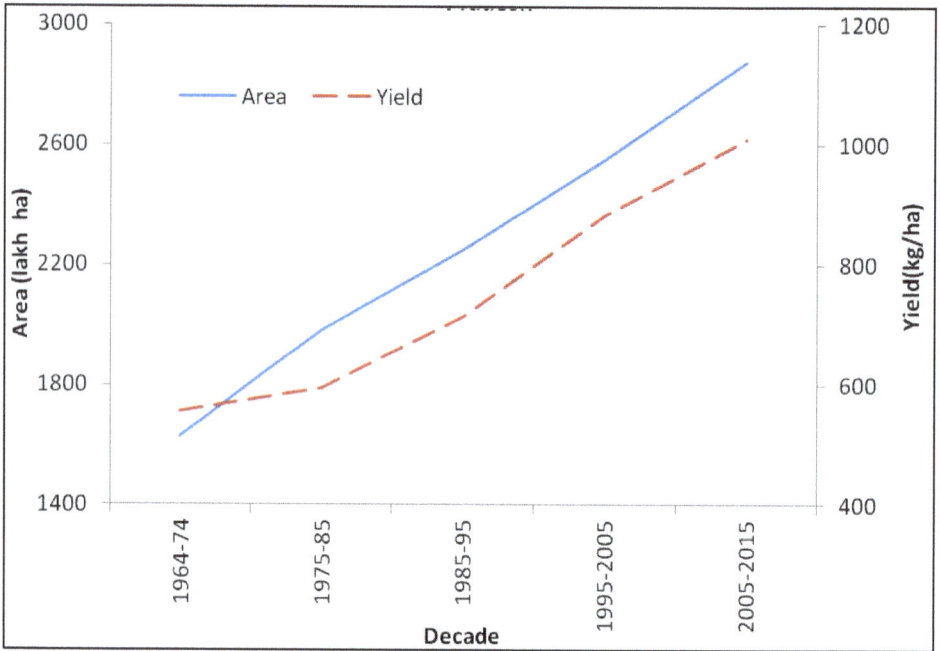

Figure 1.5: Decadal Variability of Chickpea Area and Yield in Madhya Pradesh.

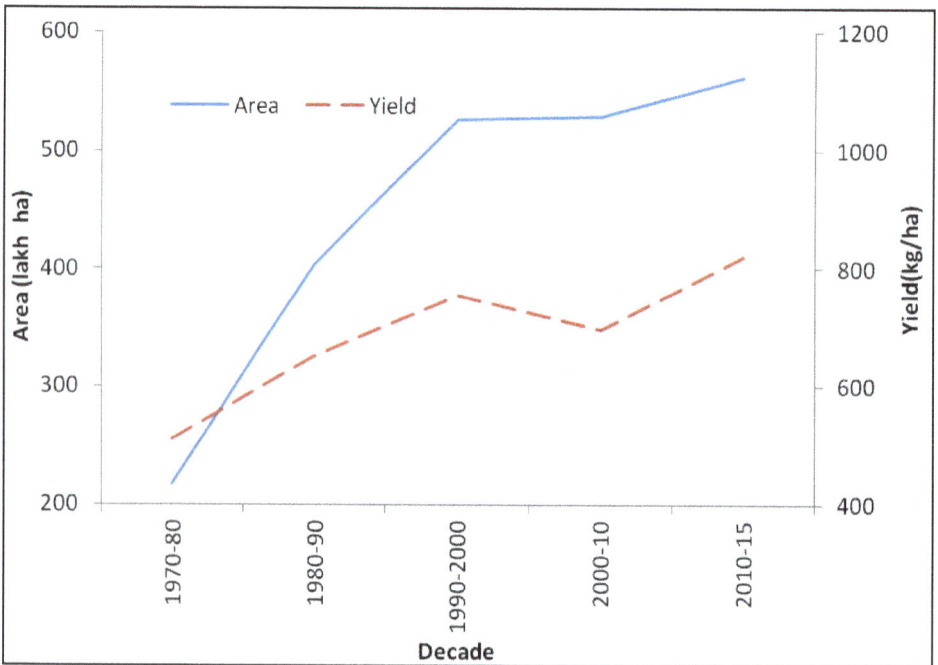

Figure 1.6: Decadal Variability of Lentil Area and Yield in Uttar Pradesh.

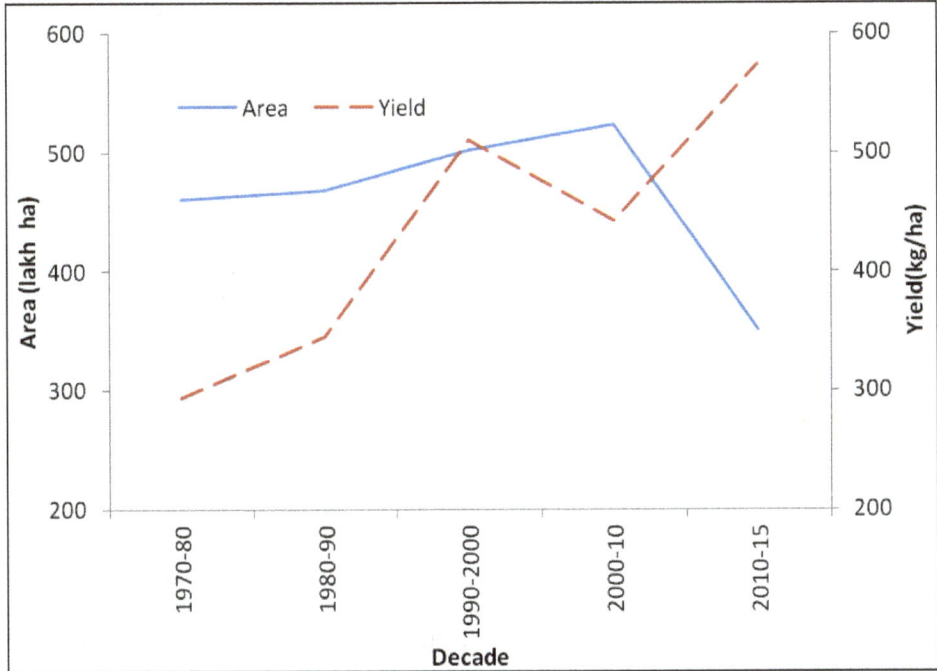

Figure 1.7: Decadal Variability of Urd Area and Yield in Maharashtra.

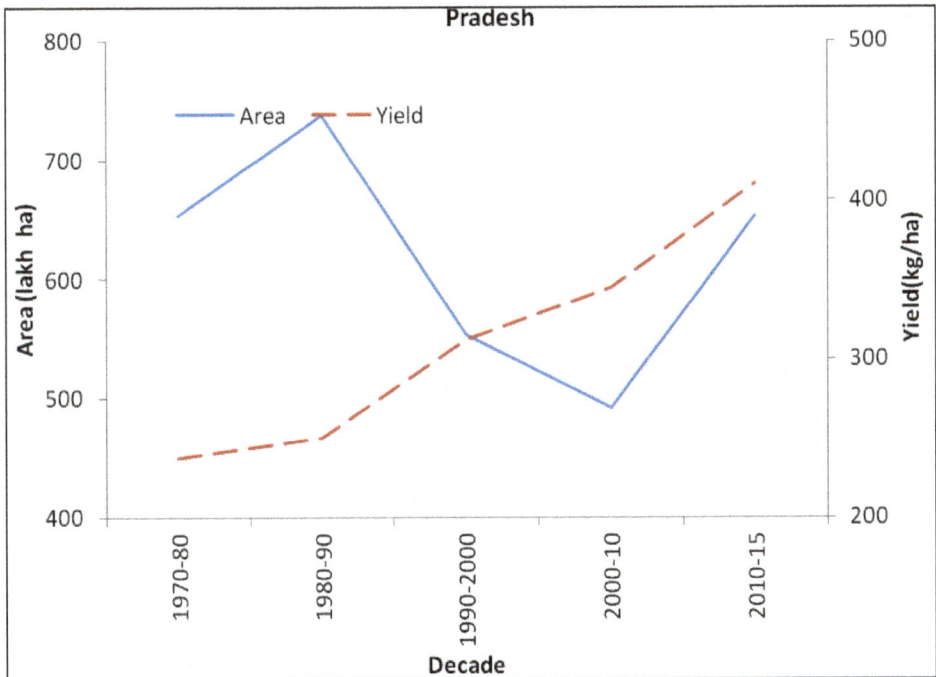

Figure 1.8: Decadal Variability of Mungbean Area and Yield in Madhya Pradesh.

Table 1.6: Major State-wise Production of Total Pulses in India
(Lakh tonnes)

States	1970-71	1980-81	1990-91	2000-01	2010-11	2011-12	2012-13	2013-14
Gujarat	2.01	5.20	6.24	2.49	7.22	7.80	5.42	7.29
Andhra Pradesh	4.50	4.15	12.49	10.54	14.40	12.30	16.23	15.51
Karnataka	4.04	4.63	5.52	9.56	15.65	11.34	12.59	16.01
Madhya Pradesh	19.92	20.11	31.04	22.75	33.62	41.62	51.66	46.44
Maharashtra	7.77	8.31	14.44	16.37	31.00	22.68	236.06	31.69
Rajasthan	17.64	11.70	17.19	7.32	32.60	24.32	19.56	24.90
Uttar Pradesh	30.69	25.23	27.72	21.60	20.37	24.03	23.32	16.97
All India	118.20	106.24	142.60	110.75	182.41	170.79	183.43	192.53

Source: Directorate of Economics and Statistics, GOI, 2017.

Export-Import and Net Availability of Pulses in India

Despite huge pulse production, India imports substantial amount of pulses from the world market for its domestic consumption since the beginning of the present millennium. Major pulse crops which are traded include pigeonpea, chickpea, lentil, peas *etc.* The quantity and value of export-import of major pulse crops are shown in Table 1.7.

Pigeopea

Though, India is a largest producer of pigeonpea in the world, but still to meet the demand, the country imports pigeonpea every year. The import of pigeonpea increased from 0.435 lakh tonne in 2000-01 to 5.75 lakh tonne in 2014-15 respectively. pigeonpea also exported. Export of pigeonpea decreased from 0.074 lakh tone to 0.012 lakh tone during 2001 to 2015.

Chickpea

As per FAO, India is the biggest producer of chickpea with a share of 68.9 per cent of globe. Other producers are Pakistan (5.4 per cent), Arharkey (4.8 per cent), Australia (4.7 per cent) and Myanmar (4.0 per cent). Australia is the biggest exporter with 37.4 per cent share followed by India (13.2 per cent) and Mexico (11.0 per cent) in 2010-11. India is the biggest importer with 18.7 per cent share of global import followed by Pakistan (13.5 per cent) and Bangladesh (13.2 per cent) in 2010-11. The import of chickpea increased from 0.639 lakh tonne in 2000-01 to 4.189 lakh tonne in 2014-15 respectively. Export of chickpea also increased from 0.026 lakh tone in 2000-01 to 1.902 lakh tonne in 2014-15.

Lentil

As per FAO, Canada was the biggest producer of lentil with a share of 38.1 per cent of globe followed by India (22.4 per cent), Arharkey (8.8 per cent) and USA (6.7 per cent) in 2011-12. Canada was the biggest exporter of lentil with a share of 63.5 per cent followed by USA (10.9 per cent) in 2010-11. Arharkey was the biggest importer with a share of 11.1 per cent of globe followed by India (9.7 per cent) and Bangladesh (7.6 per cent) in 2010-11.

Table 1.7: Export-Import of Pigeonpea and Chickpea
(Quantity: Lakh tonnes, Value: Rs. Crore, Unit: Rs/kg)

Year	Import			Export		
	Quantity	Value	Unit Value	Quantity	Value	Unit Value
Pigeonpea						
2000-01	0.435	62.78	14.45	0.074	19.24	26.00
2013-14	4.658	1755.87	377.69	0.001	0.48	48.00
2014-15	5.752	2635.85	45.82	0.012	8.82	73.50
Chickpea						
2000-01	0.639	92.75	14.50	0.026	4.56	17.55
2010-11	1.006	252.11	25.05	2.046	852.62	41.677
2011-12	2.065	772.26	37.40	1.702	1042.25	61.24
2012-13	6.976	2802.95	40.18	1.974	1297.71	66.58
2013-14	2.761	842.83	30.52	3.338	1676.75	50.23
2014-15	4.189	1334.96	31.87	1.902	1021.57	53.71

Source: Export-import Data Bank, GOI, 2014.

The import of total pulses has increased more than 20 per cent of the domestic production during the period between 2009-10 and 2012-13 (Table 1.8). In the remaining years, it has been in the range of 15-20 per cent of the domestic production, except during 2003-04 to 2005-06, where it varied to the tune of 11 per cent to 12 per cent of the domestic production. The import of total pulses has increased from 22.32 lakh tonnes in 2001-02 to 45.8 lakh tonnes in 2014-15. India's total exports of pulses have increased from 1.62 lakh tonnes in 2001- 02 to 4.48 lakh tonnes in 2005-06, declined to only 1 lakh tonnes in 2009-10 and then increased to 02.02 in lakh tonnes 2012-13.

Table 1.8: Export-Import of Total Pulses in India
(Quantity: Lakh tonnes, Value: Rs. Crore, Unit: Rs/kg)

Year	Import			Export		
	Quantity	Value	Unit Value	Quantity	Value	Unit Value
1980-81	1.73	29.776	1.72	0.001	0.35	3.21
1990-91	12.73	481.17	3.78	0.15	17.93	11.87
2000-01	3.51	500.06	14.26	2.44	573.58	22.01
2001-02	22.32	3163.72	14.17	1.62	369.20	22.84
2002-03	19.95	2741.05	13.74	1.51	351.37	23.21
2003-04	17.26	2288.28	13.26	1.54	329.12	21.36
2004-05	13.39	1777.58	13.27	2.77	6077.73	21.94
2005-06	16.96	2476.25	14.60	4.48	1116.07	24.92

Year	Import			Export		
	Quantity	Value	Unit Value	Quantity	Value	Unit Value
2006-07	22.71	3891.91	17.14	2.52	778.96	30.91
2007-08	28.35	5374.94	18.96	1.64	527.42	32.08
2008-09	24.74	6246.40	25.24	1.36	540.22	39.64
2009-10	35.10	9813.37	27.96	1.00	407.35	40.77
2010-11	26.99	7149.62	26.49	2.08	865.74	41.62
2011-12	33.65	8931.24	26.50	1.74	1065.84	41.62
2012-13	38.39	12738.64	33.18	2.0275	1285.00	61.18
2013-14	31.78	11036.75	34.73	3.4566	1749.29	50.60
2014-15	45.85	17062.94	37.22	2.2210	1218.10	54.84

Source: Export-import Data Bank, GoI., 2017.

Though, India is top pulse producing, consuming and exporting country, it has to import pulses from other countries like Canada, USA, Australia, Russia, Ukraine, France, Nepal, China Myanmar, Ethiopia, Myanmar, Tanzania *etc.* The pulses imported are dry peas, pigeon pea, chickpea, lentil, green gram, urd, rajma *etc.* Pigeon pea is imported from Myanmar, Tanzania, Mozambique, China and Malawi, while chickpea is imported from Australia, Myanmar, Tanzania, China, Russia, USA, and Canada (Table 1.9).

Table 1.9: Major Exporters of Pulses to India

Pulses	Countries
Dry Peas	Canada, USA, Australia, Russia, Ukraine, and France
Pigeon Pea (Arhar/Arhar)	Myanmar, Tanzania, Mozambique, China and Malawi
Urad and Green Gram	Myanmar, Australia, Tanzania
Lentils (Masur)	Canada, USA, Australia, Nepal and China
Chickpeas (Chana)	Australia, Myanmar, Tanzania, China, Russia, USA, Canada
Kidney Beans (Rajma)	China, Myanmar, Ethiopia, USA
Other Beans	Myanmar, China, Madagascar, Brazil
Other Dried Legumes	Myanmar, Tanzania, Mozambique, China, Malawi

Source: Faostat.fao.org.

Demand and Supply of Total Pulse

Though the major pulses are imported from across 30 countries, Canada, Myanmar, USA, Russia and Australia have been the major source of imports. Peas, lentils, chickpea continue to share a bulk of the imports. The total supply of pulses increased from 206.81 lakh tonnes in 1990-91 to 217.35 lakh tonnes in 2014-15 and that accounts for around 5 per cent increase (Table 1.10). Total availability of pulses increased from 206.66 lakh tonnes 1990-91 in to 215.13 lakh tonnes 2014-15.

Table 1.10: Demand and Supply Balance Sheet for Pulses (Lakh tonnes)

Total Pulses	1990-91	2000-01	2010-11	2011-12	2012-13	2103-14	2014-15
Production	203.68	110.75	182.41	170.89	183.42	192.53	171.50
Imports	3.13	22.32	26.99	33.65	38.39	31.78	45.85
Total supply	206.81	133.70	209.40	204.54	221.81	224.31	217.35
Export	0.15	1.62	2.08	1.74	2.03	3.46	2.22
Total availability for domestic use	206.66	131.45	207.32	202.80	219.78	202.85	215.13
Total utilization	206.81	133.70	209.40	204.54	221.81	224.31	217.35
per cent imports to production	1.54	20.15	14.80	19.69	20.93	16.51	26.73

Source: Export-import Data Bank, GoI., 2017.

Per Capita Net Availability of Cereals and Pulses

The per capita production and availability of pulses in the country has shown a sharp decline. Per capita net availability of pulses has declined from around 60 grams per day per person in the 1950s to 40 g per day in the 1980s and further it is around 35 g per day in 2000s (Table 1.11). However, in the past four years, there has been momentous increase in consumption of pulses around 50 g per day virtually because of higher production and nutritional awareness. Even though, it is rationally smaller amount in the line of achieving the WHO recommendation against of 80 g/day per person.

Table 1.11: Per Capita Net Availability of Cereals and Pulses in India

Year	Population (million)	Net Availability (million tonnes)		Per Capita Net Availability per Day (g)		
		Cereals	Pulses	Cereals	Pulses	Total
1951	363.2	43.6	8.0	334.2	60.7	349.9
1961	442.4	64.6	11.1	399.7	69.0	468.7
1971	551.3	84.0	10.3	417.6	51.2	468.7
1981	688.5	104.8	9.4	417.3	37.5	454.8
1991	851.7	145.7	12.9	468.5	41.6	510.1
2001	1033.2	145.6	11.3	386.2	30.0	416.2
2011	1201.9	180.1	18.9	410.6	43.0	453.6
2012	1213.4	181.0	18.4	408.6	41.7	450.3
2013	1228.8	160.6	19.4	358.1	43.3	401.4
2014 (P)	1244.0	201.6	21.4	444.1	47.2	491.2

Source: Directorate of Economics and Statistics, GoI, 2017.

Projected Demand of Pulses for India in 2020 and 2050

By 2050, India will need about 26.5 million tonnes pulses to meet the consumption requirements. As per the projection of Indian Institute of Pulses

Research (IIPR), India's population is expected to touch 1.68 billion by 2050. To meet this requirement, additional 3 to 5 million ha land will need to be brought under cultivation of pulses and productivity has to be increased to 1361 kg/ha.

Table 1.12: Requirement, Production and Import of Pulses in India

Year	Population	Requirement	Production	Import
	(million)	(Mt)	(Mt)	(Mt)
2000–01	1027	16.02	11.08	0.35
2004–05	1096	17.1	13.13	1.31
2009–10	1175	18.33	14.6	2.83
2020–21	1225	19.1		
2050–51	1613	26.5		

Challenges Ahead

Pulses in India are mostly grown in rain-fed areas with unstable and uncertain rainfall This increases the risk of crop failure. The major challenges being discussed ahead.

Cropping Pattern and Food Habit

The green revolution in India has changed the cropping pattern and affected the area of pulses too. Cereals based cropping has affected the food habits and people take more carbohydrate intake in their food than protein. The micronutrient malnutrition has been crept in developing countries of the world and thus become a global issue. It was reported that over 2 billion people are suffering from hidden hunger, especially micronutrient deficiencies (Judy and Joyce., 2016). Zn and magnesium deficiency is one of the most common ones prevalent in the world (Ashworth and Alloway, 2004) resulting in a high number of incidences of malnutrition and health problems; especially in children (Kutman *et al.*, 2010).

Changing Climate

The predicted changes in temperature and their associated impacts on rainfall and consequent availability of water to crops and extreme weather events are likely to affect considerably the potentials of pulse production. The most worrying part of the prediction is the estimated increase in winter and summer temperatures by 3.2° and 2.2°C respectively, by 2050. Such abnormal rises will surely have an adverse impact on pulse production in the form of a reduction in total crop-cycle duration. Most of the pulses like mungbean and urdbean are already short-duration crops (65–75 days). Any further reduction in crop duration will amount to a lower yield per unit area.

It has been observed that temperature and moisture stresses are more variable and frequently affect the crop production in general and pulse production in particular, in Central India during last two decades. A large number of studies were carried out world wide for impact assessment of climate change and global warming for food production and its supply chain with least focus on other components of food. The temperature increase and water deficit due to frequent droughts put more pressure on production environment. There is a positive impact of climate change (high CO_2 concentration) on crop production but under thermal and moisture stress conditions, the positive benefits either nullify or get converted into negative effects. Srivastava *et al.* (2016) reported that chickpea is the pulse crop whose yield was less affected under thermal and moisture stress conditions in central India. In the years to come, the nutrient value of chickpea and its production in hungry and thirsty agro-ecosystem make it a very important food item for human kind.

In the absence of information concerning the nature of increases in rainfall variability in the coming decades, *kharif* pulse yield would be affected more except *rabi* pulses yield, in the areas of high rainfall variability (CV> 30 per cent). Climate variability and extreme climate events have been shown to increase in large regions of the world because of climate warming (Coumou and Rahmstorf 2012; Hartmann *et al.*, 2013).

Minimum Support Price of Major Pulses

Price of the crop is determined by the demand and supply side factors. The supply side determinants include productivity and area used for cultivating pulses whereas the demand side determinants are consumption pattern, population, income growth, urbanization, changes in food habit and shift in trade flows and so on. The Minimum Support Price (MSP) is the highest for the pulses *i.e.* pigeonpea, followed by moong and the lowest for lentil. The domestic wholesale prices have been generally lower than the international prices during 2001-02 to 2012-13. MSP of lentil has been continuously lower than domestic wholesale prices. The MSP is Rs. 5050/- for pigeonpea in year 2016-17, whereas it is Rs. 5225/- for moong and Rs. 5000/- for urd (Table 1.13). The MSP of chickpea and lentil was Rs. 3425/- and Rs. 3325/- for the year 2015-16.

Table 1.13: Minimum Support Price of Major Pulses

Pulses	2014-15	2015-16	2016-17
Kharif			
Pigeonpea	4350	4625*	5050[§]
Moong	4600	4850*	5225[§]
Urad	4350	4625*	5000[§]
Rabi			
Chickpea	3175	3425[#]	Not announced
Lentil (Masur)	3075	3325[#]	Not announced

Source: Directorate of Economics and Statistics, GoI, 2017.

Prize stability is crucial for producers to invest in pulse production on constant basis. This is indispensable for increasing domestic production by eliminating the risks, that farmers experience while growing, processing and marketing. The policy makers must set a minimum price index with assured marketing outlets for procuring the produce, in no case the prize should go down the set index, even if there occurs higher production.

Need of Climate Smart Cultivars

ICRISAT has developed extra-early (85 to 90 days to maturity) and super-early (75 to 80 days) chickpea varieties that can escape terminal droughts. Recently, ICRISAT identified certain chickpea lines that have high levels of heat tolerance, which will enable them to be grown in areas with higher temperatures during the heat-sensitive pod filling stage.

Jawaharlal Nehru Krishi Vishwa Vidyalya (JNKVV) Jabalpur has also developed a heat tolerance chickpea cultivar JG -16 with higher yield potential under late sown conditions. Many new cultivars of major pulses are released by national as well as international organizations/institutes for abiotic and biotic stress management. The list of latest pulse wise cultivars and their specific features are given in Table 1.14.

Table 1.14: New Pulse Cultivars and their Specific Features

Major Pulse	Cultivar Name	Institute	Specific Features
Chickpea	ICCV08102	ICRISAT	Suitable for cultivation in Madhya Pradesh state
	ICCV08108	ICRISAT	Suitable for cultivation in Maharashtra state
	ICCV05106	ICRISAT	Suitable for combine harvesting. Long height
	JG 16	JNKVV	A heat tolerance and suitable for late sown conditions.
	JG 14	JNKVV	High yielding variety suitable for late sown conditions.
Pigeon pea	LRG 52	TANU	Yield potential of 2 t/ha under *rainfed* conditions and 3 t/ha under irrigated conditions. Short duration 150 days and is moderately tolerant to helicoverpa, maruca, pod fly, Fusarium wilt and Sterility mosaic diseases.
	Pusa 16	IARI	It is a semi-dwarf variety suitable for combined harvesting. Its yield levels are in the range of 15-20 quintals/ha and matured in 4 months
	ICPL 88039	ICRISAT	Short duration of 120 days duration.
	ICPH 2740	ICRISAT	Hybrid pigeonpea. High yielding 3.5t/ha.
Mungbean	OUM11-5 (Kamdev)	OUAT	Very short duration (50-55 days) suitable for sowing across the season. Yield potential of 8-10q/ha
	IPM 25	IIPR	Short duration 52-55 days.
	Smart	IIPR	13.2 q/ha. Suitable for summer sowing. Determinant growth and synchronized maturity.
	Meha	IIPR	14q/ha. Suitable for summer sowing. Determinant growth and synchronized maturity. Registant to YMV.
	Co(Gg) 8	TANU	Medium duration 55-60days. Suitable for rainfed conditions. Escape terminal drought.
Urdbean	TU-40	BARC	Resistant to YMV and moisture stress. 70-80 days duration
	TBG-104	TANU	17-18q/ha. Photo insensitive, Suitable for all season.
	OBG-32	OUAT	70-75 days duration. Resistance to YMV
Lentil	Binamasur-10	ICARDA	Drought tolerant. The variety has a yield potential of 2.0 tons per hectare. 110 days duration

Water Management

Effective irrigation measures can enhance carbon storage in soils through increased yields and residue returns. Although, harvesting of rainwater is advocated since many decades but farmers hardly adopt it at small scale merely for minimizing the impact of mid-season and terminal droughts; frequency of adoption will likely to be more in years to come. Application of nano fertilizers is also being practiced in crop management to enhance water use efficiency and enriched production.

Crop Management

Promotion of conservation agriculture practices such as zero tillage, bed planting, efficient residue management and crop rotation may be adopted. Use of climate ready crops and bright farming could be advocated. Changing crop calendars and altered intensity of pest diseases will be immediate application to deal with challenges of climate change. Introduction of new cropping sequence and crop varieties to combat the climate change is utmost required.

Selection and Breeding Approaches for Climatic Adaptation

Heat tolerance in chickpea for enhancing its productivity in warm growing conditions could be required to mitigate impact of climate change on yield. Enhancing yield and stability of pigeonpea through heterosis breeding should be on the top agenda of pigeonpea breeding programame. Likewise, enhanced lentil production for food, and nutritional security leading to improved rural livelihood should be achieved through introduction, selection and hybridization programame.

Climate Ready (Bio-engineered) crops can reduce carbon emissions and have a positive effect on climate change. California-based Arcadia recently developed a rice variety that will give the same yields but use 65 percent less nitrogen fertilizers, which directly contribute to mitigate the climate change. The adaptation of crops to certain climatic optima is of great significance in connection with the introduction of a particular variety or breed into a new area. New varieties which have better characteristics for productivity in a particular environment that have the desired characteristics for a given locality are sought in other area where the climate is analogous. However, with a combination of climate -ready varieties plus improved agronomic practices, *rainfed* farmers would be able to overcome the adversities of a warmer world.

Use of New Tools and Techniques

Biochar

Biochar is a charred carbon-enriched material intended to be used as a soil amendment to sequester carbon and enhance soil quality. Most studies showed that addition of biochar increased crop yields. Apart from carbon sequestration, there are other environmental benefits that can be derived from the application of biochar in soils which include reduction in the emission of non-CO_2 GHGs by soils.

Hydrogel

Hydrogels are cross-linked polymers with a hydrophilic group which have the capacity to absorb large quantities of water without dissolving in water. Hydrogel may prove as a practically convenient and economically feasible option to achieve the goal of agricultural productivity under conditions of water scarcity. It can be easily applied directly in the soil at the time of sowing of field crops and in the growth medium for nursery plantation. The low application rate (*i.e.* 2.5–5.0 kg/ha) of hydrogel is effective for almost all the crops in relation to soil type and climate of India. The improvement in growth and yield attributing characters and yield of different field, ornamental and vegetable crops have been reported with the application of hydrogel. Agricultural hydrogels are not only used for water saving in irrigation, but they also have tremendous potential to improve physico-chemical and biological properties of the soil. Bulk density, porosity and water holding capacity of the soil are improved with the application of hydrogel.

Boron

A recent study by ICRISAT indicated that soils in many states in India are deficient in micro-nutrients such as boron, sulfur, zinc and magnesium. Boron deficiency causes many anatomical, physiological, and biochemical changes, most of which represent secondary effects. Though, legumes fix atmospheric nitrogen but adequate boron nutrition is critical not only for high yields but also for high quality of crops.

Biotechnology for Drought Resistance

Limited success has been achieved so far through conventional breeding in pulse improvement. Genomics-assisted breeding (GAB) holds promise in enhancing the genetic gains. In the last few decades, advances in the pulse genomics are noteworthy, *e.g.* discovery of genome-wide molecular markers, high-throughput genotyping and automated sequencing facilities, high-density genetic linkage/QTL maps and, the most importantly, the availability of whole-genome sequence. Efforts are quite consistent in identifying key regulators in plant drought response through genetic, molecular, and biochemical studies using the model species *Arabidopsis thaliana*. Additionally, many signalling pathways for drought tolerance have been identified. However, only limited portion of these regulators have been explored as potential candidate genes for their application in the improvement of drought resistance in pulse crops. Based on biological functions, these genes can be grouped into the following three categories: (1) stress-responsive transcriptional regulation (*e.g.* DREB1, AREB, NF-YB); (2) post-transcriptional RNA or protein modifications such as phosphorylation/dephosphorylation (*e.g.* SnRK2, ABI1) and farnesylation (*e.g.* ERA1); and (3) osomoprotectant metabolism or molecular chaperones (*e.g.* CspB). Extensive research efforts are also focused on fine-tuning the expression of the known candidate genes for stress tolerance in specific temporal and spatial patterns to avoid negative effects in plant growth and development. Results are now apparent in the form of yield improvements in several pulse crops under a variety of water-deficit conditions. However, results are still to be materialized in the field, as most of the studies are performed under controlled growth environments creating a gap between early success in the laboratory and the application of these techniques to the elite cultivars of major pulse crops in the field.

Weather Forecasting and Early Warning System

Success in dealing with uncertainties arising out of climate change largely rests with a capacity to anticipate. Consistent agricultural productions along with maintenance of human health depend on our potential to anticipate and prepare for the uncertain future. Anticipating futuristic environmental challenges requires improved scientific understanding. A sound ecological forecasting could have an expanding role in policy and management for effective research planning and decision-making process. The spatial extent of the weather forecasting may ranges from small plots to regions to zones to continents to the globe, whereas, the time horizon can extend from a week to few decades.

An accurate estimation and its communication will determine the success of weather forecasting as well as early warning system. Early warning system attributes are ones for which uncertainty can be reduced to an extent. Technical structure of forecasts covers collection of new or augmenting existing data networks to support experimental research. A good stock of existing and new data can provide a baseline for forecasting (Carpenter, 1996; Clark *et al.*, 1999). The feedbacks' data from a crop history in response to varying climate become vital when the spatial extent of forecast is the prime objective. These data serve in developing accurate weather risk and insect-pest forewarning system to reduce the production risks. The practical demand of feedbacks' data stock is the generation of knowledge based decision support systems for translating weather information into operational management practices that encompass the tactical and strategic crop management at district and block level. The government of India has established a network of 130 agro-meteorological field units across the country and issuing bi-weekly agro-met advisory bulletins on real time basis for a number of stalk holders in every districts. These facilities may be utilized for efficient input and crop management to minimize the weather risks.

Insurance

Weather based index insurance is typically associated with abnormal changes in weather condition. The crop or weather index insurance could enhance the risk bearing ability of farmers. The insurance also could have other benefits such as providing a safety net to vulnerable households and price signals regarding the weather risks. A large array of governmental as well as social initiative has come up to facilitate weather based crop insurance to the farmers. Pradhan Mantri Fasal Bima Yojna, Modified National Agricultural Insurance Schemes are also available to the farmers at ease to enable them to bear the risk of catastrophic weather shocks. Under the crop insurance scheme, government support in the form of premium subsidies, where a fixed portion of total insurance premiums is paid to the insurance company by the government. This initiative certainly impedes adaptation by encouraging households to maintain or increase investments in unsustainable livelihoods. The insurance net and applicability may be increased and decreased respectively to increase the risk bearing ability of farmers.

Credit Facility

An easy access to sources of financing at the need is a basis for building better living conditions for farmers by ensuring better profitability for their products. The financing in farm sector has significant potential in reducing poverty among farmers by increasing their access to diversified employment opportunities in upstream and downstream agribusiness enterprises. Traditionally, the farmers engaged in pulse production are mostly small and marginal. They are deprived of banking and credit facilities. For efficient performance of agricultural sector, adequate credit facilities and proper support is essential both from government and private sector. Most of the small and medium sized farmers have inadequate capital for the purchase of costly inputs such as farm machinery, fertilizer, herbicide and pesticide which compel them to compromise with the standard and the basic need of the farm and crops. This obligation largely contributes to low production and large scale migration of the farming community. Timely provision of fund to purchase the farm inputs will definitely ameliorate this problem. Policy should be drawn for financial literacy to the farmers and to provide credit facilities to empower their ability of utilizing modern agro-chemical and machines. Government must focus on strengthening the credit facility of farmers through all bank and insurance agencies with minimal interest. The financing institutes or banks should provision easy hastle-free formalities for loan disbursement so that resource poor farmers can easily access the loan facility.

Marketing

The lack of a supporting mechanism for the procurement and marketing of pulses is a major hurdle in pulse production. There is either lack of marketing network or if available, crowded with bunch of mediators. Procurement of produce by a dedicated agency is almost lacking or in-effective. Improper access to storage and milling facility cause further risk to farmers. Also, poor market linkages lead constraints in meeting market demand. In addition, the procurement prices for pulses are low. This becomes a disincentive for farmers to grow pulses. However, to create confidence in farmers cultivating pulses, the Ministry of Agriculture through NAFED ensures to purchase pulses at MSP under price support scheme. The government of Madhya Pradesh has recently launched a scheme *viz.*, Bhavantar Yojna for farmers to give the market price of the selected crops including pulses to the farmers.

Post Harvest Processing and Value Addition

Post processing and value addition is very important. Development of post harvest machines and practices are required for increasing recovery of pulses and minimizing the losses in processing.

Most of the products available in market are of Gram. A few items are also available of Moong and Urdbeans. Therefore, there is a need to work on evolution of new items using the aforesaid and other pulses. In order to have access to the world markets and compete successfully with major pulse producing countries it is essential to add value to the commodity. The large array of snacks produced using wholegrain pulses as well as pulse flour can meet consumer acceptability in terms of texture and structural attributes if the correct extrusion conditions and feed composition are applied. Fortification and bio-fortification of some pulses can increase the bioavailability of nutrients by which demands their consumption (Anoma *et al.*, 2014). Most of the value additions in pulses are done by the ultimate consumers themselves. However, the growing health consciousness, preference for quality packaged products and shortage of labour drives the processors to use modern technology. Due to inconsistent policies and proper support, the players in the value chain of pulses are hesitant to come forward to make investment decisions including those related to research and development, marketing and input supplies. The establishment of strong linkage among producers, suppliers, consumers, processors, whole sellers, and retailers is inevitable to begin value chain.

Conclusion

India is occupying a large area for pulse production, but still its productivity is far behind from many low coverage zones of world pulse producers. A large number of constraints exist for its lower productivity including weather, price variability, vain policies and research. Integrated approach may be required to minimize or mitigate the constraints which are limiting the pulse growth and yield. Government policy initiatives as well as stalk holder's awareness and provocation activities are required. Besides, government initiative, incorporation of technological advancement in the field of research and development may be utilized and disseminated by their stalk holders. Thus the pulses sector needs a long-term policy that includes production, processing, consumption and trade. It needs adequate planning, research and investment. The government should prioritize and focus seed production, area expansion, price stabilization, widening insurance coverage and value addition for pulses to meet its future demand.

References

Anoma, A., Collins, R., and McNeil, D. (2014). The value of enhancing nutrient bioavailability of lentils: the Sri Lankan Scenario. *African Journal of Food, Agriculture, Nutrition and Development*, 14(7), 9529-9543.

Ashworth, D. J., and Alloway, B. J. (2004). Soil mobility of sewage sludge-derived dissolved organic matter, copper, nickel and zinc. *Environmental Pollution*, 127(1), 137-144.

Carpenter, S. R. (1996). Microcosm experiments have limited relevance for community and ecosystem ecology. *Ecology*, 77(3), 677-680.

Clark, J.S., Beckage, B., Camill, P., Cleveland, B., HilleRisLambers, J., Lichter, J., McLachlan, J., Mohan, J. and Wyckoff, P., 1999. Interpreting recruitment limitation in forests. *American Journal of Botany*, 86(1), pp.1-16.

Coumou, D., and Rahmstorf, S. (2012). A decade of weather extremes. *Nature climate change*, 2(7), 491-496.

Export-Import Data Bank.(2015).Indian Trade Portal. Government of India. http://www.indiantradeportal.in/

FAO 2012 FAOSTAT Online Database (available at http://faostat.fao.org/, accessed September 2017)

Food and Agriculture Organization. (2013). Food and Agriculture Organization. http://www.faostat.fao.org.tional

Gowda, C. L., Srinivasan, S., Gaur, P. M., and Saxena, K. B. (2013). Enhancing the productivity and production of pulses in India.

Hartmann, D.L., Tank, A.M.K., Rusticucci, M., Alexander, L.V., Brönnimann, S., Charabi, Y.A.R., Dentener, F.J., Dlugokencky, E.J., Easterling, D.R., Kaplan, A. and Soden, B.J. (2013). Observations: atmosphere and surface. In *Climate Change 2013 the Physical Science Basis: Working Group I Contribution to the Fifth Assessment Report of the Intergovernmental Panel on Climate Change*. Cambridge University Press.

Indian Institute of Pulses Research (2015). Indian Council of Agricultural Research, Kanpur,Vision 2050 pp. 1-38.

Judy H. P. S., and Joyce, C. T. K. (2016). Children Malnutrition in India. *Development: What Now? Past, Present and Future Challenges in International Development*, 205.

Jukanti, A. K., Gaur, P. M., Gowda, C. L. L., and Chibbar, R. N. (2012). Nutritional quality and health benefits of chickpea (*Cicer arietinum* L.): a review. *British Journal of Nutrition*, 108(S1), S11-S26.

Kutman, U. B., Yildiz, B., Ozturk, L., and Cakmak, I. (2010). Biofortification of durum wheat with zinc through soil and foliar applications of nitrogen. *Cereal Chemistry*, 87(1), 1-9.

National Sample Survey Organization. (2013). Ministry of Statistics and Programme Implementation, Government of India.

Srivastava, A. K., Silawat, S., and Agrawal, K. K. (2016). Simulating the impact of climate change on chickpea yield under rainfed and irrigated conditions in Madhya Pradesh. *Journal of Agrometeorology*, 18(1), 100-105.

Statistics, Indian Agricultural. "17, 2013-17." *Directorate of Economics and Statistics, Government of India* (2017).

World Bank 2014: Data, GDP Current, Available online at http://data.worldbank.org/indicator/NY.GDP. MKTP.CD. DoA: 26 September 2017.

2018, *Climate Risks Management: Sustainable Pulse Production*
Editors: A K Srivastava and Yogranjan
Published by: **ASTRAL INTERNATIONAL PVT. LTD., NEW DELHI**   *Pages 27–52*

Chapter 2

Agronomic Technologies for Enhancing Pulse Production Ahead of Climate Change

Nitin Gudadhe[1], A.A. Raut[2], J.D. Thanki[1], P.S. Bodke[3] and S.R. Imade[4]

[1]*Department of Agronomy, Navsari Agricultural University, Navsari, Gujarat*
[2]*ICAR-Agricultural Technology Application Research Institute, Jabalpur, M.P.*
[3]*ZARS, Igatpuri, Mahatma Phule Krishi Vidyapeeth, Rahuri, Maharashtra*
[4]*Anandniketan College of Agriculture, Warora, Dist. Chandrapur, Maharashtra*

ABSTRACT

Pulse production is mainly associated with biotic, abiotic and price risks. Pulses are mostly cultivated in rainfed areas where likelihood of drought occurrence is high. Under such scenario, pulses are obviously at higher risk. A multi-pronged strategy of using indigenous coping mechanisms, wider adoption of the existing technologies (improved, high yielding varieties and appropriate crop management practices) and or concerted research and development efforts along with policy incentives are needed for adaptation and production risk management. The present paper expresses the risks arising in agriculture from climate change and discusses smart agronomic management approaches to deal with production risks to sustain pulse production.

Keywords: Pulses, Pulse production, Climate risks, Agronomic practices.

Introduction

The year 2016 promises to be an important and exciting year for agriculture as the international community celebrates the International Year of Pulses (IYP 2016).

The idea for an international year dedicated to pulses was rooted in the fact that pulses have health benefits and can help in providing healthy nutrition, maintenance of soil fertility, resilience of global warming, sustainable development and poverty eradication (Anonymous, 2016).

Per decade increase in surface temperature of India is 0.3 °C. (Goswami *et al.*, 2006). The production potential of pulses is likely to be affected by temperature and rainfall variations. Rise in winter and summer temperatures by 3.2° and 2.2°C, respectively will be the most worrying event by year 2050. This abnormal temperature increase will reduce total crop cycle duration, which will have adverse impact on pulse production. (Ali and Gupta, 2012).

Agronomic measures need to be applied by anticipating the effect of climate change on pulses production. Use of improved varieties, changes in land management, changing planting date, planting method, sowing depth, seed priming, application of straw mulching, manipulating plant population, improved intercropping practices, advanced weed control, alterations in fertilizer application methods and use of water harvesting *etc.* are the examples of agronomic management options. These are some of the technologies, which can manage to a considerable extent, the adverse effects of global warming and climate change on grain legumes.

In order to sustain pulses production, productivity and farm income there is necessity than an option to intelligently adapt resilient technologies against the changing climate for protecting livelihoods of small and marginal farmers. (Prasad *et al.*, 2014).

Climate determine crop growth and yield, it also influence the incidence of weeds, pests and diseases. Hence growth and yield of crop will be affected if substantial climate change is happening. Parry *et al.* (2004) and IPCC (2013) predicted decrease of crop yield in developing countries in Southern Asia and Africa with hunger risk and increase in yield in developed countries. Climate change is expected to increase yields in Northern hemisphere countries and decease in nearby tropical countries. Rise in sea level of south Asian countries on cultivated land (IPCC, 1996). However, Parry *et al.* (2004) and IPCC (2007) discussed few exemptions like increase of yield due to increased monsoon intentiy or increase in moisture due to northward movement of monsoon. They also raised questions whether the climate change adoption strategies work properly or not in future. A major uncertainty is over the effect of increased concentration of CO_2 and its effect on crop growth (Parry *et al.*, 2004). It is predicted that doubling of CO_2 concentration in atmosphere will enhance photosynthesis and nitrogen fixation and which ultimately will lead to increased legume yield (Poorter and Nagel, 2000). Hungate *et al.*, 2003 predicted that increased CO_2 benefit will be available only when ample nitrogen will be available for to support growth. Zanetti *et al.*, 1996 observed those non legumes which can fix atmospheric nitrogen are stimulated by increased CO_2 concentration under controlled conditions. Hence, cool grain legume are likely to be less affected under adverse climate over non-legume crops (Andrews and Hodge, 2010).

Pulses Production Constraints

a. Constraints of Production

Biotic and abiotic stresses are major constraints for pulses production. Pod borers, wilt, rots, blight and botrytis are major biotic constraints of chickpea and pod borer, pod fly, wilt and mosaic for pigeonpea. For lentil the constraints are pod borer, aphids, cutworm, powdery mildew, rust and wilt. Insect pests and diseases are attracted to pulses due to the richness of legumes in N and P (Sinclair and Vadez, 2012). In arid and semi arid regions 50 per cent crop yield may reduce due to drought and heat stress. Soil salinity and alkalinity is another major problem in semi-arid tropics and in Indo-Gangetic plains. Temperature extremities globally are likely to reduce grain yield drastically. In the states of UP, Bihar, West Bengal, Chhattisgarh, MP and Jharkhand heavy pigeonpea yield losses are seen due to heavy rains and poor drainage, which reduce the plant population and mortality due to *phytophthora* blight.

b. Socio-Economical Constraints

India imports large amount of pulses from other countries even though India is the largest producer of pulses in the world. Pulses are treated as a secondary crop by Indian farmers. Cereals and cash crops are priority crops for farmers for investment followed by pulses. Hence, pluses are grown on poor soils with little inputs. Another fragile areas regarding pulses production are lack of post harvest innovations and poor policy support. Major constraints in increasing pulses production are lack of quality seed of high yielding varieties and other important inputs. (David *et al.*, 2002, Gowda *et al.*, 2013).

Present Status of Pulses

Pulses area has increased from 19 to 25 million ha respectively from 1950-51 to 2013-14 and production has increased from 8.41 to 19.27 million tons from same period. Pulses cultivation reduced during 1960s-70s due to adoption of high yielding varieties (HYV) of cereals during and after Green Revolution. Cultivation of pulses is an age old practice in the integrated farming system with domestic seeds and labours. Green Revolution promoted external inputs and seeds of modern varieties were used in fertile lands and hence pulses were pushed to the marginal land and rainfed areas. This has resulted in poor pulses productivity and degradation of land. (Mohanty and Satyasai, 2015).

Demand Projections of Legumes in Future

During 1980s pulses production growth rate was just 1.52 per cent and in 1990s it was 0.59 per cent, during 2001-08 it has further increased to 1.42 per cent and currently it is 0.6 per cent. During 1980s and 1990s pulses area growth rate was negative. It will require having 2.05 per cent annual pulse productivity to reach 26.5 Mt by 2050 by assuming constant area under pulses (Ali and Gupta, 2012). By 2050, domestic pulses requirement will be 26.50 Mt and this high asking rate has to be accomplished under extreme weather conditions. Reducing cost of production and improving per unit productivity are two proactive strategies are inevitable to be followed for pulses production.

Measures for Achieving Self Sufficiency in Pulses Production

Adoption of Suitable Agronomic Practices

To cope up with future climatic threat necessary agronomic practices have to be adopted. In Indian perspective for better food and nutritional security, great emphasis is needed for refinement of low–cost pulse production technologies for resource constrained Indian farmers and also from food safety and quality point of view. (Pooniya *et al.*, 2015).

However, agricultural practices have to be selected and match as per prevailing conditions of climate, soil and pulses to be cultivated. Under mechanized agriculture conditions the concept of minimum tillage is rapidly adoptable. However, deep ploughing/chiseling has to be used regularly to reduce salt accumulation. Uniform water application can leach salts in better way. Addition of organic matter keeps soil porous, improves water infiltration, increases microbial activity and enhances nutrient supply. Soil texture, type and irrigation method and crop to be grown determine soil nutrient carriers and nutrient application method. Fertilizer selection and application methods must be considered as per soil type and water salinity. Under global warming scenario, innovative sowing methods for legumes will have to be adopted. Drilling method of sowing saves water because irrigation is supplied to crop lines only. Similarly mulching can curtail evaporation and conserve soil moisture. These techniques will be beneficial in future for enhancing pulse production.

i. Tillage Management

Tillage is necessary for manipulation of soil with farm tools and implements for good seed germination conditions, seedling establishment and good crop growth (Das *et al.*, 2014). Under dryland conditions; deep ploughing results better soil moisture conservation, root proliferation and higher productivity over shallow cultivation (Vadi *et al.*, 2006). The heavy soils require one pre-shower deep ploughing followed by 2 to 3 cultivations and harrowing after early shower. All the growth parameters were significantly improved when the seed bed prepared with only one ploughing due to better tilth (Tomar and Singh, 1991). Conventional tillage is best for *tarai* region of India for higher productivity of lentil because in conventional tillage more aeration and proliferation of roots takes place, which extracts more soil moisture and nutrients from per unit area of soil (Pooniya *et al.*, 2015).

Two cropping sequences namely durum wheat-lentil-watermelon and wheat-chickpea-watermelon, lentil and chickpea do not requires deep tillage (Pala *et al.*, 2000) and promising zeal was found in shallow cultivation. Hence minimum and zero tillage practices left more water at harvesting for the succeeding crop. Faulty and impatiently done tillage practice may bring moist soil on surface which increases compaction and reduction in infiltration. If proper weed control methods are not followed in zero tillage practice, it will cause reduction in crop yields (Singh *et al.*, 2010).

ii. Planting Time and Sowing Depth

Planting time should be chalked out in such a way that pulse crop should not suffer for moisture deficiency during growth period. Plant growth, fruiting and biological nitrogen fixation are affected if planting time is delayed and which ultimately leads to forced maturity. Contrary to that sometimes off-season cultivation of green peas, beans, cowpea and other legume vegetables leads to enhanced profitability due to premium prices in the market though the yield levels are quite lower owing to less congenial climate (Rahi *et al.*, 2013). *Rabi* greengram can be sown up to end of December and this is practiced in southern part of country, where the winters are not severe. Sowing of summer greengram in first fortnight of March recorded higher yield as compared to last week of March (Patel 2003). September first week is the suitable time for horsegram sowing during *rabi* season (Kalita *et al.*, 2003). Optimum sowing depth depends on type of crop/cultivar, growing season, soil moisture, soil texture and more importantly seed size of the respective pulse crops (Dass *et al.*, 1997; Pooniya *et al.*, 2015).

Table 2.1: Influence of Dates of Sowing and Irrigation Levels on Growth, Yield Components and Yield of Kabuli Chickpea

Treatment	Plant Height (cm)	Branches/ Plant	Canopy Width (cm)	Pods/Plant	100 Seed Weight (g)	Seed Yield (kg ha⁻¹)
Irrigation levels						
Control	30.62	3.36	34.83	46.63	25.27	1454
0.4 IW/CPE ratio	35.10	3.70	38.53	49.32	25.76	1561
0.6 IW/CPE ratio	35.53	3.80	34.21	49.12	25.71	1565
0.8 IW/CPE ratio	35.57	3.73	34.80	48.65	25.67	1618
S. Em	0.13	0.03	0.50	0.59	0.11	33
C.D [5 per cent]	0.413	0.09	1.559	1.64	0.345	102
Dates of sowing						
First fortnight of October	34.66	3.70	33.66	44.50	25.57	1517
Second fortnight of October	35.34	3.74	34.43	55.69	26.65	1802
First fortnight of November	34.32	3.67	33.71	46.69	25.49	1727
Second fortnight of November	30.49	3.48	40.58	46.83	24.69	1153
S.Em	0.12	0.02	0.58	0.90	0.09	41
C.D [5 per cent]	0.35	0.06	1.60	2.50	0.25	115

Source: Mansur *et al.* (2010).

Benefit of earliest rains can be grabbed by pulse sowing, if it is followed by light showers or dry weather which leads to poor germination and weed problems.

Shifting of planting dates from spring to winter season may enhance the productivity of cool season legumes. Sowing of lentil during late April/early May use more water and increases its yield than late June/early July sowing in Australia. (Siddique *et al.,* 1998). Early maturing *kabuli* chickpea is recommended for delayed sowing for higher yield by several workers. (Trivedi and Vyas, 2000). Mansur *et al.,* 2010 observed that chickpea yield can be increased by 56 per cent if chickpea sowing is adjusted to second fortnight of October instead of second fortnight of November (Table 2.2).

Table 2.2: Legume Crops for Limited Water Supply

Scientific Name	Common Name	Degree of Tolerance
Leucaena leucacephala	Leucaena	2.0
Phaseolus vulgaris	Common Bean	1
Vigna unguiculata	Cowpea	1.5
Cajanus cajan	Pigeon Pea	2.0
Dolichos lablab	Lablab Bean	2.5
Vigna radiata	Mung Bean	2.0
Phaseolus acutifolius	Tepary Bean	2.5
Vigna aconitifolius	Mat Bean	2.5
Tylosema esculentum	Marama Bean	3.0

Source: Singh *et al.* (2014).

iii. Sowing Direction

Different directions of sowing and crop phenotypes grown in the field create various planting geometry in combination which differ in yield. The competitive ability of pulse crops can be increased by orientating crop rows at a right angle to the sun light direction, *i.e.* sow crops in an east-west direction. East-west crops more effectively shade weeds in the inter-row space than north-south crops. The shaded weeds have reduced biomass production and reduced seed set. In particularly weedy fields, the reduced weed growth leads to increased crop yield. An experiment was laid out with two row orientations *viz.,* North-South (NS) and East–West (EW) (Table 2.3). Phenology and growth parameters were observed regularly during the crop growth cycle. Line quantum sensor was used for recording photosynthetically active radiation (PAR) in all the treatments. At harvest of crop yield contributing characters and yield were recorded at crop harvest. N-S sown crop exhibited more seed yield per plant (14.8 g) plant height than E-W sown crop (13.8 g), however the heat use efficiency of the crop under N–S direction was 0.60 kg ha^{-1} day^{-1} °C which was at par with E-W directional sown crop (Singh *et al.,* 2012).

iv. Method of Sowing

Among agronomic practices, proper method of planting may considerably increase the production of pulses. In order to have maximum productivity, ideal planting geometry is important for efficient utilization of available crop growth resources. Pulses are mainly sown by broadcasting followed by planking or opening of furrow and drilling through tractor or bullock drawn seed drills.

**Table 2.3: Growth Attribute of Mungbean at Varying Dates of
Sowing, Row Direction and Spacing**

Treatment	Plant Height (cm)	Branches/ Plant	Pod/ Plant	Seeds/ Plant	100 Seed wt (g)	Yield (kg/ha)	Heat Units (day °C)	HUE (kg/ha/ day°C)
D1 : Last week of July	61.80	6.00	45.60	13.50	3.50	977.7	1493	0.67
D2 : First week of August	52.80	4.80	42.50	12.50	2.70	838.9	1526	0.55
CD (P=0.5)	3.51	1.52	2.40	0.52	0.50	-	-	-
NS: North south	60.00	5.80	44.00	14.80	3.40	912.5	1510	0.60
EW: East –West	55.00	4.50	40.10	13.80	3.20	904.2	1510	0.59
CD (P=0.5)	4.52	0.80	2.50	0.25	0.08	-	-	-
P1: 30x15cm	60.00	4.90	44.30	14.80	3.42	694.5	1510	0.46
P2:45x15cm	54.50	3.80	43.50	14.30	2.25	922.2	1510	0.62
CD (P=0.5)	3.80	0.50	0.80	NS	0.40	-	-	-

Source: Singh *et al.* (2012)

In *kharif* pulses, raised/ridge-furrow planting technique has been found very successful in draining excess water from crop root zone and increase the yield by 25-30 per cent over flat bed planting (Pramanik and Singh 2008, Das *et al.*, 2014). Ali, 1998 observed that in Ludhiana (Punjab), flat sowing recorded significantly higher pigeonpea yield over other treatments, but at Hisar and Pantnagar in North Indian conditions, raised bed with 2.7 m width recorded significantly higher yield over other sowing methods. It might be due to proper drainage of excess water from crop root zone and less incidence of insect-pest and diseases.

Application of appropriate sowing method also determines the success and productivity of crops in particular environmental, temporal and field variability regimes (Choudhary and Suri 2014). Planting by *doni* (sowing of seed in pot and transplanting them at proper time) + scooping (2-4 inches spade out dug ditches, which retained water for some periods than flat land built in between two rows) gave highest pigeonpea yield due to appropriate media addition to seeding and planted as such after hardening (Pradhan *et al.*, 2014).

v. Depth of Sowing

Soil type, crop type, emergence time and herbicide used decides the depth of sowing needed for pulse crops. In wet winter season herbicide leaching problem is observed in light texture soils. Deep sowing should be preferred in sandy soils and shallow sowing in heavy soils. Generally beans are most tolerant to deep sowing, while lentils, peas, chickpeas are intermediate and lupins are the least tolerant.

Poor seed germination may be resulted in moisture scarcity conditions when seed sowing is done in shallow depth. To harness more soil moisture recommended sowing depths for pigeon pea and chickpea is 5-8 cm and 4-5 cm for lentil in clay soil for better crop establishment than at superficial depths. If chickpea is sown at 15 cm depth the incidence of *Fusarium oxysporum* f.sp. *ciceris* is low (Dahiya *et al.*, 1988). With deeper sowing depth crop emergence may be delayed slightly but it will not affect to the plant stand. Chickpea sown at 10 cm depth, registered significant production over 7.5 cm depth (Malaviya *et al.*, 2010).

vi. Planting Geometry

Optimum spacing requirement depends on type of crop and cultivar, growing season and planting system. Most of short duration pulse varieties need narrow spacing, while long duration varieties perform well under wider spacing. An appropriate planting density in field crops and vegetables lead to better harness of the solar radiation to translate into higher crop yields (Choudhary *et al.*, 2014). In general, *kharif* sown crop requires wider spacing and less plant population compared to summer sowing due to fairly warm temperature, prolonged vegetative growth and profuse branching (Prasad 2012; Pooniya *et al.*, 2015). The sparsely sown (45 x 15 cm) mungbean crop was reported more crop yield (922 kg ha^{-1}) than narrowly sown (30 x 15 cm) crop (694 kg ha^{-1}) (Table 2.3), higher heat use efficiency (0.62 kg ha^{-1} day^{-1} °C) was observed in the higher proportions in the wider spacing (45cm × 15cm) than in closer spacing (0.46 kg ha^{-1} day^{-1} °C) of 30 cm × 10cm (Singh *et al.*, 2012).

vii. Optimum Seed Rate and Plant Population

The seed rate also varies according to weather conditions and duration of crop growth. Primarily, plant population desired per unit of land area determines the seed/seedling rate (Prasad, 2012). Spreading types of genotypes requires lesser seed rate as compared to tall and erect genotypes. Spacing between rows, spacing of plants within row, plant size and seed germination, *etc.* affect the rate of planting required to reach a particular plant population (Poehlman, 1991; Pooniya *et al.*, 2015).

viii. Priming of Seed

Germination of seed is not possible until imbibition of water takes place. Poor seed germination may result under the conditions of moisture stress. Seed germination can be improved by priming of seeds in water for given period before sowing. Seed priming can be done either by water or by adding germination inducing chemicals. Under moisture stress conditions seeds can be soaked/primed overnight, dried and can be sown in normal way in various crops like chickpea (Harris *et al.*, 1999). Seed priming enhances rapid and uniform emergence, more vigorous plants growth, moisture stress tolerance, earlier blooming and increases crop yield. Minimum period of seed soaking varies from crop to crop. However, chickpea seed should not be soaked for more than 8 hours because long duration soaking may induce seed sprouting, which can cause seed damage during sowing (Harris *et al.*, 1999). Under drought/moisture prone areas this technology could help farmers by reducing the risk of failure of crop.

ix. Seed Systems in Villages

Seed of the crop contains complete genetic information in code form from its parents. Hence genetic purity maintenance of special genotypes is necessary for having repeated good quality crop performance. Original pulses traits can be kept intact generation after generation with minimum resources and efforts by self pollination. If released improved pulses varieties not available in villages, it will be difficult to adopt such varieties and their expansion in rainfed areas.

x. Straw Mulching

Mostly crop residues are considered as a waste, these residues are burned and which loose valuable nutrients, soil flora and fauna and pollute environment. Crop residue can be used as mulch, which can conserve soil moisture. Crop residue burning enhance CO_2 emission besides depriving soil from residue incorporation. In order of cost cutting of labors in uprooting, chopping and mixing of crop residue in soil, the residues are easy to burn for farmers. Rotavator machine was introduced to discourage farmers from burning of crop residues in the NICRA villages to promote residue chopping and mixing in soil. For early seedbed preparation use of rotavator helps for early seedbed preparation soon after harvesting of *kharif* crops for sowing of *rabi* crops. Standing stubble as a mulch are left back in zero tillage practice but it may require any advance seed drill for sowing.

xi. Nutrient Management

Key of successful pulses production is balanced nutrient supply and its availability. Fertilizer management encompasses on adding right amount of nutrients at right time through an appropriate method so as to minimize nutrient losses, thereby making efficient nutrient-use for enhancing crop productivity and maintaining soil fertility (Dass *et al.*, 2014). Pulses require less amount of nitrogen as they are capable of fixing atmospheric N biologically through *Rhizobium* bacteria but need adequate phosphorus and sulphur for their root proliferation and synthesis of sulphur containing amino acids (Choudhary 2009). Pulses also require comparatively higher amount of micronutrients like molybdenum and iron, which are integral constituents of nitrogenase enzyme, essentially required for nitrogen fixation (Choudhary *et al.*, 2014). Application of sulphur significantly increases the grain (9.1 per cent) and straw yield (9.6 per cent) and protein content with application of 20 kg sulphur ha^{-1}. Combined application of *Rhizobium* + phosphorus solubilizing bacteria alone or its combination with N and P given higher economic returns due to low cost biofertiliers (Kushwaha 2007). Use of micronutrient like $ZnSO_4$ @ 25 kg ha^{-1} has increased pigeonpea yield positively (Puste and Jana 1988). Phosphorus solubilisation, transformation and its economy with enhanced green peas productivity is possible with application of VAM fungi. (Kumar *et al.*, 2014, Yadav *et al.*, 2015) (Pooniya *et al.*, 2015).

Farmers does not get enough time for basal fertilizer application in utera method of sowing of pulses after rice crops harvest. Therefore application of KCl @ 1 per cent and DAP @ 2 per cent foliar spray is recommended at 45 and 30 DAS. Palta *et al.* (2005) concluded that in chickpea the problem of terminal drought can be managed by application of foliar spray of urea @ 30 kg N ha^{-1}, which also enhances grain and protein yield. In anticipation of drought in pulses where it is a regular character should be sprayed with urea as a foliar spray. Ali *et al.* (2002) focused on different genotypes have different feature for absorption of different nutrients. *e.g.* drought tolerant chickpea cultivar having longer roots can absorb more phosphorus than other cultivars. Pulses green manuring usually perform multiple functions that include soil improvement w.r.t. physico-chemical and biological properties as well as enhancement of soil microbial biomass and enzymatic activity.

xiii. Water Management Practices

Due to more intensification in agriculture and industries in future, the demand for groundwater use will rise. Hence, not only water saving planning is needed but also innovative irrigation practices need to be adopted to minimize the use of saline/sodic water. Irrigation methods like drip, sprinkler and subsurface will have to be accepted instead of traditional flooding irrigation. To avoid under/over irrigation a warning system needed to be developed. Alternate use of saline and good quality water is a good option to avoid the problems of salinity to crop and soil. But while application of saline water the knowledge of salinity sensitive critical growth stages of pulses is necessary to avoid the damage to the crops.

Nearly 90 per cent of chickpea is cultivated under *rainfed* conditions grown under residual soil moisture during *rabi*. Use of furrow irrigation with raised-bed systems improve the irrigation water-use efficiency under permanent raised-bed seeding where, tillage is done on top of the beds. Study of anti-transpirants can be a new area of research to be explored in other legumes, so that pulse crops to enhance their production under rainfed areas (Pooniya *et al.*, 2015). Water harvesting techniques like macro and micro catchments can be used. The harvested rain water can be used, when crop expose to drought at critical growth stages. In most of the legume crops pod initiation is the most important stage as far as moisture stress is concerned. Hence, under water scarcity conditions supply of life saving irrigation will enhance crop yield under critical stages pod initiation/formation (Singh *et al.*, 2010).

Different irrigation levels at various stages of crop showed significant response on growth and production parameters (Mansur *et al.*, 2010) (Table 2.2). IW/CPE ratio irrigation @ 0.8 given significantly higher crop yield over other ratios. In conditions like dry spell, salinity, temperature tolerant crops should be selected. As per tolerance ability to abiotic stress following crops are shown in Table 2.5 (Singh *et al.*, 2014). Based on degree of tolerance to water scarcity the categorization of crops can be done. Among all types of crops grasses and beans are more hardy and tolerant in nature. Rice followed by late drilled chickpea needs irrigation as compared to normal drilled crop because of poor root growth in late sown chickpea.

xiii. Improved Weed Management

It is known fact that in a CO_2 elevated environment weeds convey higher carbon to rhizome/roots and towards shoots. Hence, traditional weed control techniques can't manage such stronger rooting weeds which defoliate the plants only and no strong impact on roots. Earlier weeds were not able to migrate to temperate areas but with increased temperature conditions now they can migrate to new area.

Crop yield reduces because weeds competes with plant for space, light, nutrients and moisture. Weed quantum, weed flora present, crop weed competition

Table 2.4: Promising Pulse Varieties in India and their Salient Features

Pulse Crop	Features	Pulse Crop	Features
Redgram (Cajanus cajan)		*Horsegram (Macrotyloma uniflorum)*	
UPAS 120	Extra early, suitable for double cropping, escapes drought	PHG 9	Powdery mildew and anthracnose resistant
Narendra Arhar 1	Resistant to sterility mosaic and tolerant to wilt, pod borer	VL Gahat 10	Yellow mosaic resistant
PPH 4	Short duration	*Mothbean (Phaseolus aconitifolia)*	
Durga	Short duration and determinate plant type	Jawala	Suitable for western part of India
Greengram (Vigna radiata)		FMM 96	Resistant to yellow mosaic
Varsha	Synchronous and early maturity	*Chickpea (Cicer arietinum)*	
PM 5	Bold seeded, extra early and resistant to yellow mosaic virus	Pusa 256	Resistant to wilt, tolerant to *Ascochyta* blight, suitable for late planting
RMG 492	Suitable for spring season and yellow mosaic resistant	Anubhav	Suitable for rainfed condition
TARM 1	Suitable for *rabi* season	Karnal chana 1	Suitable for saline situation
Co 6 (COGG 902)	Suitable for all seasons and yellow mosaic resistant	GPF 2	Plant grow erect with thick stems resistant to lodging
Blackgram (Phaseolus mungo)		*Field pea (Pisum sativum)*	
PDU 1	Suitable for spring season	Pant P 13 and 14	Resistant to powdery mildew, tolerant to rust, semi dwarf
NDU 99-3	Suitable for *kharif* and resistant to yellow mosaic virus	Pusa Prabhat	Extra early maturity
VBN 5	Suitable for all seasons and yellow mosaic and powdery mildew resistant	*Lentil (Lens esculenta)*	
Cowpea (Vigna unguiculata)		Pant L 5	Resistant to wilt, rust and blight, bold seeded
Swarna	High protein cultivar	HUL 57	Small seed, tolerant to rust and wilt
Gujrat Cowpea 2	Drought tolerant	*Frenchbean (Phaseolus vulgaris)*	
Pusa Sukomal	Suitable for *Kharif* and summer, resistant to golden yellow mosaic virus and leaf spot disease (*Pseudocercospora cruenta*).	IPR 98-5 (Utkarsh)	Cold tolerance, attractive seed colour
UPC 628	Forage cowpea, resistance to yellow mosaic virus, anthracnose, collar/root rot, aphid, flea beetle and root-knot nematode	IIPR 96-4	Resistant to bean common mosaic virus

Source: Prasad (2012); Bana *et al.* (2014) and Pooniya *et al.* (2015).

and stage decide the reduction of crop yield. Among various pulse crops, critical time of crop and weed competitions varies. Hand weeding, herbicides and cultural practices are the common methods of weed control. For first 20-30 days duration the plants should be weed free. First and second weeding should be done after 20-25 and 34-50 days after sowing, respectively. Integrated weed management is a practice of effective control of weeds through chemical, biological, cultural and mechanical methods that can be used economically by the farmers. Choudhary (2013) obtained higher grain yield with the pendimethalin application @ 1 kg a. i. ha^{-1} as a pre-emergence application in various *kharif* and *rabi* season pulses. Application of fluchloralin @ 1.0 litre a. i. as a pre plant application before sowing or pendimethalin @ 750 ml to 1.0 liter *a.i.* as pre-emergence application within 72 hrs after sowing provide good control. Quizalophop @ 40-50 ml a. i. for control of monocot 20 to 25 days after sowing can be used (Anonymous, 2012).

Table 2.5: Yield Advantage (per cent) from Improved Technologies of Pulse Crops under Frontline Demonstrations (2006-09)

Technology	Chickpea	Pigeon-pea	Lentil	Mung-bean	Urdbean	Fieldpea
Improved varieties	22.4	24.7	23.6	23.3	21.9	20.0
Sulphur application	15.4	17.4	20.3	19.0	19.9	24.1
Rhizobium inoculation	13.4	13.5	21.0	11.1	14.2	13.2
Weed management	40.0	30.0	24.7	29.6	18.3	26.4
IPM	19.9	28.1	13.1	20.4	17.6	20.2
Full package	24.9	34.6	41.9	33.9	27.8	40.1
Number of demonstrations	3480	3773	1454	1640	1098	686

Source: Ali and Gupta (2012).

Highest seed yield of chickpea was recorded at row spacing of 30 cm, while among weed control measures, hand weeding gave significantly higher seed yield over other treatments and it remained at par with chickpea + paddy straw mulching (Pooniya *et al.*, 2009). Gajera *et al.* (1998) reported that mulching with sugarcane trash @ 8 tonnes ha^{-1} is effective for control of weeds and equally important in increasing yield, conservation of soil moisture and moderation of soil temperature in pigeonpea.

xv. Plant Protection

Pulses are susceptible to many insect-pests and diseases. The losses in yield due to lack of plant protection measures vary from 46-96 per cent depending on the crop and varieties. Use of resistant varieties, crop rotation *etc.* are the components of integrated pest management practice. Intercropping of gram + linseed/mustard or gram + coriander encourages natural enemies of pod borers. Use of bioinsecticide NPV @ 250-500 LE ha^{-1} controls pod borers. Use of neem seed kernel extract (5 per cent) is also helpful for control of pod borers. Use of sex pheromone trap is also helpful in controlling pod borers. Use of all management practices for control of diseases is known integrated disease management (IDM) which can curtail disease pressures below economic injury threshold. It does not recommend only chemical application, but supports the integration of chemical, biological, physical and cultural control strategies to prevent disease outbreak. IDM includes deep summer ploughing and field sanitation, growing resistant varieties, seed treatment with fungicides, crop rotations with sorghum and tobacco, soil solarization and soil treatment with formaldehyde, captan and vapam *etc.* Various fungicides and bio-agents are tried as seed treatment to control pulse diseases. Application of carbendazim + thiram and bio-agent (*Trichoderma viride*) in combination with vitavax are best for reducing wilt incidence in pulses.

Selection of Suitable Varieties/Cultivars

Nowadays various improved varieties are available for higher yield, disease and pest resistance, short duration, synchronous maturity and short stature, *etc.* suitable to varied agro-climatic and soil conditions. Development of high yielding, disease resistant, short duration varieties has made a platform for replacing less productive pulses/cereals. This has given opportunity to utilize rice fallows for *rabi* pulses. Breeding work is also underway to produce short duration varieties and hybrids for almost all pulse growing areas throughout country. A list of some promising pulse cultivars for different production zones in India as well as suitable pulses and their cultivars specifically are furnished in Table 2.6 (Pooniya *et al.*, 2015). A pulse revolutionary opportunity has came by developing a new variety of hybrid pigeon pea by joint venture of Indian Council of Agriculture Research and International Crops Research Institute for the Semi-arid Tropics (ICRISAT) (Saxena and Nadarajan, 2010).

Table 2.6: Means, Range and F Value for Agronomic Traits of IPM 205-7 and IPM 409-4 alongwith Eight Check Varieties at Kanpur (India) during Summer, 2011

Genotype	Days to 50 per cent Flowering	Days to 75 per cent Maturity	Seed Yield (kg ha^{-1})
IPM 205-7	33.33	47.00	960.00
IPM 409-4	32.00	45.00	876.67
PDM 139	36.00	56.67	1066.67
IPM 99-125	36.00	61.33	1010.67
IPM 2-14	38.67	58.67	830.00
IPM 02-3	37.33	59.33	1000.00
Pant M 2	36.67	61.00	996.67
Sona Yellow	36.33	59.00	1166.67
Pusa 9531	38.00	57.00	1026.67
Pusa Vishal	36.33	61.00	1216.67
Range	32.00-38.67	45.00-61.33	876.67-1216.67
Mean SEm	36.07±0.74	55.76±0.94	987.33±52.45
CD at 1 per cent level	3.01	3.78	211.01
CV (per cent)	3.55	2.92	9.2
F Value at 1 per cent	ns	**	**

Source: Pratap *et al.* (2013).

If newly developed high yielding varieties are giving good production then such varieties will be accepted by farmers for future. 'Seeds for Needs' campaign is looking for successful genotypes that will adopt to vulnerable and changing climate. It was initiated in India by Biodiversity in collaboration with several important institutes like NBPGR, IARI, small and local organization and Humana People to People India (HPPI).

Bridging Yield Gap by Use of Existing Technology

Wheat, rice and vegetables are the crops of prime importance for farmers and apply fertilizers, irrigations and insecticides at very little rates to pulses after satisfying the needs of above crops. Hence, there is wide difference of yield between the farmers field, frontline demonstration plots and experimental results. New technologies developed through on farm demonstration during last five years on large scale have shown its superiority over local practices. Productivity of pulses can be increased by 13 to 42 per cent by adoption of these technologies in the country (Table 2.5). Productivity of pulses can be increased by 20 to 25 per cent by use of improved varieties, however improved technology package like high yielding varieties, integrated management of nutrients, pest and diseases has shown 25 to 42 per cent production advantage over farmers practices. After successful experiments and broad level frontline demonstration the use of ridge furrow and broad bed furrow planting, seed treatment with biofertilizer, sulphur application @ 20 kg ha^{-1}, urea spray @ 2 per cent, pendimethalin 1 to 1.25 kg ha^{-1} and various IPM modules are advocated for higher pulse productivity. Farmers' participation is necessary in on farm developmental research for complex rainfed areas. Technology refinement and adoption constraints elimination are possible by farmers participatory research. KVKs, SAUs and ICARs involvement is desperately needed for demonstration on farmers field on large scale which can enhance average productivity from 637 to 737 kg ha^{-1} and production of 17.69 Mt by 2020 (Ali and Gupta, 2012). Paudela *et al.* (2014) observed that selection of the legume in a millet + legume intercrop was found to be important for the feasibility of conservation agriculture system.

Pulses Cultivation in Rice Fallows

Four countries namely India, Nepal, Bangladesh and Pakistan comprising whole Indo Gangetic plains (IGP) of South Asia and is most important area from agriculture point of view. IGP covers an area of 14.3 million ha under *kharif* rice and remain fallow during *rabi* season which provide huge scope for expansion of *rabi* pulses area *e.g.* chickpea, grasspea and lentil. Just after harvest of rice the cultivation of short duration varieties of pulses can increase production level 1-2.5 q ha^{-1} is observed on large scale on farm trials by Agricultural Universities of Madhya Pradesh, Odisha, West Bengal, Chhattisgarh and Jharkhand. In these states *Utera* or *Paira* double cropping system is adopted in which pulses or oilseed are sown in standing rice crop before its maturity stage. Cultivation of low yielding and late maturing rice varieties followed by lathyrus as a paira or utera cropping without fertilizer is a common tradition in above states (Anonymous, 2012). *Kabuli* varieties of chickpea like ICCV 2, KAK 2 and JGK 1 receive good market prices and hence they are preferred by farmers and short duration *desi* varieties are also preferred up to certain extent. In rice fallow areas under late sown conditions, chickpea variety JG14 a heat tolerant variety is found highly adoptable (Gowda *et al.*, 2013).

New Cropping Systems, Intercropping and Crop Diversification

In order to increase food productivity of India the cereal-cereal based cropping system became the most dominating cropping system of India. The prerequisite character of any genotype is early maturing; so as to fit well in multiple cropping system and crop rotation. Early pigeonpea was introduced in crop development programmes to address such issues. Srivastava *et al.* (2012) developed elite x elite crosses in pigeonpea for improvement of traits like test weight, grain yield and early maturity. Thermo/photo sensitivity issue is associated with pigeonpea and that can be resolved by recently developed super-early lines IIPR, Kanpur developed two extra short duration green gram genotypes, which mature in 45 to 48 days during summer and *kharif* season (Pratap *et al.*, 2013) which have shown resistance against mungbean yellow mosaic India virus. New genotypes IPM 409-4 and IPM 205-7 took 47 days for maturity over check PDM 139, which took 60 days for maturity; hence these two genotypes had advantage of 13 days for maturity (Table 2.6). Promising economic returns from pulses have potential for replacement of upland paddy. Non conventional cropping systems require involving a thrust for enhancing pulses production and productivity (Table 2.7) (Pooniya *et al.*, 2015). Gudadhe *et al.* (2015) convinced that cotton-chickpea is a good sequence for higher returns to farmers.

Table 2.7: New Options in Pulse Based Intercropping and Cropping Systems in India

Intervention	New Options in Pulse Based Intercropping and Cropping Systems in India
Short duration pulse cultivation	☆ Introduction of short duration pigeonpea varieties into irrigated cropping system in northern and central India in sequence with wheat;
	☆ Introduction of *rabi* pigeonpea in Bihar, West Bengal, Odisha, eastern Uttar Pradesh, Gujarat, Andhra Pradesh.
	☆ Introduction of *rabi* kidneybean in Uttar Pradesh, Bihar, Odisha, Madhya Pradesh, Maharashtra and West Bengal.
Pulses as summer crops	☆ Introduction of summer pulses (blackgram, greengram, cowpea) in irrigated areas after the harvest of *rabi* crops.
	☆ Introduction of greengram and blackgram in fields vacated by potato, mustard, sugarcane and wheat in Andhra Pradesh, Bihar, Haryana, Madhya Pradesh, Odisha, Punjab, Rajasthan and Uttar Pradesh.
Shift to pulses from other crops	☆ Replacement of high water duty crops by low water intensive crops like pulses in command areas in order to make irrigation water available at critical stages of crop growth through effective water scheduling.
	☆ Substitution of upland crops like rice, sorghum, maize, pearl-millet and diverting these areas to short duration pulses in eastern and southern states.
Intercropping	☆ Intercropping of pigeonpea, greengram and blackgram with sorghum, pearlmillet, maize, cotton, groundnut, soybean, *etc.*

Source: Pooniya *et al.* (2015).

Growing of two/more crops in different rows on same area in same season is called as intercropping. This practice is followed to have security if one crop fails due to adverse weather conditions. In Indian subcontinent chickpea + mustard is a beneficial intercropping system (Arya *et al.*, 2007). Marginal and small farmers, who execute different operations of field manually are generally practicing intercropping. Both crops in intercropping will not suffer if some management practices like irrigation, weed control, insect pest control and harvesting are specifically designed. Intercropping wheat + chickpea has land equivalent ratio > 1 and is found advantageous over mono cropping (Gunes *et al.*, 2007). Hence, cultivation of short duration pulses as a inter and relay crop during *kharif* and *rabi* season can expand acreage and production of pulses. In anticipation of the consequences of climate change necessary measures for building resilience has to be taken for future agricultural production. Crop diversification is rational and its economic feasibility is one of the way in this strategy. But production selective monocrops for economic interest and use of biotechnology strategy for the same creates hindrance in adoption of this strategy (Lin, 2011).

In rice wheat cropping system cultivation of oilseeds and pulses is good for diversification which can enhance productivity and soil fertility. As pulses are fixing atmospheric nitrogen, black and green gram are mainly promoted as a intercrop in rice wheat cropping system. In between harvest of wheat and transplanting of rice, farmers are growing green gram and sesbania as a *in-situ* green manuring crop.

Modeling and Agronomic Practices

Weather conditions determine plant growth. Crop manipulate physiological processes in their life with change in weather parameters like wind speed, humidity, radiation, soil temperature and air (Monteith, 1981). Without much time and money such type of models can anticipate plant growth and yield with confidence (McKenzie and Andrews 2010). The effect of climate change on productivity of groundnut in different groundnut producing regions of India can be quantified by CROPGRO-Groundnut model. This model can quantify the feasibility of different agronomic technologies and adaptable option individually/with combinations for increasing groundnut productivity ahead of climate change. From this experiment it is concluded that COPGRO-Groundnut model was useful to identify the use of various agronomic practices regionwise for higher groundnut yield under climate change (Singh *et al.*, 2014).

Mustak (2015) studied suitability of Seonath basin through Land Suitability Modeling (LSM). The study reveals that the degree of suitability increases indicated the degree of suitable factors increases and degree of limitations decreases and hence the spatial distribution of land suitability varied in the area which sharing 13.83 per cent High Suitability (S1) and 18.22 per cent unsuitable due to the spatial variation of the suitable factors and limitations for the growth and production of arhar pulse in the area. Vishwajith *et al.* (2014) observed that both ARIMA (*Autoregressive Integrated Moving Average*) and GARCH (*Generalised Autoregressive Conditional Heteroscedastic*) models can be used for modelling pulses production in India; inclusion of the factors like fertilizer and rainfall increases the accuracy of the model. The simulation effect of agronomic management, weather, soil, pest attack, nitrogen, water, soil carbon and green house gas emission can be studied by model InfoCrop, a generic crop model (Aggarwal *et al.*, 2006). Assessment of climate change can't be studied by all available models. Those model who are studying the effect of CO_2 concentration and temperature on radiation use and photosynthesis are suitable for such assessment.

Abiotic Stress Studies

Inheritance of environmental and genetic interactions helps to tolerate abiotic stresses. Chickpea genotypes which are drought tolerant are known for higher absorption and uptake of macro and micro nutrients like N, P, K, Ca, Zn, Mn and B (Gunes *et al.*, 2006). Under abiotic stress situations there is need to design cultivars, which can absorb and uptake higher quantity of nutrients; so that yield will not be limited at least due to nutrient deficiency. During drought situation the process of nitrogen fixation and nodulation are hampered though legume crop can fix atmospheric nitrogen. In faba bean, application of potassium increases production of dry matter and nitrogen fixation. Inoculation of arbuscular mycorrhiza in chickpea increases plant nutrients levels and enhanced growth than no application of arbuscular mycorrhiza (Anilkumar and Kurup, 2003). Under drought stress scenario use of biofertilizers can improve fixation of nitrogen for smart growth and yield of legume crops. Cycocel at the rate 100 ppm can mitigate the effect of drought through root development. To curtail flower dropping due to drought condition, foliar spray of NAA is recommended during flowering and at 15 days interval thereafter. Quest for germplasm as a gene against stress is the prime step in plant breeding programme for stress tolerance. Early flowering genes support chickpea against yield loss and escape the conditions of terminal drought. 'efl-1' a recessive gene was reported as a gene responsible for early flowering (Kumar and van Rheenen 2000), and a flowering within 24 days in super early genotype ICCV 96029 had supported to this fact at ICRISAT (Kumar and Rao, 1996). A field study during the 2014-15 investigated the genotypic variation for heat tolerance of 50 lentil genotypes. Four genotypes have been identified (72578, 70549, 71457 and 73838) to have improved tolerance to high temperature and absolute yield equivalent to current commercial cultivars (Delahunty *et al.*, 2015). This provides the opportunity for breeding programs to improve the tolerance of lentil to heat stress, leading to better yield stability and profitability for growers.

Throughout world the foremost abiotic factor for retarding agricultural production is frost. During reproductive period field pea, fababean, lentil and chickpea are the crops susceptible to injury by radiant frost. Flowering and pod formation are more susceptible stages for frost. An inverse relation is seen between plant age and frost tolerance. Frost tolerance is more during plants vegetative growth period and further it is found decreasing in the field conditions. At cellular level adaptation to function and structure can tolerate freezing temperature (Margesin *et al.*, 2007; Maqbool *et al.*, 2010).

Agro-advisory and Pulses Production

In Indian conditions, weather has its role from crop sowing to harvesting. Weather forecast technologies changed with advancement of time and its accuracy percentage has been improved in recent years. Planning and management of cropping is immaterial without weather forcast, which is issued by India

Meteorological Department (IMD). The weekly information on crop management practices *e.g.* variety selection, irrigation date, pest and disease control measures, sowing and fertilizer application time *etc.* are provided under agroadvisory services of IMD. The information on weather conditions like large temperature variation, heavy rain and strong winds is also provided by IMD as an early warning system.

How to Minimize the Losses

Among various agronomic practices timely sowing is one of the important practice, which minimizes the yield losses. Aphid attack can be avoided by early sowing. High flower drop and abortion of pod may be encountered in weather conditions like high temperature which further reduces seed weight, enhance early maturity and reduces crop yield. Under cloudy weather, polyphagous pest namely gram pod borer reproduce faster. For fast multiplication of pest the maximum, minimum, average temperature and relative humidity requirement is 21.7-28.4, 10.0-16.8, 17.8-21.8°C and 85-88 per cent respectively. Hence losses can be curtailed and productivity of pulses can be enhanced by using weather based agro advisory services (Anonymous, 2012).

Pulse Policy Issue

Pulses productivity can be raised by governments' positive policy, institutional support and complementary attempts of scientists. Pulses procurement is not so effective as of cereals through government frequently announces pulses minimum support prices. It is reported that the import prices of chickpea are comparatively lesser than shipping of chickpea from one state to other. Pulses productivity and profitability can be increased if rice fallow land can be brought under cultivation of pulses and if cheap pulses shipping facility available from government. No proper policy is available to compensate crop failure due to biotic and abiotic constraints. Farmers prefer to cultivate cereals over risky pulse crops except the cases where pulses are profitable over cereals (Pandey *et al.*, 2012).

During February, 2011 a network project NICRA governed by ICAR was launched for long term strategic research to assess the impact of climate change on agriculture in India and to save Indian farmers from climate variability by demonstrating best existing package of practices. Under the programme of research germplasm of cereals and pulses was exposed to various biotic and abiotic stress and selected germplasm lines were used to develop stress resistant high yielding varieties. A district level vulnerability atlas has been prepared in India for giving economic priority to vulnerable localities. The project has demonstrated best management practices such as resource conservation technology, farm mechanization, use of tolerant varieties, water harvesting *etc.* in 100 vulnerable districts of India to immune farmers from climatic variability, encouraging high yields (NAAS, 2013).

Conclusion

Production risks of pulses in view of climate change and vulnerabilities of rainfed agriculture could be marginalized by selection of suitable technological options across the region. All these would go a long way in preparing a nation to deal with climate change impacts on pulse production. With technological supplementation, India needs to produce the required quantity, but also remain competitive to protect indigenous pulses production. Advanced agronomic management will play an important role in adoption of climate smart technologies. May technological options are available, which have shown capability to enhance high pulse production. Use of weather forecast based agro-advisory does not only minimize input cost but also maximizes the input use efficiency. Search of new genotypes and adoption of demonstrated technological options (*e.g.* IPM, IDM, and IWM) would certainly increase the overall pulse production. A regional and location specific adaptation research must be carried out through agricultural universities and other regional research centers. In order to make India, a self sufficient country in pulses production, integrated efforts at national level are necessary *e.g.* new policy initiatives, mitigation options which are cost effective and global cooperation. Whatever initiatives and efforts are being put forward by government are bearing fruits now, and if this momentum is continued further, the country is believed to be self sufficient in pulses production.

REFERENCES

Aggarwal, P. K., Banerjee, B., Daryaei, M. G., Bhatia, A., Bala, A., Rani, S., Chander, S., Pathak, H. and Kalra, N. 2006. InfoCrop: A dynamic simulation model for the assessment of crop yields, losses due to pests and environmental impact of agro-ecosystems in tropical environments. II. Performance of the model. *Agricultural Systems* **89**: 47–67.

Ali, M. 1998. Consolidated report on *kharif* and *rabi* pulses, Agronomy 1997–1998, All India Pulse Improvement Project, DPR, Kanpur.

Ali, M. and Gupta, S. 2012. Carrying capacity of Indian agriculture: Pulse crops. *Current Science* **102** (6): 874–81.

Ali, M., Mishra, J. P. and Chauhan, Y. S. 1998. Effective management of legume for maximizing biological N fixation and other benefits. (*In*) *Residual effect of legume in rice and wheat cropping system in the Indo-Gangetic plains* pp. 127–8.

Ali, M.Y., Krishnamurthy, L., Saxena, N. P., Rupela, O. P., Kumar, J. and Johansen, C. 2002. Scope for genetic manipulation of mineral acquisition in chickpea. *Plant and Soil* **245**: 123–134.

Andrews, M. and Hodge, S. 2010. Climate change, a challenge for cool season grain legume crop production ISBN 978-90-481-3708-4 (Eds. Yadav, S. S., McNeil, D. L., Redden, R., and Patil S. A.) Published by Springer Dordrecht Heidelberg London New York pp. 1-10.

Anilkumar, K. K. and Kurup, G.M. 2003. Effect of vesicular arbuscular mycorrhizal fungi inoculation on growth and nutrient uptake of leguminous plants under stress conditions. *Journal of Mycology and Plant Pathology* **33**: 33–36.

Annoymus. 2013. Report of expert group on pulses. Department of Agriculture and Cooperation Government of India, Ministry of Agriculture. p. 139.

Anonymous, 2012. Model training course on production techniques in rabi pulses, Directorate of Extension Services Indira Gandhi Krishi Vishwavidhyalaya Raipur, 492012, Chhattisgarh. 09-16 January, 2012.

Anonymous 2016. International year of pulses.*WFO E-magazine Formletter* Issue No. 47, January 2016 pp. 2.

Arya, R. L., Varshney, J. G., and Kumar, L. 2007. Effect of integrated nutrient application in chickpea + mustard intercropping system in the semi-arid tropics of North India. *Communications in Soil Science and Plant Analysis* **38**: 229–240.

Bana, R. S., Pooniya, V., Choudhary, A. K. and Rana, K. S. 2014. Agronomic interventions for sustainability of major cropping systems of India. *Technical Bulletin* (ICN: 137/2014), Indian Agricultural Research Institute, New Delhi, p 34.

Birthal, P. S., Khan, Md. T., Negi, D. S. and S. Agarwal 2014. Impact of climate change on yields of major food crops in India: Implications for food security. *Agricultural Economics Research Review* **27** (2): 145-155.

Choudhary, A. K., Yadav, D. S. and Singh, A. 2009. Technological and extension yield gaps in oilseeds in Mandi district of Himachal Pradesh. *Indian Journal of Soil Conservation* **37** (3): 224–229.

Choudhary, A. K. 2013. Technological and extension yield gaps in pulses in Mandi district of Himachal Pradesh. *Indian Journal of Soil Conservation* **41** (1): 88–97.

Choudhary, A. K. and Suri, V. K. 2014. Scaling up of pulses production under frontline demonstrations technology programme in Himachal Himalayas, India. *Communication in Soil Science and Plant Analysis* **45** (14): 934–48.

Choudhary, A. K., Pooniya, V., Bana, R. S., Kumar, A. and Singh, U. 2014. Mitigating pulse productivity constraints through phosphorus fertilization – A review. *Agricultural Reviews* **35**(4): 314–9.

Choudhary, A. K., Pooniya, V., Rana, D. S., Bana, R. S. and Rana, K. S. 2014. Planting geometry and integrated nutrient management schedules for late–season cauliflower and cabbage for higher productivity and profitability in irrigated Indo–Gangetic plains region. (*In*) *Proceedings of National Symposium on Crop Diversification for Sustainable Livelihood and Environmental Security*, held during 18-20 November 2014 at PAU, Ludhiana, pp. 74–6.

Dahiya, S. S., Sharma, S. and Faroda, A. S. 1988. Effect of date and depth of sowing on the incidence of chickpea wilt in Haryana, India. *International Chickpea Newsletter* **18**: 24–25.

Das, T. K., Choudhary, A. K., Sepat, S., Vyas, A, K., Das, A., Bana, R. S. and Pooniya, V. 2014. Conservation agriculture: A sustainable alternative to enhance agricultural productivity and resources use- efficiency. Technical Extension Folder, IARI, New Delhi.

Dass, A., Kharwara, P. C. and Rana, S. S. 1997. Response of gram varieties to sowing dates and phosphorus level under on-farm conditions. *Himachal Journal of Agricultural Research* **23** (1 and 2): 112–5.

Dass, A., Suri, V. K., Choudhary, A. K. 2014. Site-specific nutrient management approaches for enhanced nutrient-use efficiency in agricultural crops. *Research and Reviews: Journal of Crop Science and Technology* 3 (3): 1–6.

David, S., Mukandala, L. and Mafuru, J. 2002. Seed availability, an ignored factor in crop varietal adoption studies: a case study of beans in Tanzania. *Journal of Sustainable Agriculture* 21: 5-20.

Delahunty, A., Nuttall, J., Nicolas, M., Brand, J. 2015. Genotypic heat tolerance in lentil *"Building Productive, Diverse and Sustainable Landscapes"* Proceedings of the 17th ASA Conference, 20 – 24, September 2015, Hobart, Australia.

ESI. 2015. The Economic Survey 2014–15. The Economic Survey of India, New Delhi.

Gajera, M.S., Ahlawat, R. P. S. and Ardeshna, R. B. 1998. Effect of irrigation schedule, tillage depth and mulch on growth and yield of winter pigeonpea. *Indian Journal of Agronomy* 43 (4): 689–93.

Goswami, B. N., Venugopal V., Sengupta D., Madhusoodanan, M. S. and Xavier Prince K. 2006. Increasing trend of extreme rain events over India in a warming environment, *Science*, 314: 1442-1445.

Gowda, Laxmipathi, C. L., Srinivasan, S., Gaur, P. M. and Saxena, K. B. 2013. Enhancing the productivity and production of pulses in India. In: Climate change and sustainable food security (ISBN: 978-81-87663-76-8) (Eds. Shetty, P.K., Ayyappan, S. and Swaminathan, M. S.) published by National Institute of Advanced Studies, Bangalore and Indian Council of Agricultural Research, New Delhi. pp. 145-159.

Gudadhe, Nitin, M. B. Dhonde and N. A. Hirwe 2015. Effect of integrated nutrient management on soil properties under cotton-chickpea cropping sequence in vertisols of Deccan plateau of India. *Indian Journal of Agricultural Research* 49 (3), 207-214.

Gunes, A., Cicek, N., Inal, A., Alpaslan, M., Eraslan, F., Guneri, E. and Guzelordu, T. 2006. Genotypic response of chickpea (*Cicer arietinum* L.) cultivars to drought stress implemented at pre- and post-anthesis stages and its relations with nutrient uptake and efficiency. *Plant Soil Environment* 52: 368–376.

Gunes, A., Inal, A., Adak, M. S., Alpaslan, M., Bagci, E. G., Erol, T. and Pilbeam, D. J. 2007. Mineral nutrition of wheat, chickpea and lentil as affected by mixed cropping and soil moisture. *Nutrient Cycling in Agroecosystem*78: 83–96.

Harris, D., Joshi, A., Khan, P. A., Gothkar, P. and Sodhi, P. S. 1999. On-farm seed priming in Hungate, B. A., Dukes, J. S., Shaw, M. R., Luo, Y. Q. and Field, C. B. 2003. Nitrogen and climate change. *Science* 302: 1512–1513.

IPCC (InterGovermental Panel on Climate Change) (2007). Summary for policymakers. In: Solomon S., Qin D., Manning M., Chen Z., Marquis M., Averyt K.B., Tignor M., and Miller H.L. (eds.), Climate change 2007: The Physical Science basis. Contribution of Working Group 1 to the Fourth Assessment Report of the Intergovernmental Panel on Climate Change. Cambridge University Press, Cambridge and New York.

IPCC (Intergovernmental Panel on Climate Change) (1996). Climate change 1995: Impacts adaptationsand mitigation of climate change: Scientific-technical analyses contribution of workinggroup II to the second assessment report of its intergovernmental panel on climate change. In: Watson, R.T., Zinyowera M. C., Moss R. H. (eds.), Cambridge University Press, Cambridge/NewYork.

IPCC (Intergovernmental Panel on Climate Change) 2013. Working Group 1, Fifth Assessment Report on *Climate Change 2013: The Physical Science Basis*. Geneva, Switzerland.

Kalita, U., Suhrawardy, J. and Das, J. R. 2003. Response of horsegram (*Dolichos biflorus*) to different seed rates and dates of sowing under rainfed upland situations. *Crop Research* **26** (3): 443–445.

Kumar, A., Suri, V. K. and Choudhary, A. K. 2014. Influence of inorganic phosphorus, VAM fungi and irrigation regimes on crop productivity and phosphorus transformations in okra (*Abelmoschus esculentus* L.)–pea (*Pisum sativum* L.) cropping system in an acid Alfisol. *Communications in Soil Science and Plant Analysis* **45** (7): 953–67.

Kumar, J. and B. V. Rao, 1996. Super early chickpea developed at ICRISAT Asia Center. *International Chickpea Pigeonpea Newsletter* **3**: 17-18.

Kumar, J., and Rheenen van, H. A., 2000. A major gene for time of flowering in chickpea. *Journal of Heredity* **91**: 67-68.

Kushwaha, H. S. 2007. Response of chickpea to nitrogen and phosphorus fertilization under rainfed condition. *Journal of Food Legumes* **20** (2): 179–81.

Lin, B. 2011. Resilience in agriculture through crop diversification: adaptive management for environmental change. *BioScience* 61(3): 183-193.

Malviya, Deepak, Verma, H.D., Nawange D. D. and Verma, Hemlata 2010. Effect of seed priming, depth of sowing and seed. *Agricultural Science Digest* treatment on productivity of chickpea (*Cicer arietinum* L.) under rainfed condition **30** (2): 145 – 147.

Mansur, C. P., Palled, Y. B., Halikatti, S. I., Chetti, M. B. and Salimath, P. M. 2010. Effect of dates of sowing and irrigation levels on growth, yield parameters, yield and economics of kabuli chickpea. *Karnataka Journal Agricultural Sciences* **23**(3): 461-463.

Maqbool, A., Shafiq, S. and Lake, L. 2010. Radiant frost tolerance in pulse crops—a review. *Euphytica* **172**: 1–12.

Margesin, R., Neuner, G., Storey, K. B. 2007. Cold-loving microbes, plants, and animals—fundamental and applied aspects. *Naturwissenschaften* **94**: 77–99.

McKenzie, B. A. and Andrews, M. 2010. Modelling climate change effects on legume crops: lenmod, a case study ISBN 978-90-481-3708-4 (Eds. Yadav, S. S., McNeil, D. L., Redden, R., and Patil S. A.) Published by Springer Dordrecht Heidelberg London/New York pp. 11-22.

Mohanty, S., Satyasai K. J. 2015. Feeling the pulse *NABARD Rural Pulse* Issue X-July-August, 2015 pp. 1-4.

Monteith, J. L. 1981. Coupling of plants to the atmosphere. In: Grace, J., Ford, E. D., Jarvis, P.G. (eds.), Plants and their atmospheric environment. Blackwell, London.

Mustak, S. L., Baghmar, N. K. and Singh, S. K. 2015. Land suitability modeling for gram crop using remote sensing and GIS: A case study of Seonath basin, India *Bulletin of Environmental and Scientific Research* 4 (3): 6-17.

NAAS, 2013. Climate resilient agriculture in India. Policy Paper No. 65, National Academy of Agricultural Sciences, New Delhi: 20 p.

Pala, M., Harris, H. C., Ryan, J., Makboul, R. and Dozom, S. 2000. Tillage systems and stubble management in a Mediterranean-type environment in relation to crop yield and soil moisture. *Expermental Agriculture* 36: 223–242.

Palta, J. A., Nandwal, A. S., Kumari, S. and Turner, N. C. 2005. Foliar nitrogen applications increase the seed yield and protein content in chickpea (*Cicer arietinum* L.) subject to terminal drought. *Australian Journal of Agricultural Research* 56: 105–112.

Pande, S and Sharma, M and Ghosh, R 2012. *Role of pulses in sustaining agricultural productivity in the rainfed rice-fallow lands of india in changing climatic scenario.* In: Climate Change and Food Security in India (Proceedings of the National Symposium on Food Security in Context of Changing Climate), 30 October –01 November 2010, CSAUAT, Kanpur, India.

Parry, M. L., Rosenzweig, C., Iglesias, A., Livermore, M. and Fischer, G. 2004. Effects of climate change on global food production under SRES emissions and socio-economic scenarios. *Global Environmental Change* 14: 53–67.

Patel, J. J., Mevada, K. D. and Chotaliya, R. L. 2003. Response of summer mungbean to date of sowing and level of fertilizers. *Indian Journal of Pulses Research* 16 (2): 122–124.

Paudela, B., Theodore, J. K., Radovichb, Catherine Chan-Halbrendta, Crowa S., Tamangc, B. B., Halbrendta, J. and Thapac, K. 2014. Effect of conservation agriculture on maize-based farming system in the mid-hills of Nepal. *Procedia Engineering* 78: 327-336.

Poehlman, J. M. 1991. *The Mungbean*, p. 375. Oxford and IBH Publishing Co. Pvt. Ltd., New Delhi.

Pooniya, V., Choudhary, A. K., Dass, A., Bana, R. S., Rana, K. S., Rana, D. S., Tyagi, V. K. and Puniya, M. M. 2015. Improved crop management practices for sustainable pulse production: An Indian perspective. *Indian Journal of Agricultural Sciences* 85 (6): 747–58.

Pooniya, V., Rai, B. and Jat, R. K. 2009.Yield and yield attributes of chickpea as influenced by various row spacings and weed control. *Indian Journal of Weed Science* 41 (3 and 4): 222–223.

Poorter, H. and Nagel, O. 2000. The role of biomass allocation in the growth response of plants to different levels of light, CO_2, nutrients and water: A quantitative review. *Australian Journal of Plant Physiology* **27**: 595–607.

Pradhan, A., Thakur, A., Sao, A. and Patel, D. P. 2014. A new planting technique of Arhar (*Cajnus cajan*) for higher production under rain-fed condition. *International Journal of Current Microbiology and Applied Sciences* **3**(1): 666-669.

Pramanik, S. C. and Singh, N. B. 2000. Boost pulse production through new planting techniques. *Indian Farming* **58** (1): 4–6.

Prasad, R. 2012. *Textbook of Field Crops Production-Food Grain Crops*, Vol I, pp. 248–319.

Prasad, Y. G., Maheswari, M., Dixit, S., Srinivasarao, Ch., Sikka, A. K., Venkateswarlu, B., Sudhakar, N., Prabhu Kumar, S., Singh, A. K., Gogoi, A. K., Singh, A. K., Singh, Y. V. and Mishra, A. 2014. Smart practices and technologies for climate resilient agriculture. Central Research Institute for Dryland Agriculture (ICAR), Hyderabad. 76 p.

Pratap, A., Gupta, D. S., Singh, B. B. and Kumar S. 2013. Development of super early genotypes in greengram [*Vigna radiata* (L.) Wilczek]. *Legume Research* **36** (2): 105-110.

Puste, A. M. and Jana, P. K. 1988. Effect of phosphorous and zinc on pigeonpea varieties grown during winter. *Indian Journal of Agronomy* **33** (4): 399-404.

Rahi, S, Thakur, S. K. and Choudhary, A. K. 2013. Off-season pea cultivation: An income enhancement venture in Mandi district of Himachal Pradesh. (*In*) *Proceedings of National Seminar on Indian Agriculture: Present Situation, Challenges, Remedies and Road Map*, held at CSK HPKV, Palampur during 4-5 Aug. 2012, CSK HPKV Publication, pp. 47–8.

Ram, H., Singh, G., Aggarwal N., and J., Kaur 2011. Soybean (*Glycine max*) growth, productivity and water use under different sowing methods and seeding rates in Punjab. *Indian Journal of Agronomy* **56** (4): 377-380.

Saxena, K. B. and Nadarajan, N. 2010. Prospects of pigeonpea hybrids in Indian agriculture. *Electronic Journal of Plant Breeding* **1**(4): 1107-1117.

Siddique, K. H. M. and Loss, S. P. 1999. Studies on sowing depth for chickpea, faba bean and lentil in a mediterranean type environment of south western Australia. *Journal Agronomy and Crop Science* **182**: 105-112.

Sinclair, T. R. and Vadez V. 2012. The future of grain legumes in cropping systems. Crop and Pasture Science. http://dx.doi.org/10.1071/CP12128.

Singh, A. K., Sangle, U. R. and Bhatt, B. P. 2012. Mitigation of imminent climate change and enhancement of agricultural system productivity through efficient carbon sequestration and improved productiontechnologies. *Indian Farming*, *61*(10), 5-9.

Singh, A. K., Singh, K. M. and Bhatt, B. P. 2014. Efficient water management: way forward to climate smart grain legumes production. MPRA Paper No. 59316 pp. 1-16. Online at https://mpra.ub.uni-muenchen.de/59316/.

Singh, A. K., Singh, S. S., Prakash, V., Kumar, S. and Dwivedi, S. K. 2015. Pulses production in India: Present Status, Bottleneck and Way Forward. *Journal of Agrisearch* **2**(2): 75-83.

Singh, G., Ram, H., and Aggarwal, N. 2010. Agronomic approaches to stress management. in: climate change and management of cool season grain legume crops ISBN 978-90-481-3708-4 (Eds. Yadav, S. S., McNeil, D. L., Redden, R., and Patil, S. A.) Published by Springer Dordrecht Heidelberg London New York pp. 141-154.

Singh, P., Singh, N. P., Boote, K. J., Nedumaran, S., Srinivas, K. and Bantilan, M. C. 2014 Management options to increase groundnut productivity under climate change at selected sites in India. *Journal of Agrometeorology* 16 (1): 52-59.

Singh, R.B. 2011. *Towards an Evergreen Revolution — The Road Map.* National Academy of Agricultural Sciences (NAAS), New Delhi.

Singh, S. P., Sandhu, S. K., Dhaliwal, L. K. and Singh, I. 2012. Effect of planting geometry on microclimate, growth and yield of mungbean (*Vigna radiata* L.). *Journal of Agricultural Physics* **12**(1): 70-73.

Singh, U. P., Singh, Y, Singh, R.G., Gupta, R. K. 2011. Opportunities for increasing food legume production through Conservation Agriculture based resource conserving technologies in rice-wheat System Resilient Food Systems for a Changing World: Proceedings of the 5th World Congress on Conservation Agriculture Incorporating 3rd Farming Systems Design Conference 25 – 29th September 2011, Brisbane, Australia.

Srivastava, R. K., Vales, M. I., Sultana, R., Saxena, K. B., Kumar, R. V., Thanki, H. P., Sandhu, J. S. and Chaudhari, K. N. 2012. Development of 'super-early' pigeonpeas with good yield potential from early x early crosses. *An Open Access Journal published by ICRISAT* **10**: 1-6.

Tomar, S. P. and Singh, R. R. 1991. Effect of tillage, seed rates and irrigation on the growth, yield and quality of lentil. *Indian Journal of Agronomy* **36** (2): 143–147.

Trivedi, K. K. and Vyas, M. D., 2000, Response of kabuli chickpea cultivars under different dates of sowing and row spacing. *Crop Research-Hissar* **20**(1): 52-55.

Vadi, H. D., Kachot, N. A., Polara, J. V., Sekh, M. A. and Kikani, V. L. 2006. Effect of tillage and mulching on yield and yield attributing chatacters of pigeonpea. *Advances in Plant Sciences* **19** (2): 497–499.

Vishwajith, K. P, Dhekale, B. S., Sahu, P. K., Mishra, P. and Noman, M. D. 2014. Time series modeling and forecasting of pulses production in India. *Journal of Crop and Weed,* **10**(2): 147-154.

Yadav, A., Suri, V. K., Kumar, A., Choudhary, A. K. and Meena, A L. 2015. Enhancing plant water relations, quality and productivity of pea (*Pisum sativum* L.) through AM fungi, inorganic phosphorus and irrigation regimes in an Himalayan acid Alfisol. *Communications in Soil Science and Plant Analysis* **46** (1): 80–93.

Zanetti, S., Hartwig, U.A., Lüscher, A., Hebeisen, T., Frehner, M.,. Fischer, B.U, Hendrey, G.R., Blum, H. and Nösberger, J. 1996. Stimulation of symbiotic N_2 fixation in *Trifolium repens* L. under elevated atmospheric pCO_2 in a grassland ecosystem. *Plant Physiology* **112**: 575–583.

Section B

Climatic Risks and Yield Maximization

2018, *Climate Risks Management: Sustainable Pulse Production*
Editors: *A K Srivastava and Yogranjan*
Published by: **ASTRAL INTERNATIONAL PVT. LTD., NEW DELHI**   *Pages 53–95*

Chapter 3

Climate Risks and its Management for Maximizing Chickpea Productivity in Central India

A K Srivastava

College of Agriculture, Jawaharlal Nehru Agricultural University,
Tikamgarh – 472 001, M.P.

ABSTRACT

Climate change and variability are affecting the crop productivity worldwide. The different futuristic climate change scenarios of national and state levels are presented here. Climate variability of five agro climatic zone of Madhya Pradesh was analyzed and related with chickpea yield and acreage variability. Impact of different climate change scenario on chickpea productivity were examined and discussed. It was found that there are some increasing and decreasing trends in maximum and minimum temperatures, however these changes were not affecting the chickpea productivity. Though, the rainfall variability has shown a little impact on chickpea acreage. The simulation studies suggest that chickpea productivity could be benefitted under projected climate change scenario. Chickpea yield would be sustainable and could be geared up by adopting new technological options. Climate risk losses, if any, may be transferred through insurance and can be minimized by use of real time forecast based tailored decisions.

Keywords: Climate risks, Climate change, Chickpea yield variability and Yield sustainability.

Agriculture is affected by many external factors, which are beyond the control of human and hence; it is the most risky business. There is much greater inherent risk in agriculture because of weather, pest, disease, price shocks, and policies. Risk can be defined as the possibility of adverse outcomes due to uncertainty and imperfect knowledge in decision making.

Risks in Agriculture

Risk is an inevitable aspect of the farming business. The uncertainties inherent in weather, yields, prices, Government policies, global markets, and other factors that impact farming can cause wide swings in farm income. Risks are generally categorized into five types, *viz;* production risk, price risk, financial risk, institutional risk, and human or personal risk.

Production risk is associated with the losses caused by climate, weather, disease, pests, and other factors. Climate risks result from climate change and mainly of two types *i.e.* direct or physical and indirect risks.

Direct risks are continuous increase of temperature, melting of ice cap, sea level rise, changes of the planet's ecology *etc.* whereas Indirect climatic risks are regulational risks, litigational risks, competition risks and production risks *etc.*

Agriculture production is often associated with production risk. Unlike most other entrepreneurs, agricultural producers cannot predict with certainty the amount of output.

Climate Change

Climate change and extreme weather events could significantly impact food production in the coming decades. The annual, maximum and minimum temperature have significantly increased by 0.2°C per decade has been observed across India in recent decades (Kothawale and Rupa Kumar, 2005). Several areas including frequently drought prone regions of the country have been recognized as being predominantly climate risk prone. However, these changes, in shorter timescale are very crucial in context of its socio-economic impacts and engross millions of small scale farmers, fishers, milk producers and forest-dependent people into its agony.

Many pulses are also resilient to adverse climate such as drought and heat, and grown in the dryland regions of the world. This makes them important food crops that adapt easily to the rising temperatures and increasingly frequent droughts under the changing climate. Climate variability and extreme climate events have been shown to increase in large regions of the world because of climate warming (Coumou and Rahmstorf 2012; Hartmann *et al.*, 2013). An increased frequency of extreme weather events such as heat waves, prolonged droughts and flooding has been observed in recent decades (Alexander *et al.*, 2006; Coumou and Rahmstorf 2012).India has a high rate of climate variability on seasonal, inter annual and decadal time scales due to its monsoon climate.

Climate Variability at Macro and Micro Level

Climate projections were depicted for national and state level. In India, Madhya Pradesh, Maharashtra, Rajasthan, Uttar Pradesh and Andhra Pradesh are major pulse producing states. Madhya Pradesh has been leading pulse state in terms of area and production of total pulses. Therefore the state Madhya Pradesh is taken for climate change and variability analysis. In Madhya Pradesh among 11 agroclimatic zone Bundelkhand Agroclimatic zone is most vulnerable for climate change and variability (SKMCC, 2014). Hence the Bundelkhand Agroclimatic zone of Madhya Pradesh has been focused to see the impact of climate change and variability impact on chickpea yield and area in this agroclimatic zone.

National Scenario

Estimates indicate that there may be several location specific uncertainties in the temperatures increase and have adverse effect the crop productivity in different agroclimatic zone of the country. Lal (2001) reported that annual mean air temperature will be ranged between 3.5 and 5.5°C over the region by 2080 (Table 3.1). Particularly, higher surface air temperature change was predicted in *rabi* season for North India. These projected climate changes will have both beneficial and adverse effects on crop production and socio-economic set up. For example, the growth of C_3 plants improve and favor biomass at the cost of grain biomass; thus leading to decrease in seed production. To combat the impact of climate change, careful analysis of crop-weather data, crop acreage and yield changes, is required. The impact of climate change on agriculture is quite location specific that includes soil type, crop and even the socio-economic conditions of the farmers. In view of this, it is important to study the impact of climate change on regional basis.

Table 3.1: Climate Change Projection for India

Year	Season	Temperature Change (°C)		Rainfall Change (per cent)	
		Lowest	Highest	Lowest	Highest
2020	Annual	1.00	1.41	2.16	5.97
	Rabi	1.08	1.54	-1.95	4.36
	Kharif	0.87	1.17	1.81	5.10
2050	Annual	2.23	2.87	5.36	9.34
	Rabi	2.54	3.18	-9.22	3.82
	Kharif	1.81	2.37	7.18	10.52
2080	Annual	3.53	5.55	7.48	9.90
	Rabi	4.14	6.31	-24.83	-4.50
	Kharif	2.91	4.62	10.10	15.18

Source: Lal (2001).

The NATCOM I reported that by 2050s maximum temperature is expected to rise by 2-4°C over south India and by more than 4°C over north. Minimum temperature is expected to rise by more than 4°C all over India over the same period. Total monsoon rainfall is expected to be relatively unchanged through to the 2050s. A decrease in the number of rainy days is, however, expected.

The importance of understanding the likelihood of climate change on agriculture is often either underestimated or overestimated. Though, there may be many uncertainties in projected climate change scenarios for a region.

State (Madhya Pradesh) Scenario

The average surface daily maximum and minimum temperatures are projected to rise from 1.8 to 2.0°C and between 2.0 and 2.4°C respectively throughout Madhya Pradesh upto 2030s. By 2080s, the maximum and minimum temperatures are projected to rise between 3.4 and 4.4°C throughout the r Madhya Pradesh (Table 3.2).

Table 3.2: Projected Climate Changes Parameters in Madhya Pradesh (PRECIS model for A1B scenarios)

Projected Change in Climate	2021-2050	2071-2100
Daily Maximum Temperatures	1.8-2.0 °C increase	3.4-4.4 °C increase
Daily Minimum Temperatures	2.0-2.4 °C increase	>4.4 °C increase
Monsoon Precipitation	Increase in precipitation by 1.25 times the current observed rainfall (1970) in most parts of Madhya Pradesh.	More than 1.35 times increase in precipitation with respect to observed climate in most parts of Madhya Pradesh. The extreme northern and western part of the state will also experience excess rainfall but less than most of the other areas.
Winter Precipitation	Decrease in precipitation	Substantial increase in precipitation in Central and South western part of Madhya Pradesh increasing from between 1.45 to 1.85 times.

Adapted from: Madhya Pradesh state action plan on climate change,State knowledge Management Centre on Climate change,EPCO, Housing and Environment Department, Govt. of M.P., 2014 pp-21

In general; winter rainfall is likely to be decrease in Madhya Pradesh for the period 2021 to 2050. In the Monsoon period, there is a slight increase in rainfall all over Madhya Pradesh (the increase being 1.25 times the rainfall observed in the current climate).

Climate and Yield Variability

Globally, climate variability accounts for roughly a one third (32–39 per cent) of the observed yield variability (Ray *et al.*, 2015). Ray *et al.*(2015) reported averaged globally yield reduction over areas ranged from 32 to 39 per cent and significant relationships exits between year to year yield (maize, rice, wheat and soybean) variability and explained by climate variation. The yield variability shows distinct spatial patterns in the relative effects of temperature, precipitation and their interaction within and across the regions. The expected increase in climate variability in many regions can increase the need for early warning systems to support agricultural decision makers (Dury *et al.*, 2011).

Rainfall and air temperature have a predominant role in crop growth and yield. Analysis of the observed climate records globally, has revealed increase in global mean surface air temperature of 0.4 to 0.8°C since the late 19th century. During the last decade or so, global annual mean surface temperatures have been among the warmest on the available records and in the line' year 2017 was recorded the warmest year of the recent past. It has been reported that C_3 crop are more responsive to climate change and the climate change would be more pronounced below the 25°N latitude. Considering these the districts having latitude below 25°N are selected for this study.

To find out change in climatic parameters, long period daily weather data of six districts namely; Jabalpur, Rewa, Indore, Chhindwara, Tikamgarh and Hoshangabad along with crop acreage and yield data of Chickpea have been collected and analyzed. The annual, seasonal and decadal variability has also been studied and presented. Chickpea is a *rabi* season crop and grows well in humid, semi-arid and arid conditions. It is best suited to areas having low to moderate rainfall and cool weather. The impacts of climate variability and change on chickpea yield were also analyzed and presented.

Climate Variability in Madhya Pradesh

To see short term (decadal) and long term (30 years) climate variability; an analysis of available weather date set of five agroclimatic zones of Madhya Pradesh was carried out. Among the 11 agroclimatic zones, 5 agroclimatic-zones were selected for the present study. Six districts representing the 5 agro climatic zone were selected and their latitude and normal rainfall and temperatures are given in Table 3.3.

Table 3.3: Agro Climatic Zone and Climate of Six Districts

Sl.No.	Agro-climatic Zone	Latitude	Selected Districts	Normal Rainfall (m.m.)	Temperature (°C)	
					Maximum	Minimum
1	Kymore Plateau and Satpura Hills	24.53°N	Rewa	1014.0	31.7	31.6
		23.16°N	Jabalpur	1377.0	18.7	18.4
2	Central Narmada Valley	22.83°N	Hoshangabad	1201.0	31.9	18.5
3	Bundelkhand	24.75°N	Tikamgarh	958.0	32.5	18.2
4	Satpura Plateau	22.50°N	Chhindwara	950.0	30.0	18.9
5	Malwa Plateau	22.71°N	Indore	978.0	31.7	18.2

The decadal, annual and seasonal variability were analyzed. Any change in weather parameters adversely affects the crop productivity and thereby livelihood of the farmers. The change in crop season's weather parameters may affect the crop much more than the annual and monthly variability. Only *rabi* (October- February) season is considered and mean, standard deviation and coefficient of variation were estimated and are presented here.

Decadal Variability

The daily weather data collected and annual mean were calculated and then the decadal mean, standard deviation and C.V. were estimated and are presented in Tables 3.4–3.9 and presented below:

Jabalpur

The Jabalpur is sub-humid climate and received the highest annual rainfall among the selected districts. The coefficient of variation of annual rainfall is ranged between 18 and 31 per cent at Jabalpur. The monsoon rainfall was observed to be normal in 83 per cent of the occasion (Table 3.4). The highest coefficient of variation was observed for decade 1971-80 to 1981-90 (30 per cent). The amount of rainfall was increased in the recent decade at Jabalpur by 4 per cent. There is no change observed in maximum and minimum temperature at Jabalpur over the decades.

Table 3.4: Decadal Climate Variability at Jabalpur

Decade	Maximum Temperature (°C)			Minimum Temperature (°C)			Rainfall (mm)		
	Mean	SD	CV	Mean	SD	CV	Mean	SD	CV
1951-1960							1380.9	250.3	18.1
1961-1970	31.4	0.6	2.0	18.7	0.6	3.0	1208.2	326.6	27.0
1971-1980	31.2	0.8	2.6	18.9	0.5	2.4	1394.4	426.5	30.6
1981-1990	31.6	0.4	1.3	18.3	0.3	1.9	1301.1	399.3	30.7
1991-2000	31.5	0.4	1.4	18.4	0.3	1.8	1368.4	376.3	27.5
2001-2010	31.9	0.6	2.0	18.3	0.7	3.8	1437.0	261.3	18.2

Rewa

The Rewa is semi-arid climate and received nearly 970 mm annual rainfall (Table 3.5). The perusal of the above data reveals that variability in rainfall was decreased over the decades and variability of maximum temperature was increased in recent decade. There was remarkable decrease (1.2°C from decadal mean) in the value of minimum temperature observed in the district.

Table 3.5: Decadal Climate Variability at Rewa

Decade	Maximum Temperature (°C)			Minimum Temperature (°C)			Rainfall (mm)		
	Mean	SD	CV	Mean	SD	CV	Mean	SD	CV
1961-1970							909.6	288.4	31.7
1971-1980	31.3	0.6	1.9	19.3	0.6	3.2	1066.1	289.6	27.2
1981-1990	31.4	0.6	1.8	19.3	0.5	2.4	929.4	187.4	20.2
1991-2000	31.3	1.1	3.4	19.1	1.3	7.0	1031.4	176.8	17.1
2001-2010	32.6	1.9	5.8	17.7	0.5	2.7	917.5	255.6	27.9

Indore

The larger climate variability was observed at Indore as compared to other districts. There was sharp increase in minimum temperature over the decades and compared to the decadal mean value (18.3°C). The increased is 0.5°C in last two decades (Table 3.6). The total rainfall amount over the decades was decreased compared to its decadal mean value (950mm). Rainfall has decreased by an amount of 79mm/decade during the last decade. Though there was slight increase in maximum temperature value during the past two decades. Increase in maximum and minimum temperature and decrease in rainfall might have impact on chickpea yield.

Table 3.6: Decadal Climate Variability at Indore

Decade	Maximum Temperature (°C)			Minimum Temperature (°C)			Rainfall (mm)		
	Mean	SD	CV	Mean	SD	CV	Mean	SD	CV
1961-1970							956.7	212.8	22.2
1971-1980	31.0	0.6	2.0	17.3	0.6	3.5	1093.2	412.4	37.7
1981-1990	31.8	1.4	4.5	18.3	0.9	5.1	882.9	151.5	17.2
1991-2000	32.5	0.7	2.1	18.8	0.6	3.3	947.6	307.4	32.4
2001-2010	32.0	1.0	3.0	18.8	0.5	2.8	871.3	182.1	20.9

Chhindwara

The Chhindwara is a semi-arid climate and received 935mm of annual rainfall (Table 3.7). There was increase in the value of decadal minimum temperature during last two decades and the increase was noted above 0.6°C/decade at Chhindwara. Rainfall value has decreased during last decade by 80mm from its decadal normal value. It is interesting to note that the standard deviation of every decadal was found to be similar for maximum and minimum temperature. Rainfall standard deviation value was decreased in last decade as compared to its decadal mean.

Table 3.7: Decadal Climate Variability at Chhindwara

Decade	Maximum Temperature (°C)			Minimum Temperature (°C)			Rainfall (mm)		
	Mean	SD	CV	Mean	SD	CV	Mean	SD	CV
1971-1980	31.2	0.7	2.3	17.9	0.7	3.9	972.0	174.6	18.0
1981-1990	29.6	0.8	3.8	18.1	0.8	4.3	969.3	264.5	27.3
1991-2000	28.9	0.8	3.5	20.1	0.8	4.0	943.2	273.6	29.0
2001-2010	30.4	1.1	4.4	19.5	1.1	5.8	855.2	185.1	21.7

Tikamgarh

The Tikamgarh is semi-arid climate and received about 970mm rainfall (Table 3.8). The coefficient of variation of minimum temperature ranged between 2.7 and 17.3 per cent at Tikamgarh, The variability was observed to be very high in last decade. There was decrease in rainfall and minimum temperature over the decades (Table 3.8). The amount of rainfall was decreased in the recent decade by 144mm from decadal normal rainfall. Decrease in minimum temperature by 0.4°C was observed from its mean decadal value. Increase in maximum temperature was observed during last two decades, though the coefficient of variation was found almost constant over the decades.

Table 3.8: Decadal Climate Variability at Tikamgarh

Decade	Maximum Temperature (°C)			Minimum Temperature (°C)			Rainfall (mm)		
	Mean	SD	CV	Mean	SD	CV	Mean	SD	CV
1971-1980	32.2	0.5	1.6	18.3	0.5	2.7	978.4	281.7	28.8
1981-1990	32.4	0.6	1.9	18.4	0.4	2.0	1018.7	328.5	32.2
1991-2000	32.5	0.5	1.5	18.4	0.4	2.0	921.5	171.0	18.6
2001-2010	33.0	0.5	1.4	17.8	3.1	17.3	780.6	301.7	38.7

Hoshangabad

The Hoshangabad is sub-humid climate and received about 1200mm rainfall (Table 3.9). The coefficient of variation of maximum temperature ranged between 1.6 and 31.4 per cent the variability and noted to be very high (31 per cent) in last decade. Mean rainfall has increased over the decades (Table 3.9). No change was observed in maximum and minimum temperature values over the decades.

Table 3.9: Decadal Climate Variability at Hoshangabad

Decade	Maximum Temperature (°C)			Minimum Temperature (°C)			Rainfall (mm)		
	Mean	SD	CV	Mean	SD	CV	Mean	SD	CV
1951-1960							1076.4	316.8	29.4
1961-1970							1275.5	243.4	19.1
1971-1980							1282.6	275.7	21.5
1981-1990	32.5	0.5	1.6	18.9	0.6	2.9	1224.4	227.7	18.6
1991-2000	31.1	1.2	3.8	19.2	0.4	2.1	1023.1	343.5	33.6
2001-2010	29.6	9.3	31.4	18.7	1.4	7.5	1200.7	375.2	31.3

Annual Variability

The annual variability and trend analysis have been made and plotted in Figures 3.1 to 3.18 with trend and its equations.

Jabalpur

No change was observed in maximum temperature (MaxT) at Jabalpur during last 50 years (Figure 3.1). The annual minimum temperature was observed to be slightly more variable in recent decade and has negative trend but found to be non-significant (Figure 3.2). The trend analysis of monsoon rainfall indicates no-significant trend (Figure 3.3) during last 68 years of rainfall data analysis.

Rewa

Decreasing trend in minimum temperature in 43 years of data analysis was found at Rewa (Figure 3.5). The decrease in minimum temperature at Rewa was 0.06°C/year in annual minimum temperature. No trend was found in maximum temperature (Figure 3.4) and rainfall parameters at Rewa (Figure 3.6).

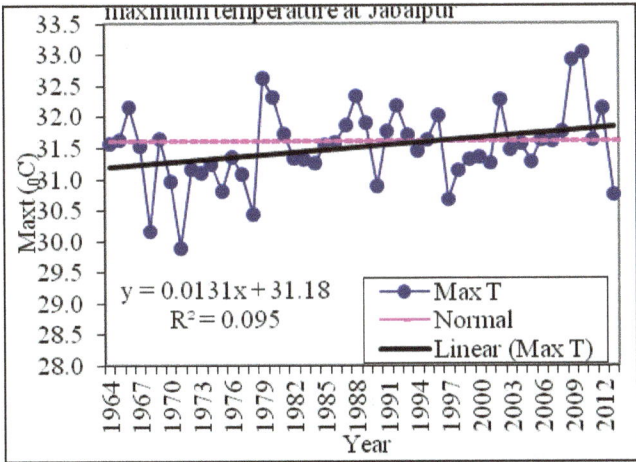

Figure 3.1: Annual Variability and Trend of Maximum Temperature at Jabalpur.

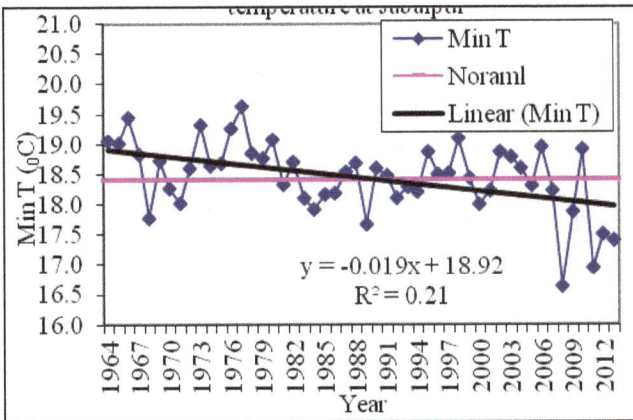

Figure 3.2: Annual Variability and Trend of Minimum Temperature at Jabalpur.

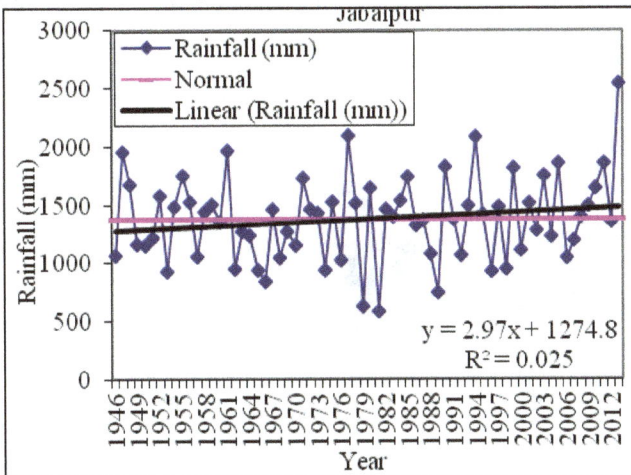

Figure 3.3: Annual Variability and Trend of Rainfall at Jabalpur.

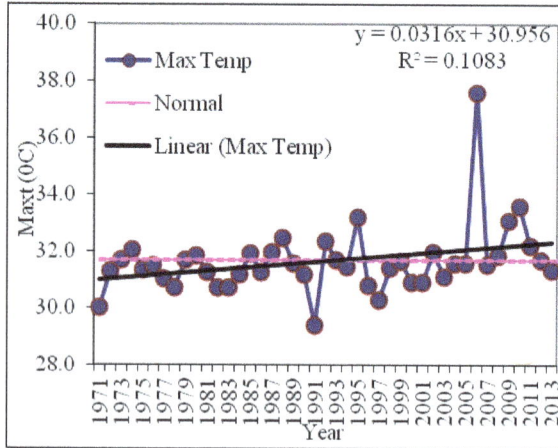

Figure 3.4: Annual Variability and Trend of Maximum Temperature at Rewa.

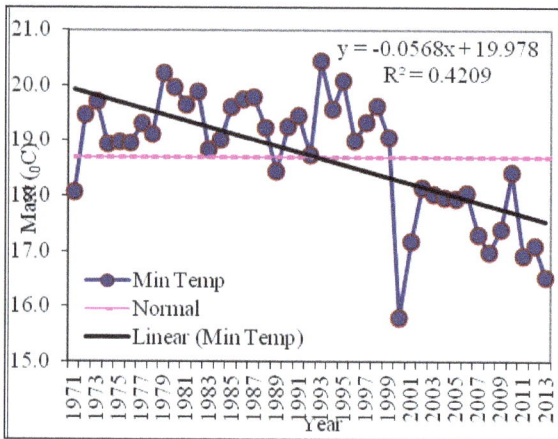

Figure 3.5: Annual Variability and Trend of Minimum Temperature at Rewa.

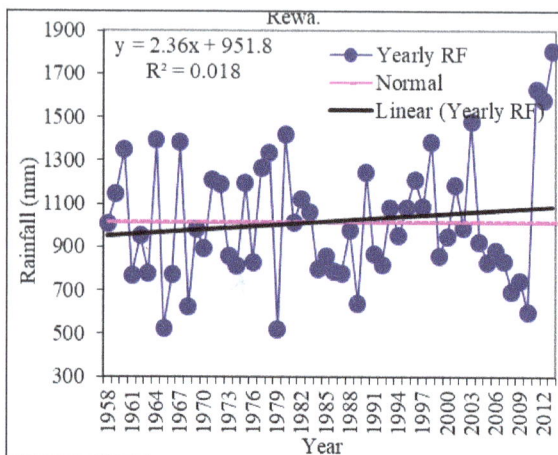

Figure 3.6: Annual Variability and Trend of Rainfall at Rewa.

Indore

Increasing trend in minimum temperature in 43 years of data analysis was found at Indore (Figure 3.8). The increase in minimum temperature was of the order of 0.03°C/year. No change in maximum temperature (Figure 3.7) and rainfall (Figure 3.9) was observed at the station.

Chhindwara

Increasing trend in minimum temperature in 43 years of data analysis was found at Chhindwara (Figure 3.11). The increase in minimum temperature was of the order of 0.04°C/yearand found non-significant. No change in maximum temperature (Figure 3.10) and rainfall (Figure 3.12) was observed at the station.

Tikamgarh

No significant change was observed in maximum, minimum and rainfall at Tikamgarh (Figures 3.13–3.15).

Hoshangabad

No change was observed in annual maximum (day) and minimum (night) temperatures at Hoshangabad during last 50 years (Figures 3.16–3.18). The annual rainfall was observed to be slightly more variable in recent years and has not shown any trend (Figure 3.18).

Seasonal Variability

The *rabi* season's maximum, minimum temperatures and rainfall variability and their trends were shown in figures (Figures 3.192–36).

Jabalpur

It was observed that there is higher variability in maximum temperature and rainfall but both the trends are found to be non-significant.

Rewa

No trend was observed in maximum temperature and rainfall at Rewa (Figures 3.22 and 3.24). But a decreasing trend (-0.04°C/year) in minimum temperature during *rabi* was observed. The above finding contradicts the earlier projection of the climate change as reported by other researchers.

Indore

Increasing trend in maximum and minimum temperature was observed in the order of 0.06 and 0.07 per year respectively at Indore (Figures 3.25 and 3.26). There was no any trend was observed in rainfall (Figure 3.27).

Chhindwara

An increasing trend (0.05°C/year) in minimum temperature during *rabi* season was observed (Figure 3.29) but was found to be non-significant. No trend was observed in maximum temperature and rainfall (Figures 3.28 and 3.30).

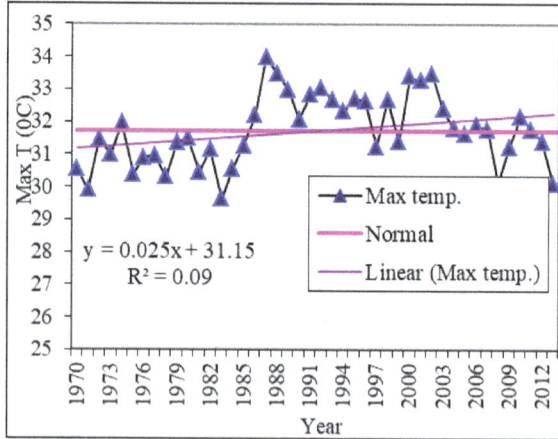

Figure 3.7: Annual Variability and Trend of Maximum Temperature at Indore

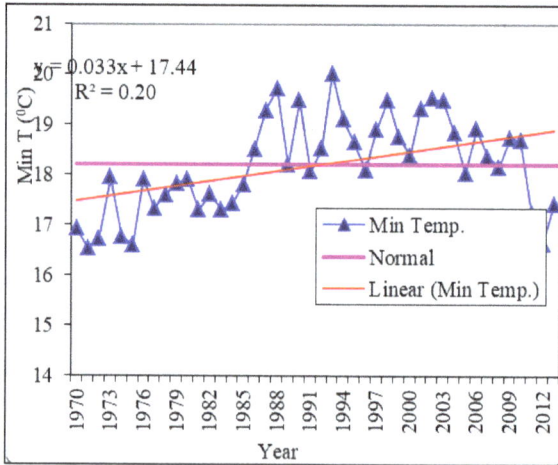

Figure 3.8: Annual Variability and Trend of Minimum Temperature at Indore

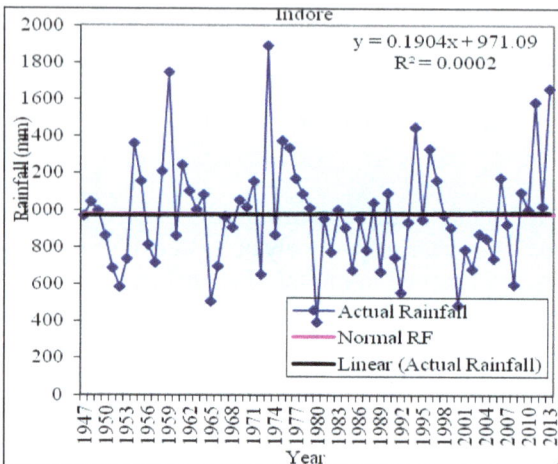

Figure 3.9: Annual Variability and Trend of Rainfall at Indore.

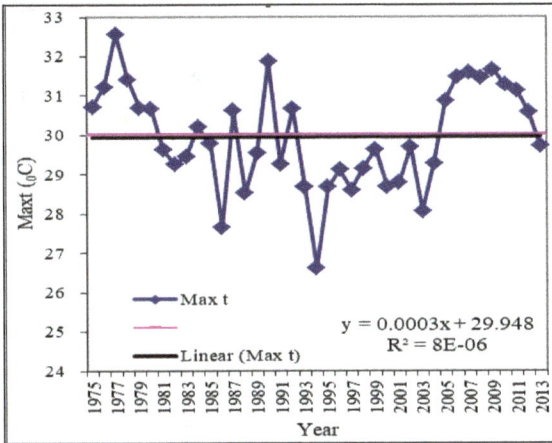

Figure 3.10: Annual Variability and Trend of Maximum Temperature at Chindwara.

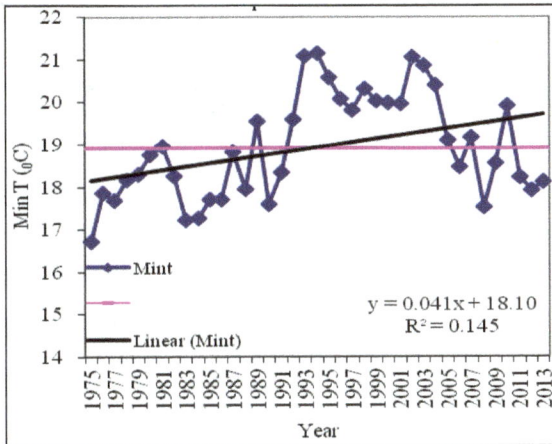

Figure 3.11: Annual Variability and Trend of Minimum Temperature at Chindwara.

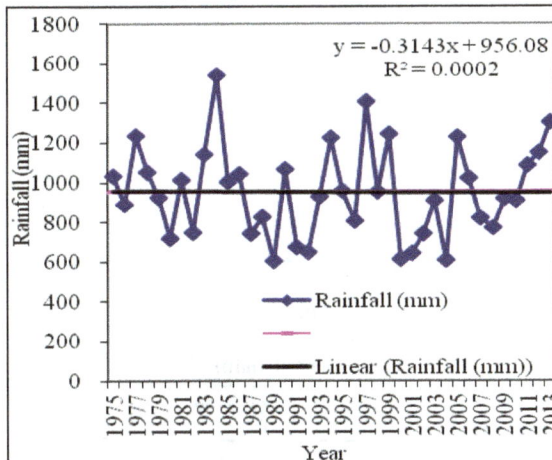

Figure 3.12: Annual Variability and Trend of Rainfall at Chindwara.

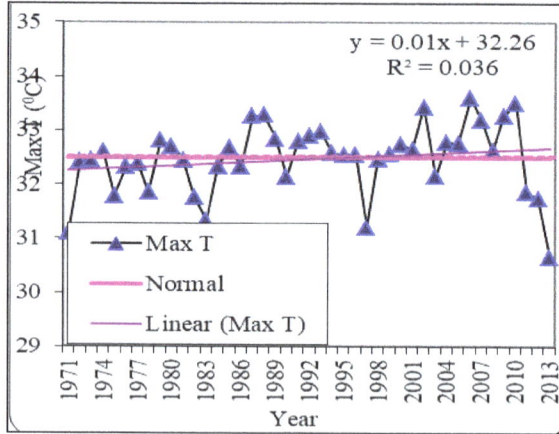

Figure 3.13: Annual Variability and Trend of Maximum Temperature at Tikamgarh.

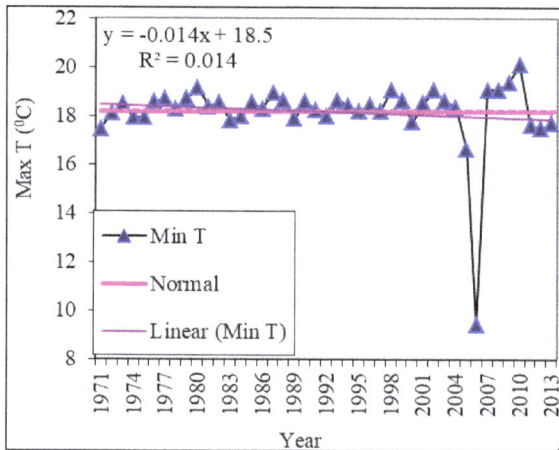

Figure 3.14: Annual Variability and Trend of Minimum Temperature at Tikamgarh.

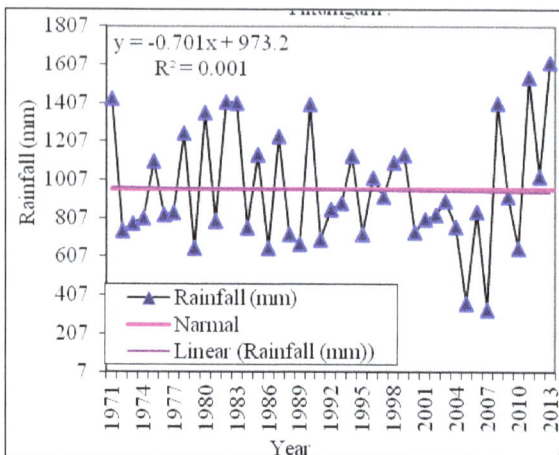

Figure 3.15: Annual Variability and Trend of Rainfall at Tikamgarh.

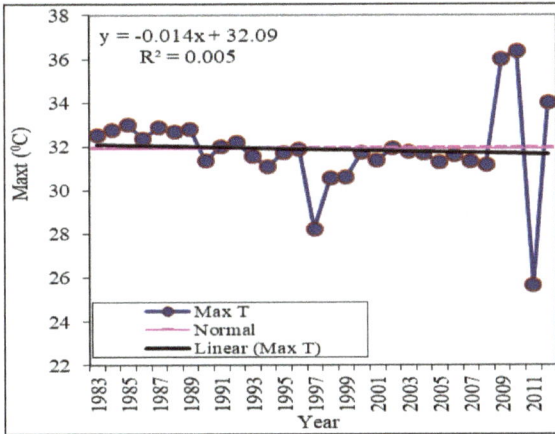

Figure 3.16: Annual Variability and Trend of Maximum Temperature at Hoshangabad.

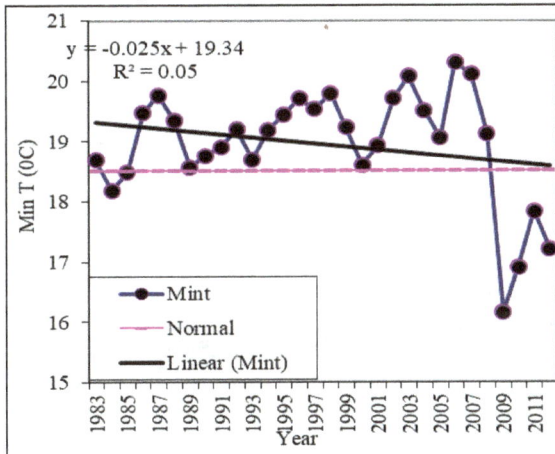

Figure 3.17: Annual Variability and Trend of Minimum Temperature at Hoshangabad.

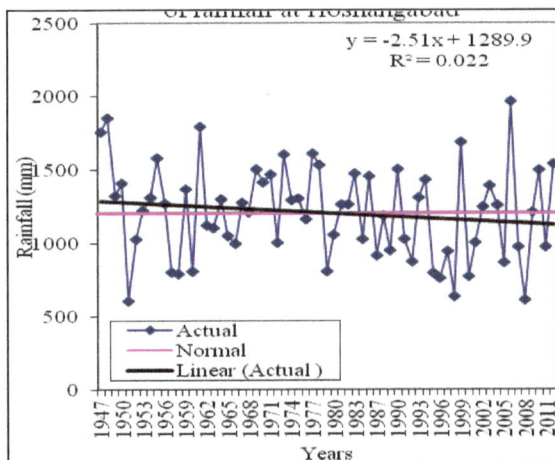

Figure 3.18: Annual Variability and Trend of Rainfall at Hoshangabad.

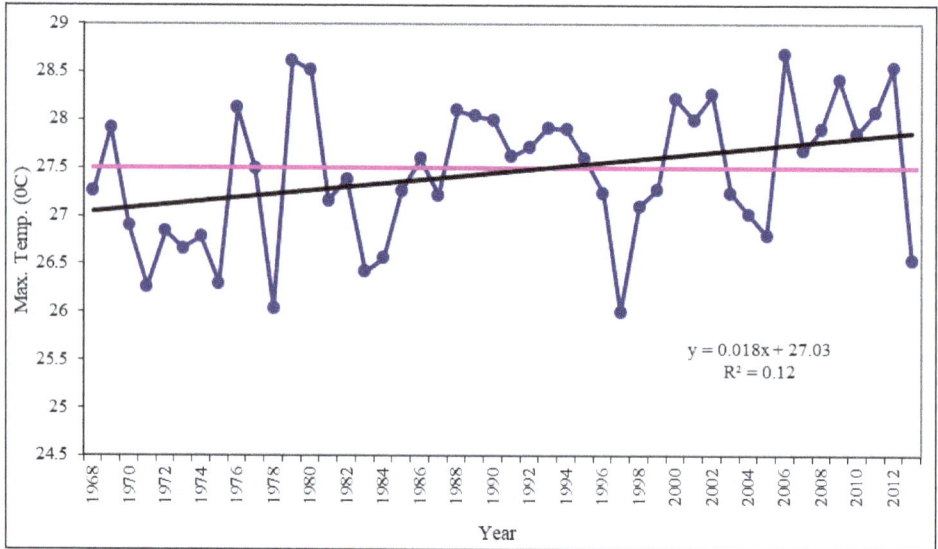

Figure 3.19: Maximum Temperature Trend in Rabi Season at Jabalpur.

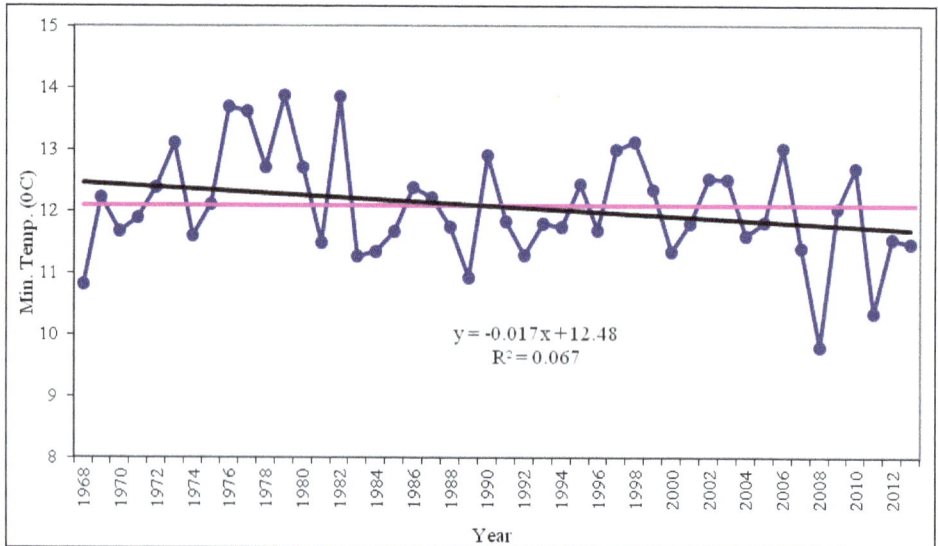

Figure 3.20: Minimum Temperature Trend in Rabi Season at Jabalpur.

Tikamgarh

No trend was observed in maximum, minimum temperature and rainfall (Figures 3.31–3.33).

Hoshangabad

The decreasing trend was observed in minimum temperature during *rabi* (Figure 3.35). This decrease in *rabi* minimum temperature was of the order

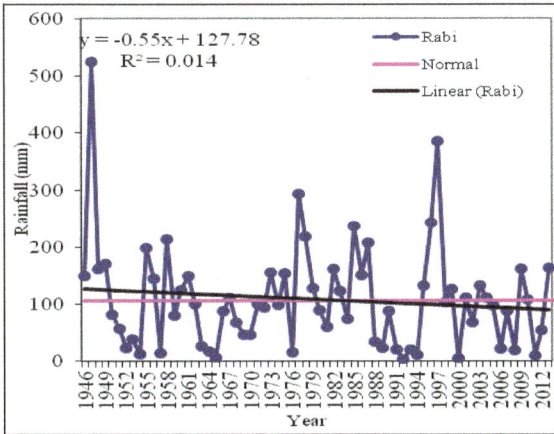

Figure 3.21: Rainfall Trend in Rabi Season at Jabalpur.

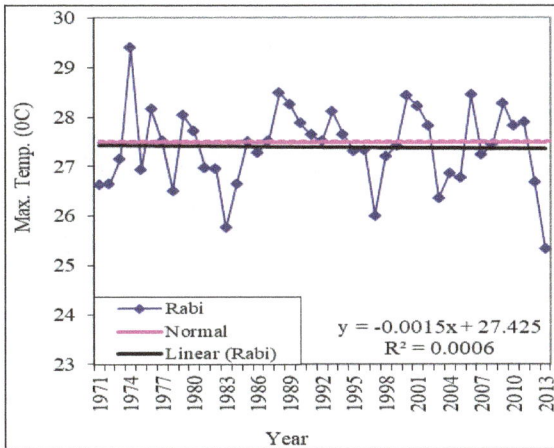

Figure 3.22: Maximum Temperature Trend in Rabi Season at Rewa.

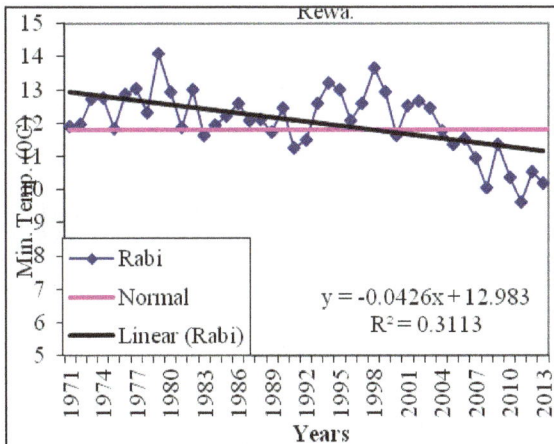

Figure 3.23: Minimum Temperature Trend in Rabi Season at Rewa.

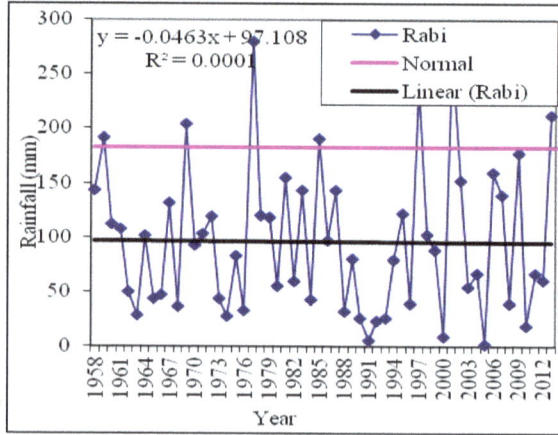

Figure 3.24: Rainfall Trend in Rabi Season at Rewa.

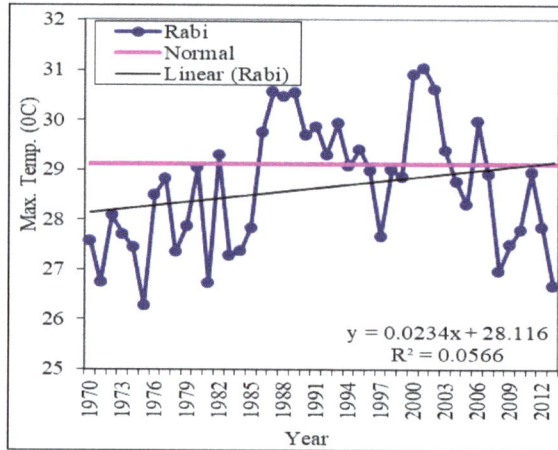

Figure 3.25: Maximum Temperature Trend in Rabi Season at Indore.

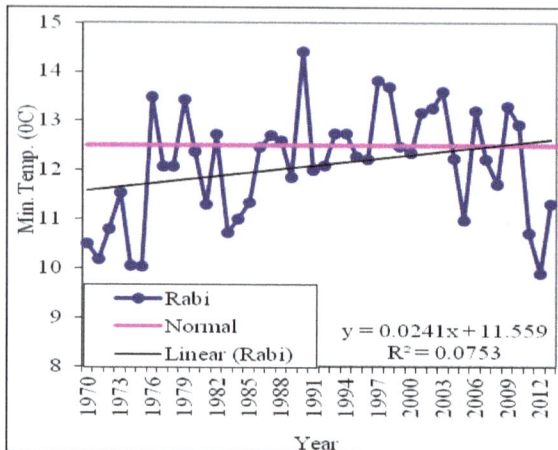

Figure 3.26: Minimum Temperature Trend in Rabi Season at Indore.

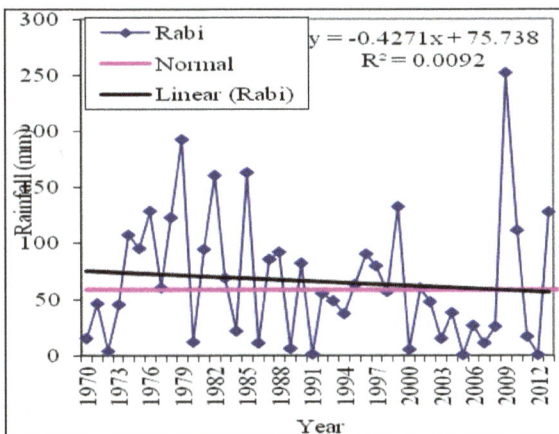

Figure 3.27: Rainfall Trend in Rabi Season at Indore.

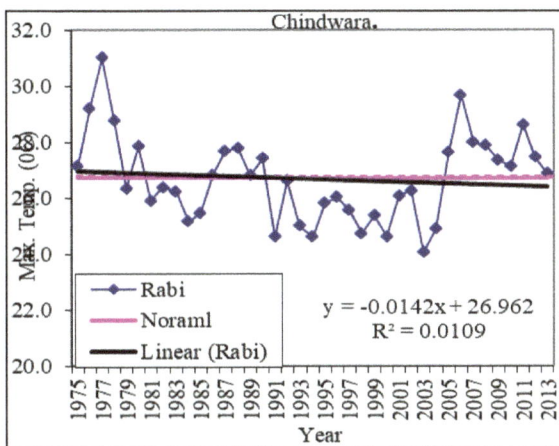

Figure 3.28: Maximum Temperature Trend in Rabi Season at Chindwara.

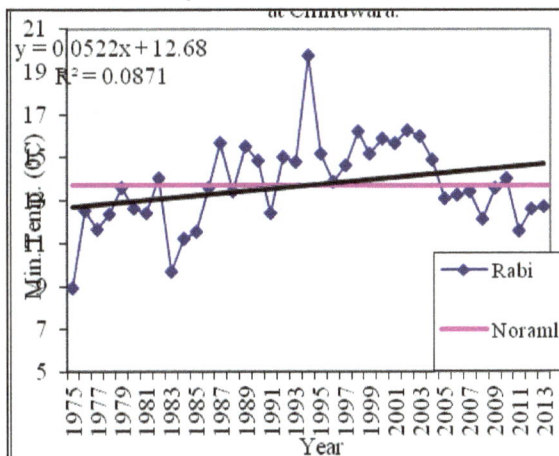

Figure 3.29: Minimum Temperature Trend in Rabi Season at Chindwara.

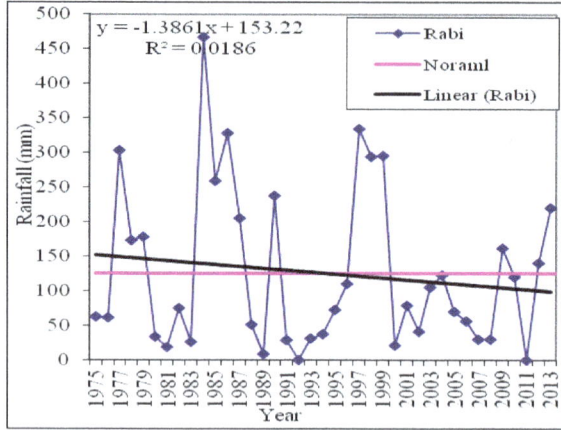

Figure 3.30: Rainfall Trend in Rabi Season at Chindwara.

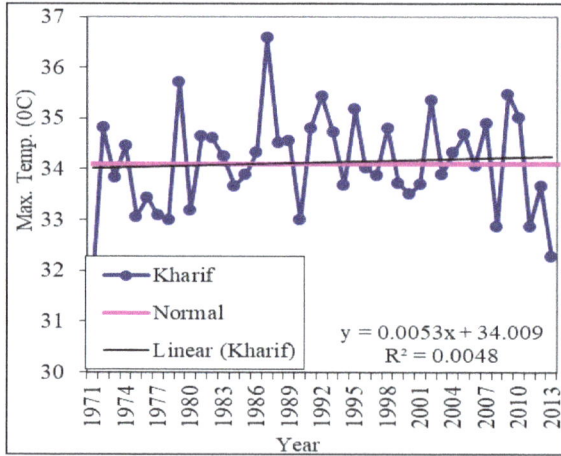

Figure 3.31: Maximum Temperature Trend in Rabi Season at Tikamgarh.

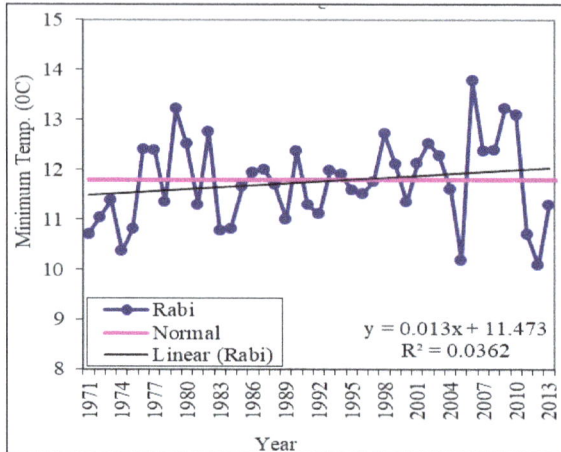

Figure 3.32: Minimum Temperature Trend in Rabi Season at Tikamgarh.

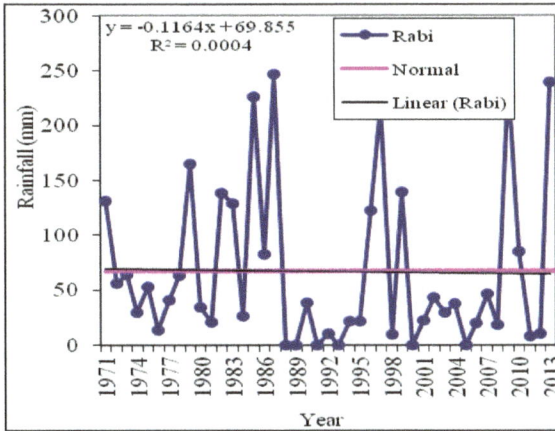

Figure 3.33: Rainfall Trend in Rabi Season at Tikamgarh.

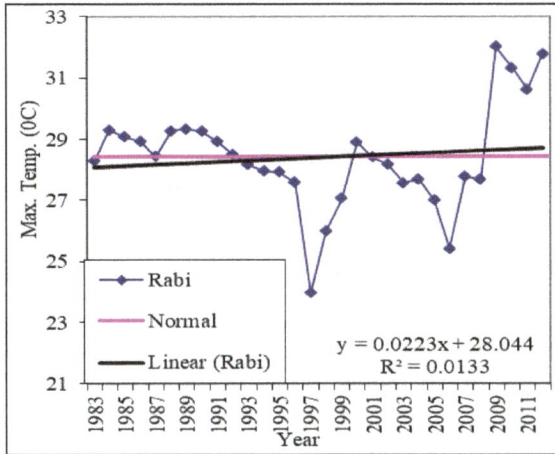

Figure 3.34: Maximum Temperature Trend in Rabi Season at Hoshangabad.

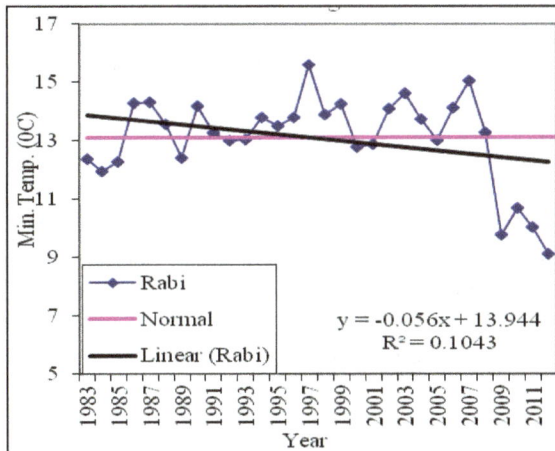

Figure 3.35: Minimum Temperature Trend in Rabi Season at Hoshangabad.

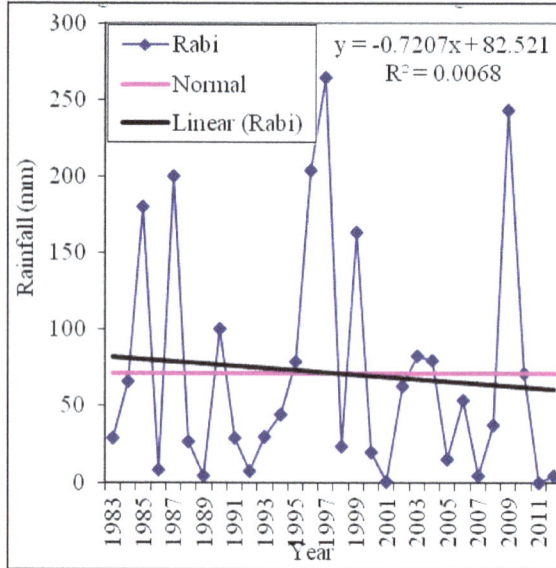

Figure 3.36: Rainfall Trend in Rabi Season at Hoshangabad.

of -0.06 °C/year. No trend was observed in maximum temperature and rainfall (Figures 3.34 and 3.36).

Climate Variability and Extreme Events in Bundelkhand Agroclimatic Zone

This study has focused on three districts *viz;* Datia, Tikamgarh and Chhatrapur of Bundelkhand Agro-climatic Zone (BACZ) of Madhya Pradesh. This agroclimatic zone is drought prone and affected by extreme weather events. Agriculture in this zone is more risky and vulnerable; mainly *rainfed* crops are gown in this zone.

Extreme Weather Events

Extreme weather events like hot, frost, heavy rainfall, and drought was calculated for all the three districts of Bundelkhand agroclimatic zone of Madhya Pradesh and presented district and decade wise extreme weather events is shown in Tables 3.14–3.16..

Tikamgarh District

The number of hot days, number frost days and number of drought year were increased over the decade in this district. A sharp decrease was observed in the number of heavy rainy days as well as frost day over the decade. The decrease in number of heavy rainy and rainy days is very crucial for sustainable *kharif* crop production in this district (Table 3.14).

Chhattarpur District

The number of hot days and number of drought year were increased over the decade in this district. A sharp increase was observed in the number of hot days over the decade. This increase in number of hot days were very high which may adversely affect the crop. The decrease in number of heavy rainy was not very high and significant in Chhatarpur district. The decade wise extreme weather events are shown in Table 3.15.

Table 3.14: Extreme Weather Events at Tikamgarh District during Past Four Decades

Decade	Number of Hot days	Number of Frost days	Number of Heavy Rainy Days	Number of Drought Year
1971-80	33	13	47	01
1981-90	53	03	49	01
1991-00	65	04	48	00
2001-10	40	00	19	03

Table 3.15: Extreme Weather Events at Chhattarpur District during Past Four Decades

Decade	Number of Hot days	Number of Frost days	Number of Heavy Rainy Days	Number of Drought Year
1971-80	66	07	58	3
1981-90	58	02	59	1
1991-00	111	01	58	1
2001-10	112	02	51	4

Datia District

The number of hot days, and number of drought years were increased over the decade in this district. A very high and alarming jump was observed in the number of hot days over the decade. The decrease in number of heavy rainy and rainy days is very crucial for sustainable *kharif* crop production in this district. Number of frost day has also increased many folds over the decade in this district. The decade wise extreme weather events is shown in Table 3.16.

Table 3.16: Extreme Weather Events at Datia District during Past Four Decades

Decade	Number of Hot days	Number of Frost days	Number of Heavy Rainy Days	Number of Drought Year
1971-80	-	-	-	1
1981-90	43	00	43	1
1991-00	78	01	37	1
2001-10	137	03	29	3

Decadal Variability

The Bundelkhand Agroclimatic Zone of Madhya Pradesh has witnessed extreme weather events in recent decades like excessive hotness, dryness, coldness, and number of successive drought years (lowest rainfall in *kharif* 2007). Decreasing number of rainy days in this zone was also highlighted by the media. All these necessitated examining the fluctuations of extreme weather events at spatial and temporal scale. Daily temperatures (maximum and minimum) and rainfall data of three districts *viz*; Tikamgarh, Chhatarpur and Datia (Bundelkhand Agroclimatic Zone) during the period 1971 - 2010 and their normals were collected and screened for extreme weather events on decadal scale. The number of hot (daily maximum temperature>45°C), frost (daily minimum temperature <2°C) days, heavy rainfall (daily rainfall>50mm) and rainy (daily rainfall =2.5mm) days were analyzed and compared on decadal scale to examine the climatic fluctuations. There was sharp increase (more than 100 per cent) in numbers of hot days during last decades (2001-10) in Chhatarpur and Datia districts. The highest daily maximum temperature (48.1°C) was recorded on 28th May 1998 at Chhatarpur. The number of heavy rainfall events has decreased from 47 days to 19 day in Tikamgarh and from 43 days to 29 days in Datia. The very heavy (133.8mm) daily rainfall event was recorded at Tikamgarh on 27th November 1979. The number of rainy day was almost constant in Datia and Chhatarpur but decreased in Tikamgarh (from 484 to 355 days). Strong trends in reduction in event of frost (occurrence of a 2°C screen temperature or lower) events over the decades have been found in this region. The lowest daily minimum temperature (-0.6°C) was recorded on 7th February 1974 at Tikamgarh. If these decadal variability are likely to continue, it might not be advantageous for winter crop although other *rabi* crops like mustard and vegetable would be benefited in this zone. The decrease in number of heavy rainfall days, will affect the crop area also.

Chickpea Yield and Acreage Variability

To analyze the variability of chickpea yield and acreage data (1968-2011) along with their trend and with regression equations of the districts Jabalpur, Rewa, Indore, Chhindwara, Tikamgarh and Hoshangabad were plotted and have been shown in the Figure 3.37–3.48.

Annual Variability

The figures reveal that chickpea yield and acreage are more variable than the climate parameters in the selected districts. chickpea yield was found to be more variable at districts like Jabalpur (35.3 per cent) Chhindwara (48.8 per cent) and Hoshangabad (39.2 per cent) (Figures 3.37, 3.40 and 3.42). The yield has shown an increasing trend in all districts. The highest increase in yield was found for Hoshangabad, Chhindwara and Tikamgarh districts. This increase was found to be the order of 25kg/ha/year at Hoshangabad, 24kg/ha/year at Chhindwara and 16Kg/ha/year at Tikamgarh. Though highest mean yield (991kg/ha) was found for Tikamgarh followed by Hoshangabad (903kg/ha) and by Chhindwara (896kg/ha), this yield scenario has been changed during last decade and Hoshangabad (1325kg/ha) yield was highest followed by Chhindwara (1155kg/ha) and Tikamgarh (1136kg/ha).

The area of chickpea was found to be more variable at Hoshangabad (46 per cent), Tikamgarh (36.1 per cent) and Chhindwara (30 per cent).The acreage of

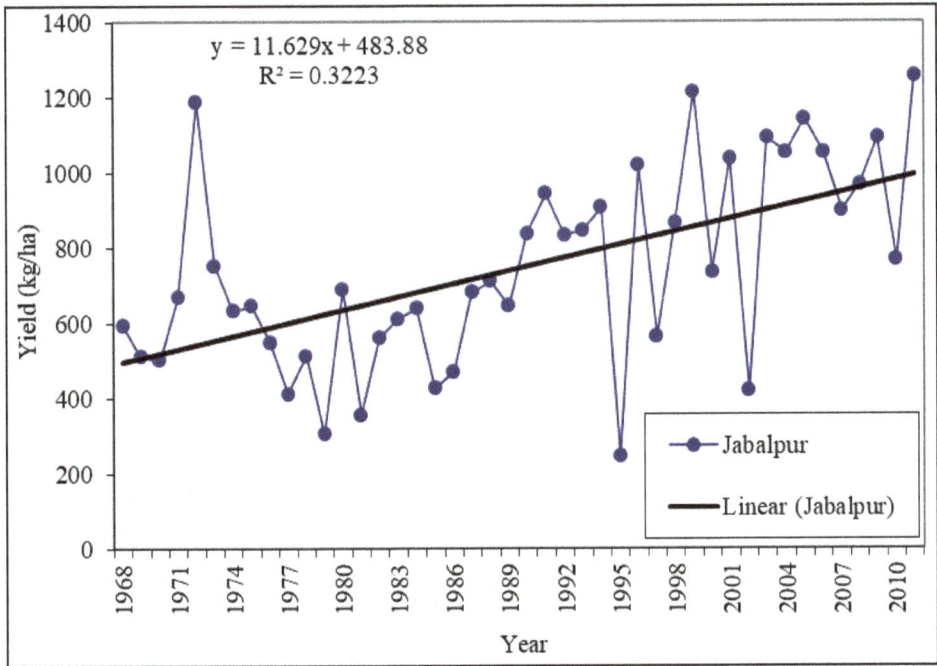

Figure 3.37: Chickpea Yield Variability and Trend at Jabalpur.

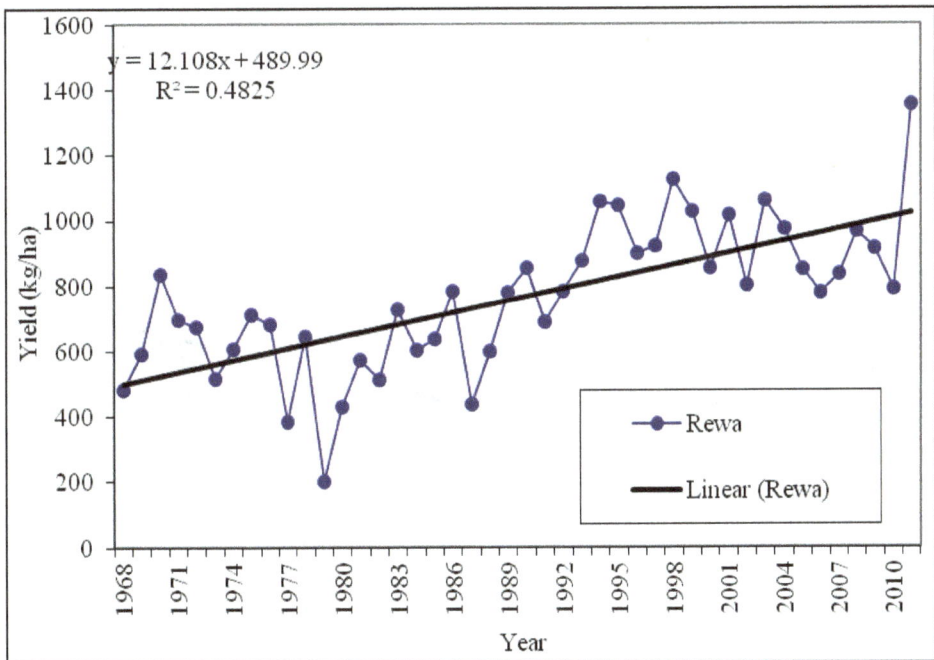

Figure 3.38: Chickpea Yield Variability and Trend at Rewa.

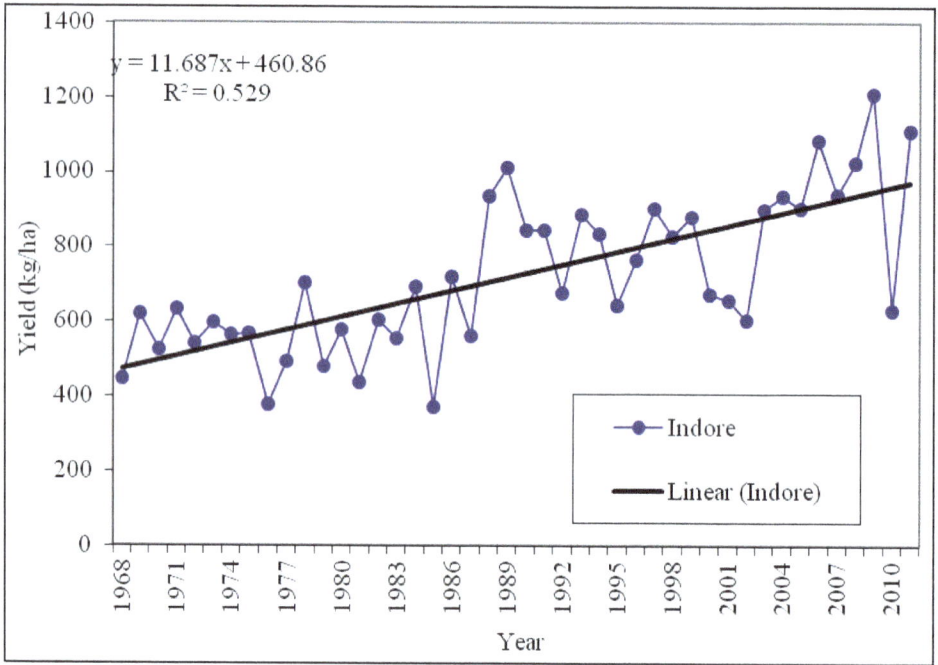

$$y = 11.687x + 460.86$$
$$R^2 = 0.529$$

Figure 3.39: Chickpea Yield Variability and Trend at Indore.

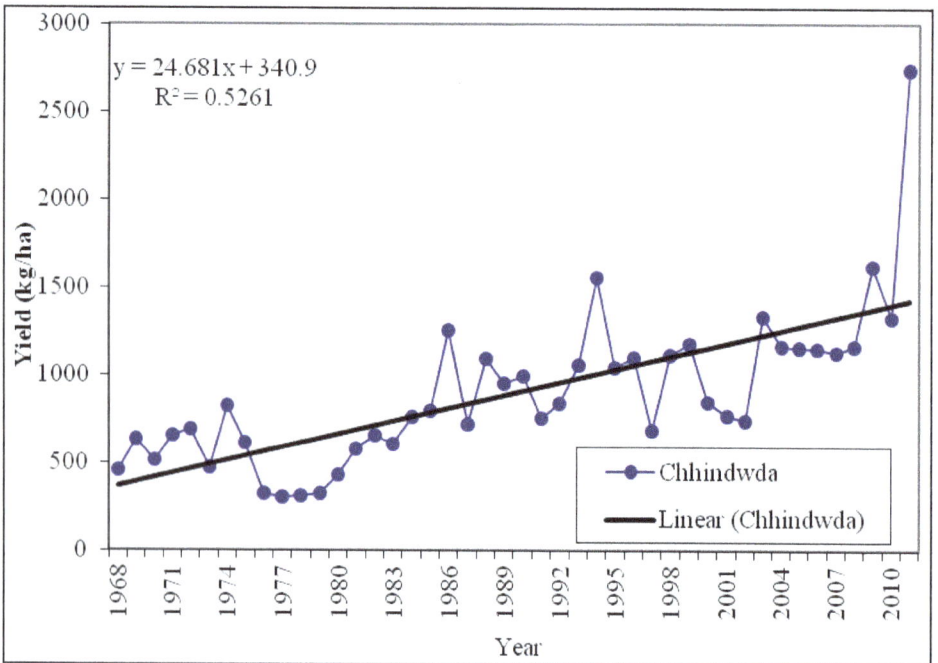

$$y = 24.681x + 340.9$$
$$R^2 = 0.5261$$

Figure 3.40: Chickpea Yield Variability and Trend at Chhindwara.

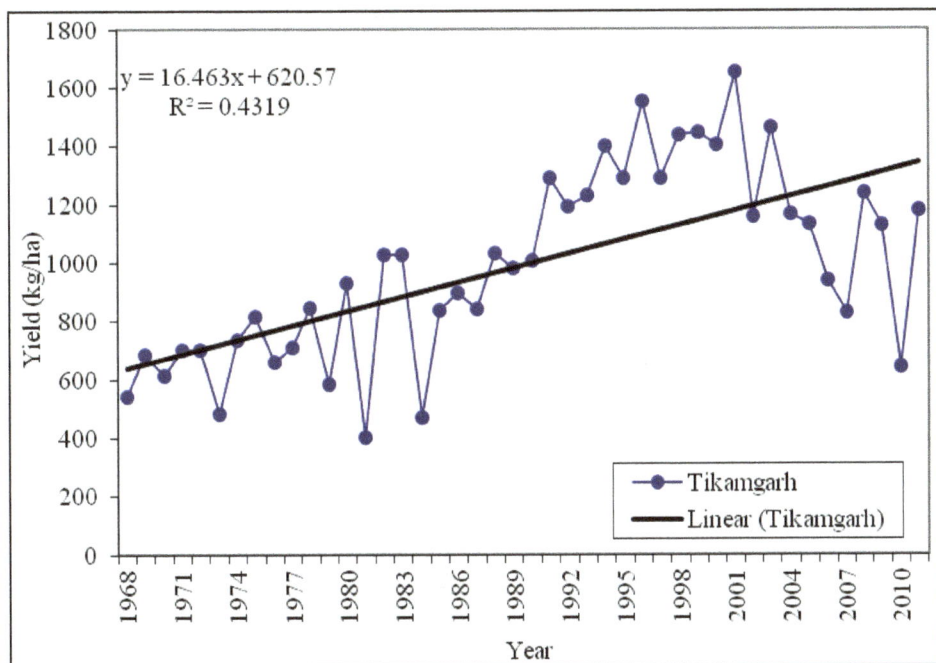

Figure 3.41: Chickpea Yield Variability and Trend at Tikamgarh.

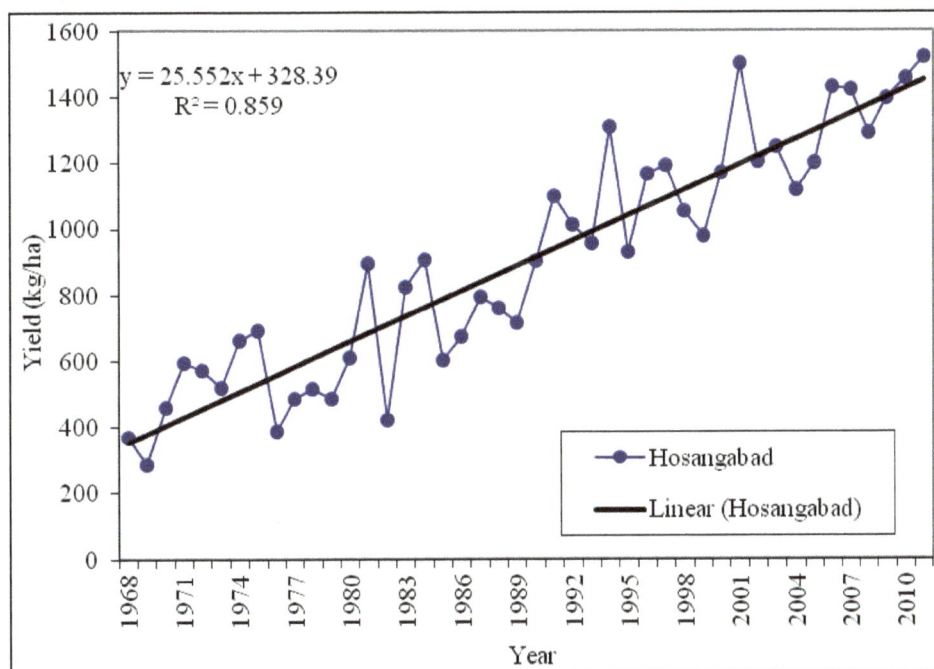

Figure 3.42: Chickpea Yield Variability and Trend at Hoshangabad.

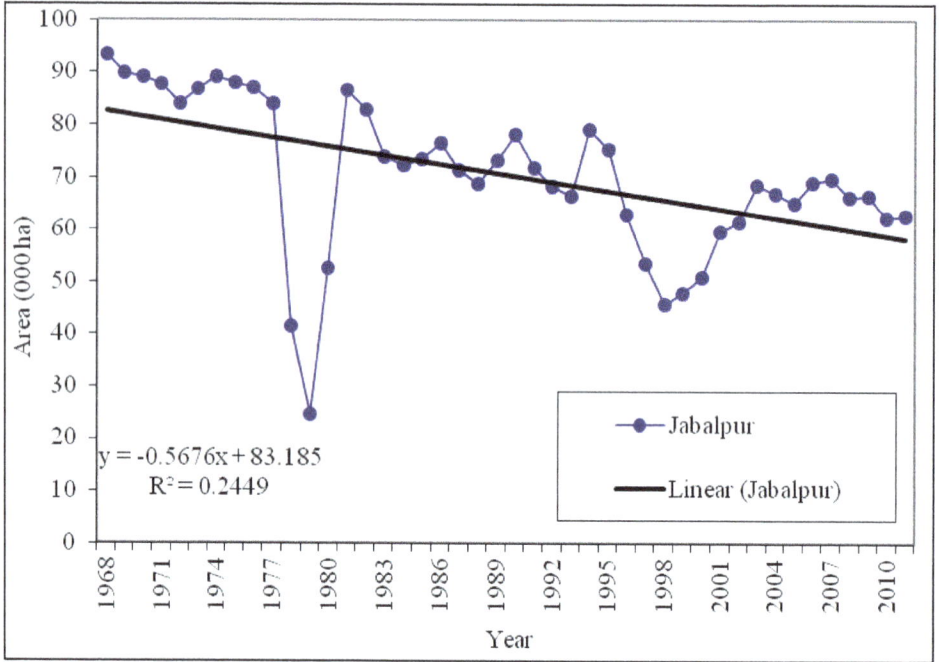

Figure 3.43: Chickpea Acerage Variability and Trend at Jabalpur.

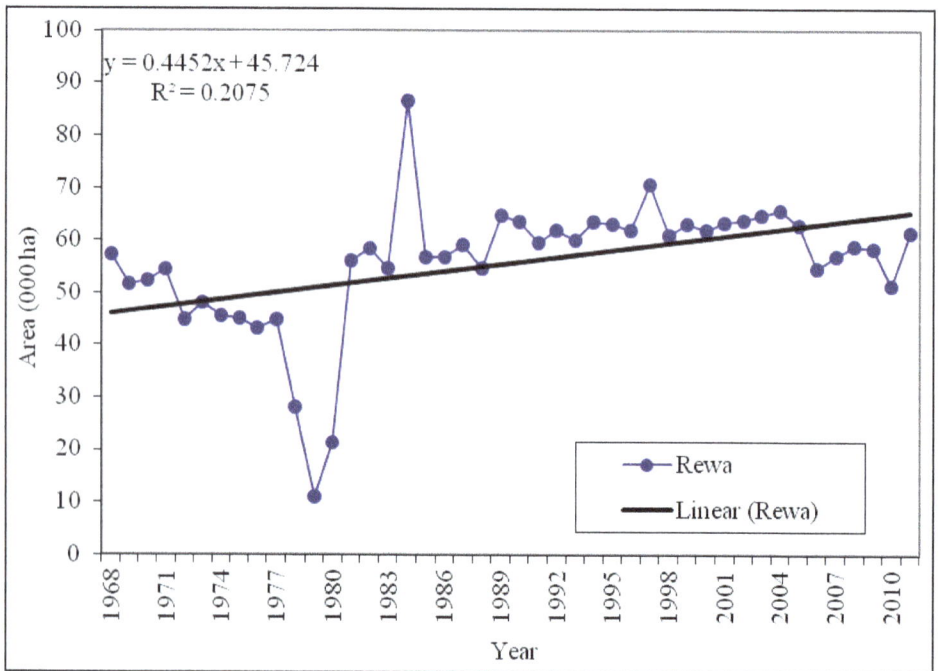

Figure 3.44: Chickpea Acerage Variability and Trend at Rewa.

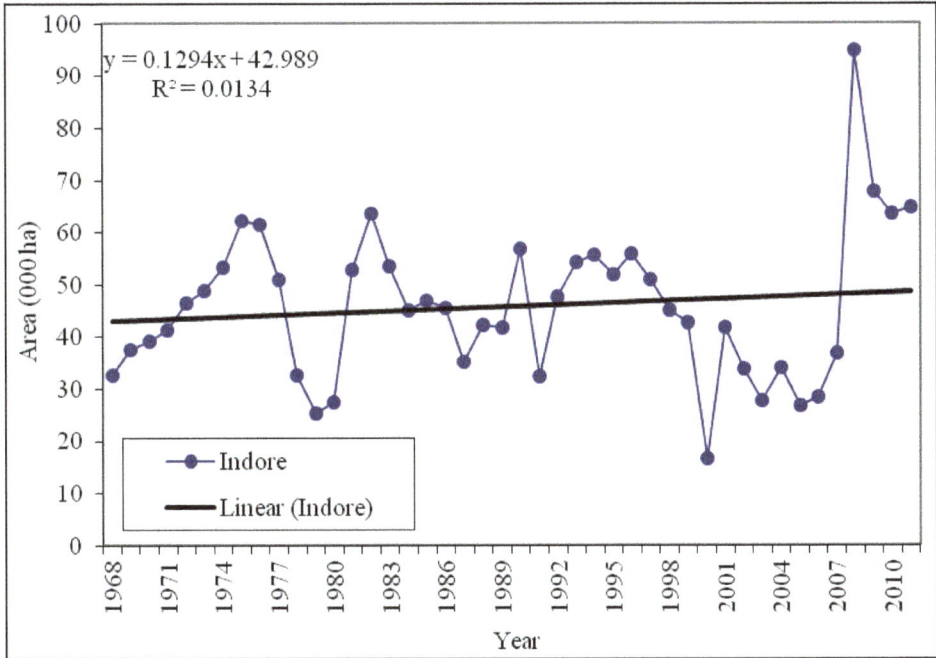

Figure 3.45: Chickpea Acerage Variability and Trend at Indore.

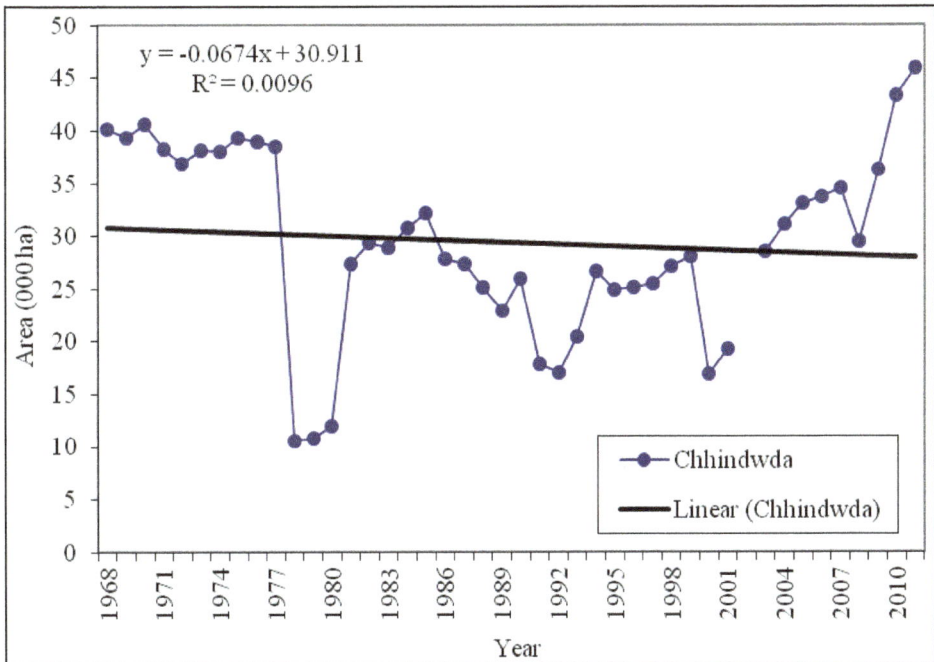

Figure 3.46: Chickpea Acerage Variability and Trend at Chhindwara.

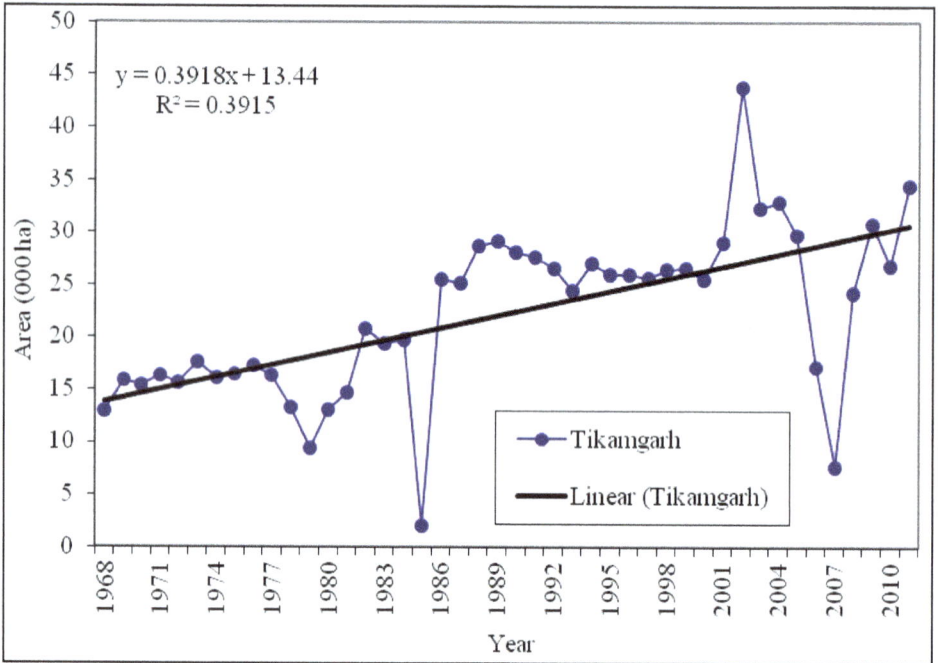

Figure 3.47: Chickpea Acerage Variability and Trend at Tikamgargh.

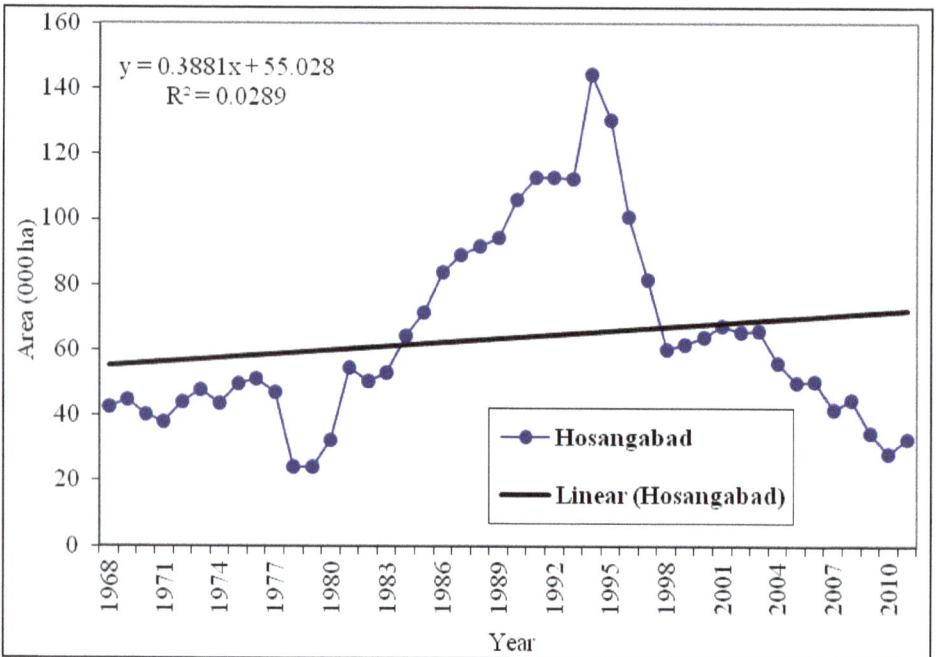

Figure 3.48: Chickpea Acerage Variability and Trend at Hoshangabad.

chickpea was found to be increasing trend at Tikamgarh (0.4 thousand ha/year) and Rewa (0.4 thousand ha/year).

Decadal Variability

The decadal variability of yield and area are shown in Figures 3.49 and 3.50. From these figures it was found that yield has increased over the decades at Hoshangabad, Indore, Chhindwara and Jabalpur. The yield over the decades was decreased at Tikamgarh. The area of Chickpea has decreased at very high rate at Hoshangabad and slightly at Jabalpur over the decade. The area of chickpea at Hoshangabad decrease of 51 per cent from 98.1 thousand ha (1991-200) decreased to 50.4 thousand ha (Figure 3.50).

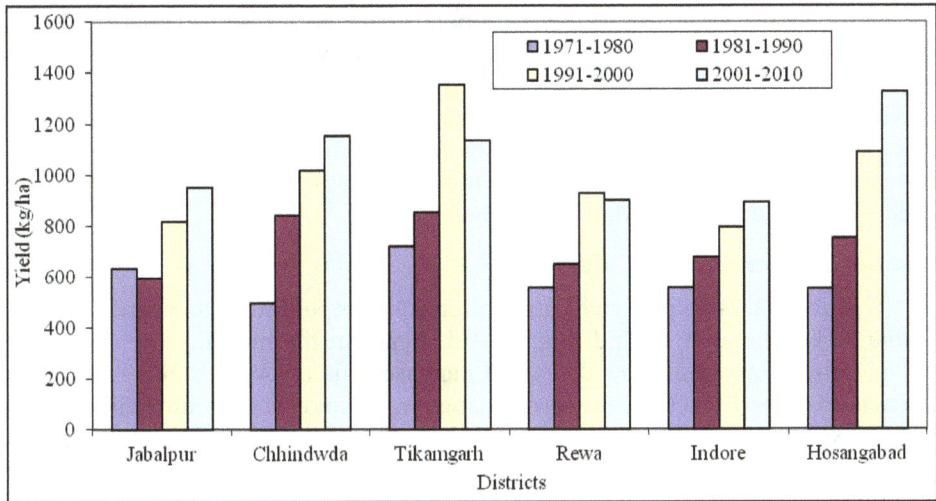

Figure 3.49: Decadal Yield Variability of Chickpea.

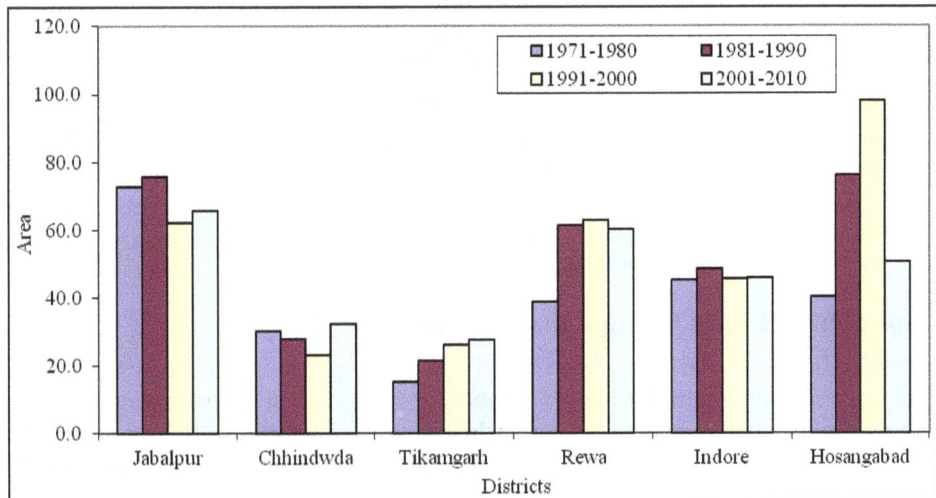

Figure 3.50: Decadal Acerage Variability of Chickpea.

Impact of Temperatures Variability Chickpea Yield

Chickpea is sensitive to both high and low temperatures during its growth period, particularly during flowering and pod filling stages. The optimum temperature for growing chickpeas is around 22.5°C. The vegetative growth takes place in cool days and warm nights (18-22°C). During the growing season of chickpea, high temperature above 34°C may limit the photosynthesis of chickpea. A night temperature of 10°C and a day temperature of 30°C are sub and supra optimal respectively. To examine the chickpea yield and acreage variability the change variability of temperatures, following analysis was carried out and presented below:

Maximum Yemperature and Yield

The chickpea decadal yield and temperatures were calculated and presented in Table 6.10. It was found that maximum temperature slightly increased over the decades at Jabalpur, Rewa and Tikamgarh districts and also the chickpea yield. The maximum temperature was found to be decreased over the decades and chickpea yield has found to be increased at Hoshangabad. Thus the slight changes either increase or decrease in the maximum temperature has not influence the chickpea yield in the six districts of Madhya Pradesh.

inimum Temperature and Yield

The decadal chickpea yield and minimum temperature has been analyzed (Table 3.11). The analysis indicates that decrease in the minimum temperature at Jabalpur, Tikamgarah and Rewa has increased the chickpea yields over these districts. In other districts there was no change was found in minimum temperature, but chickpea yield has increased in all these districts. The chickpea yield has increased over the decades irrespective of the slight increase or decrease in the minimum temperature over the decades.

The recent study across many countries and a variety of crops indicate that climate change has not so far seriously affected the yield and gross production (Jayaraman, 2010). There has been an overall rise in agriculture production (Hafner, 2003). According to Hafner (2003), cereal yields must grow at a minimum rate of 33.1kg/ha/yr in order to maintain current per-capita production level in 2050.

Similar findings have been reported in India by various researchers. Lal *et al.*(1999) reported that soybean yield was increased by 50 per cent with doubling of CO_2, effect. Increase in yield with combined effect of CO_2 and temperature change (maximum temperature is increased by 1°C and minimum temperature increased by 1.5°C) was restricted to 35 per cent only.

Impact of Climate Change on Chickpea Yield

Srivastava *et al.* (2016) has used CROPGRO-chickpea model to study the impact of climate change on chickpea yield at Jabalpur (irrigated) and Tikamgarh (*rainfed*). Under climate change scenarios (increasing maximum temperature by +1 to +3 °C, minimum temperature by+0.5 to 2.5°C and CO_2 from 400 to 600ppm (Table 3.12).

The per cent change in seed yield of chickpea cultivars were simulated under all the five climate change scenarios are presented in Table 3.12. It may be seen that

Table 3.10: Decadal Chickpea Yield and Maximum Temperature

Decade	Jabalpur		Chhindwada		Tikamgarh		Rewa		Indore		Hoshangabad	
	Yield	Max. Temp.	Yield	Max. Temp.	Yield	Max. Temp.	Yield	Max. Temp.	Yield	Max. Temp.	Yield	Max. Temp.
1971-80	634	31.2	494.4	31.2	717.1	32.2	553.5	31.3	555	31	552.1	
1981-90	592.4	31.6	841.9	29.6	852.3	32.4	649.9	31.4	674.1	31.8	749.2	32.5
1991-00	816.8	31.5	1016.5	28.9	1353.4	32.5	927.2	31.3	794.8	32.5	1085.2	31.1
2001-10	951.2	31.9	1155	30.4	1135.5	33	898.3	32.6	890.1	32	1324.7	29.6

Table 3.11: Decadal Chickpea Yield and Minimum Temperature

Decade	Jabalpur		Chhindwada		Tikamgarh		Rewa		Indore		Hoshangabad	
	Yield	Min. Temp.	Yield	Min. Temp.	Yield	Min. Temp.	Yield	Min. Temp.	Yield	Min. Temp.	Yield	Min. Temp.
1971-80	634	18.9	494.4	17.9	717.1	18.3	553.5	19.3	555	17.3	552.1	
1981-90	592.4	18.3	841.9	18.1	852.3	18.4	649.9	19.3	674.1	18.3	749.2	18.9
1991-00	816.8	18.4	1016.5	20.1	1353.4	18.4	927.2	19.1	794.8	18.8	1085.2	19.2
2001-10	951.2	18.3	1155	19.5	1135.5	17.8	898.3	17.7	890.1	18.8	1324.7	18.7

the seed yield of chickpea was found to increase under all climate change scenarios. The yield increased with increase in CO_2 concentration as well as with temperature (Figures 3.51 and 3.52). The variety JG-11 was found to have more (87 to 108 per cent) beneficial effect than variety JG-315 (61 to 97 per cent).

Table 3.12: Climate Change Scenario Selected for the Study (IPCC, 2013)

Projected Climate Change Scenarios	Maximum Temperature (°C)	Minimum Temperature (°C)	CO_2 Concentration (ppm)
S1	+1.0	+0.5	400
S2	+1.5	+1.0	400
S3	+2.0	+1.5	450
S4	+2.5	+2.0	500
S5	+3.0	+2.5	600

Srivastava (2003), has also reported a high impact (40-50 per cent increase in yield) CO_2 concentration on the productivity of chickpea at Raipur. Vanaja *et al.*

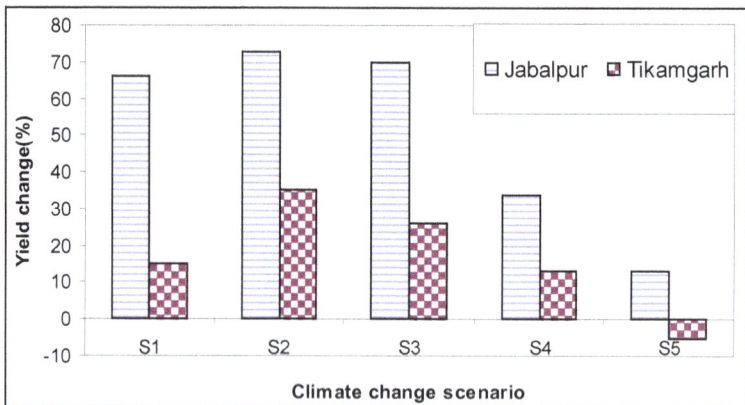

Figure 3.51: Yield Change in JG-315 under different Climate Change Scenarios.

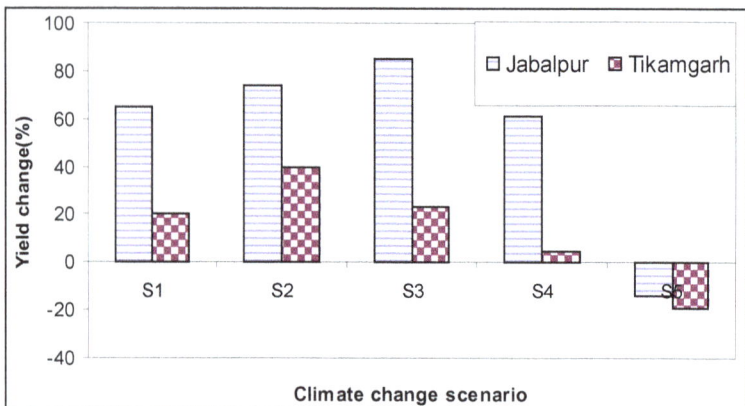

Figure 3.52: Yield Change in JG-11 under different Climate Change Scenarios.

(2011) reported that seed yield of pigeonpea improved from 22.8 g plant^{-1} at ambient to 42.4 g plant^{-1} at 700 ppm, thereby showing an increment of 85.9 per cent with enhanced CO_2. In blackgram, the seed yield recorded 1.74 g plant^{-1} at ambient and improved to 3.99 g plant^{-1} at elevated CO_2.

Date-wise percentage changes in yields were simulated and their mean values were presented in Table 3.13. Among the different dates of sowings, the impact was found to be beneficial in all except in extremely late condition (Dec.26). The highest seed yield increase (+200 to 256 per cent) was obtained under D1. Under very late sown condition (D6), the impact of climate change under S1, S2 and S3 scenarios was positive while under S4 and S5 scenarios it was negative. Thus under irrigated conditions the beneficial effect of climate change was observed. This result indicated that for maximization of chickpea yield, sowing date may be crucial under climate change conditions. Seed yield variability was increased in case of climate change and large variability was noted for *rainfed* condition as compared to irrigated condition. Hajarpoor *et al.* (2014) was simulated the impact of changing climate on chickpea at four major producing dry areas of Iran with different sowing dates. They used increase CO_2 concentration from 350 to 700 ppm and temperature (+2, +4 and

Table 3.13: Percentage Change in Yield under different Climate Change Scenarios in Irrigated and Rainfed

Treatments	S1	S2	S3	S4	S5
a) Irrigated Variety					
JG-315	+102.	+110.9	8 +121.3	+129.4	+137.5
JG 11	+138.5	+150.6	+165.9	+172.1	+187.6
Date of sowing					
D1	+208.6	+217.9	+231.5	+242.0	+256.6
D2	+158.4	+184.7	+217.0	+240.9	+265.3
D3	+160.0	+165.4	+173.7	+184.8	+200.7
D4	+83.2	+105.7	+118.3	+127.1	+137.7
D5	+91.6	+100.1	+119.5	+116.6	+129.2
D6	+22.0	+10.6	+1.6	-7.1	-14.1
b) Rainfed Variety					
JG 315	+61.8	+66.6	+73.2	+77.2	+96.7
JG -11	+87.3	+91.6	97.2	+100.6	+108.3
Date of sowing					
D1	+177.1	+183.5	+195.0	+204.1	+215.45
D2	+127.5	+143.5 +	165.5	+183.2	+205.20
D3	+57.6	+60.8	+64.3	+70.8	+80.50
D4	+52.0	+59.2	+60.3	+62.5	+65.45
D5	+34.4	+35.7	+39.8	+33.8	+35.10
D6	-1.5	-8.1	-13.7	-21.2	-13.40

D1: October 11, D2: October 26, D3: November 11, D4: November 26, D5: December 11, D6: December 26.

+6°C) with reduction in precipitation by 2 and 10 percent from current weather situations. They reported that chickpea yield would be raised between 37-89 per cent in *rainfed* conditions under the future climate in all sites. The increase in CO_2 concentration and maximum and minimum temperature had positive effects on other yield components as well.

Impact of Ozone Change on Chickpea Yield

The ozone with present concentration of 30 ppb is considered as one of the most serious environmental stresses for agro ecosystems (Agathokleous *et al.*, 2015). Singh *et al.* (2017) conducted experiments on chickpea yield under elevated carbon dioxide (550±10 ppm) and elevated ozone (70 ±10 ppb) and combination both gases in free air carbon enrichment and ozone experiments at IARI, New Delhi and reported that. They reported that elevated ozone has negative impact on yield whereas elevated carbon dioxide has positive impact on yield of chickpea and when both are combined the negative impact of elevated ozone were counteracted by elevated carbon dioxide. Chickpea yield is significantly increased under elevated CO_2 but significantly decreased under elevated O_3 treatment.

Chickpea Yield Sustainability

To see the sustainability of chickpea yield in the most vulnerable agro-climatic zone *i.e.* Bundelkhand agro-climatic zone of Madhya Pradesh, yield sustainability was calculated for four decades and presented in the Table 3.14. It is found that chickpea yield is highly sustainable for districts Chhatarpur and Tikamgarh.

Table 3.14: Chickpea Yield Sustainability in Bundelkhand Agro-climatic Zone of M.P.

Decade	Tikamgarh		Chharatpur		Datia	
	SYI	Status	SYI	Status	SYI	Status
1971-80	0.63	S	0.51	S	0.55	S
1981-90	0.6	S	0.69	S	0.54	S
1991-00	0.8	HS	0.72	HS	0.7	S
2001-10	0.51	S	0.74	HS	0.67	S

US: Unsustainable; S: Sustainable; HS: Highly sustainable

Risk Management

Risk management involves choosing among alternatives that could reduce financial losses resulting from uncertainties. Climate risk management (CRM) is an approach for climate-sensitive decision making. The approach seeks to promote sustainable development by reducing the vulnerability associated with climate risk. CRM involves strategies aimed at maximizing positive and minimizing negative outcomes for agriculture, food security, water resources, and health. It includes early-response systems, strategic diversification, dynamic resource-allocation rules, financial instruments, infrastructure design and capacity building.

Managing Weather Risk in Agriculture

There are two primary production risks *viz.,* weather and price risks. The risk management options are given in Table 3.15.

Table 3.15: Risk Management Strategies in Agriculture

	Informal Mechanisms	Formal Mechanisms	
		Market based	Publicly Provided
On-farm	Avoiding exposure to risk		Agricultural extension
	Crop diversification and Inter-cropping.		Integrated Pest management systems
	Diversification of income source		Irrigation systems
	Buffer stock accumulation of crops or liquid assets		
	Adoption of advanced cropping techniques (fertilization, irrigation, resistant varieties)		
Sharing risk with others	Crop sharing	Contract marketing and futures contracts	
	Informal risk pool	Insurance coverage net	
Coping with shocks	Sale of assets	Credit	Social assistance
	Reallocation of labor		Social funds
	Mutual aid		Cash transfer

Source: Adamenko 2004, World Bank 2001.

Index Insurance

Instead of multiple-peril farm-level crop insurance products, weather based index insurance products are based on an independent weather variable and highly correlated with farm-level yield or revenue outcomes. The independent weather variables such as temperature or rainfall being used mostly for farm-level insurance in India. Many index based insurance such as rainfall insurance, deviation in temperature insurance for crops are provided by many insurance companies in India. To transfer the climate risks losses, multiple peril farm insurance like Pradhan Mantri Fasal Beema Yojna or parametric insurance may be utilized. Crop/weather insurance should be promoted to reduce the impact of climate change and achieving stability in from income.

Price Risk Management

The traditional way to manage price variability used to be the pre harvest agreements between growers and purchases through entering into a pre-determined specific price for future delivery. The forward contracts, contract farming and assured minimum support price would be the ways to manage the market price fluctuations. Bhavantar Yojna, a recently launched scheme by the government of Madhya Prdesh is an ambitious step towards price risk management.

Crop Management Strategies

Management Options to Climate Variability

The conservation agriculture based crop management technologies including zero tillage with residue recycling, direct drilling into the residue, direct seeding of rice, brown manuring with *sesbania*, raised bed planting and integrated approach for water, nutrient, pest-disease and weed management technologies may have potential to combat the climate change and its variability impact in future. It has been demonstrated that mitigation of GHG emission is possible through sowing of rice and wheat with direct drill seeded rice and wheat on beds or with zero tillage.

Diversification in cropping system and cultivation of C_4 crops like maize, sorghum and bajra in semi-arid climate may be adopted to minimize the ill effect of climate variability Changes in land use pattern may be adopted to minimize the GHGs emission and energy balance. A brief of the crop management actions is given in Table 3.16.

Table 3.16: Crop Management and Early Warning Strategies for Climate Risk Management

Sections	Drought	Extreme High Temperature	Heavy Rainfall	Higher Temperature difference
Crops	Drought tolerant cultivars	Heat tolerant cultivars and bright farming	Adoption of ridge and furrow system, Drainage	Shifting of sowing window
Weather forecasting/ Early warning	Fortnightly forecasting of dry days	Excessive hot day and cold day forecasting	Forecasting of very heavy to heavy rainfall forecasting	Forecasted at weekly interval

Adaptation and Mitigation Options

There are several agricultural practices which can be fined tuned to reduce the emission of green house gases from the agricultural fields. The approaches that best reduce emissions depend on local conditions, and therefore, vary from region to region.

Conclusion

Variability and trends were found in climatic parameters in the selected districts of the five agroclimatic zones of Madhya Pradesh, however these climate variability and trends are far below as observed over other parts of the world. The long-period (around 50 years) crop–yield and climate data analysis suggest that rainfall variability and its trend influence chickpea yield and acreage. The change in maximum and minimum temperatures has not influenced the yields. The major findings are presented below:

☆ The rainfall was found to be decreasing at Tikamgarh, Chhindwara, Indore and increasing over decades at Jabalpur and Hoshangabad.

☆ The amount of rainfall during the last decade was decreased at Tikamgarh by 144mm, at Chhindwara by 80mm, at Indore by79mm from its decadal mean rainfall value.

☆ Increase in the value of mean minimum temperature 0.6°C/decade during last two decades at Chhindwara was observed. Increase in minimum temperature over the decades was also at Indore and it has increased by 0.5°C in last two decades. But at Tikamgarh district a decrease in mean minimum temperature was observed and this was 0.4°C from its mean decadal value during last decade.

☆ The decrease in annual minimum temperature at Rewa was 0.06°C/year. The increase in minimum temperature was in the order of 0.03°C/year at Indore, by 0.04°C/year at Chhindwara and this increase was found to be non-significant.

☆ The maximum temperature slightly increased over the decades at Jabalpur, Rewa and Tikamgarh districts and increased the chickpea yield.

☆ The decrease in the minimum temperature at Jabalpur, Tikamgarah and Rewa has increased the Chickpea yields over these districts. In other districts, no change was found in minimum temperature, but Chickpea yield has increased in all these districts selected under the study.

☆ The impact of climate change on chickpea yield was found to be favourable. The cultivar JG-11 would be more benefited than JG-315. The early sown (Oct.) crop would be maximum benefited.

☆ Development of heat and drought tolerant crop cultivars and adaptation of new technology would further minimize the projected climate change and variability impact on crop growth and yield.

☆ Develop knowledge based decision support systems for translating weather information into operational management practices at district and block levels.

☆ Improvement in water harvesting and its efficient utilization will also be required to be adopted for sustainable crop yield under *rainfed* conditions.

☆ However, with a combination of climate -ready varieties plus improved agronomic practices, *rainfed* farmers would be able to overcome the adversities of a warmer world.

Under adaptive strategy, new technologies will have to be developed to cope with future impacts and also to reduce the adaptation costs. Efficient crop cultivars of lower water demand, hot loving and higher harvest index may be developed. However, the agriculture is currently not proactively managing these fluctuations, even though changes in farming practices are suggestive of the fact that farmers of this zone have already responded to these observed climate fluctuations. Chickpea may be a sustainable pulse crop for the future in India, whose yield could be maximized by adaptation of new tools and techniques.

REFERENCES

Adamenko, T.(2004). "Agroclimatic Conditions and Assessment of Weather Risks for Growing Winter Wheat in Kherson Oblast." The World Bank Commodity Risk Management Group (CRMG) and International Finance Corporation Partnership Enterprise Projects (IFC-PEP), unpublished report from the Ukrainian Hydrometeorological Centre, Kiev, July.

Agathokleous, E., Saitanis, C. J., and Koike, T. (2015). Tropospheric O_3, the nightmare of wild plants: a review study. *J. Agric.Meteorol.*, 71(2), 142-152.

Hafner Sasha 2003. Sensitivity of evapo-transpiration to global warming: A case study of arid zone of Rajasthan (India). *Agricultural Water Management* 69: 1-11.

Hajarpoor Amir, Soltani Afshin, Zeinali Ebrahim and Sayyedi Faramarz (2014). Simulating climate change impacts on production of chickpea under water-limited conditions. *Agric. Sci. Dev.*, 3(6.): 209-217.

Hajarpoor Amir, Soltani Afshin, Zeinali Ebrahim and Sayyedi Faramarz (2014). Simulating climate change impacts on production of chickpea under water-limited conditions. *Agric. Sci. Dev.*, 3(6.): 209-217.

IPCC, (2013): Summary for Policymakers. In: Climate Change (2013). The Physical Science Basis. Contribution of Working Group I to the Fifth Assessment Report of the Intergovernmental Panel on Climate Change [Stocker, T.F., D. Qin, G.-K. Plattner, M. Tignor, S.K. Allen, J. Boschung, A. Nauels, Y. Xia, V. Bex and P.M. Midgley (eds.)]. Cambridge University Press, Cambridge, United Kingdom and New York, NY, USA.

IPCC, (2013): Summary for Policymakers. In: Climate Change (2013). The Physical Science Basis. Contribution of Working Group I to the Fifth Assessment Report of the Intergovernmental Panel on Climate Change [Stocker, T.F., D. Qin, G.-K. Plattner, M. Tignor, S.K. Allen, J. Boschung, A. Nauels, Y. Xia, V. Bex and P.M. Midgley (eds.)]. Cambridge University Press, Cambridge, United Kingdom and New York, NY, USA.

Jayaraman.T. 2010. Climate change and agriculture: A review article with special reference to India.

Kothawale, D.K. and Rupa Kumar, 2005. On the recent changes in surface temperature trends over India; *Geophys Res. Lett.* 32 L18714,doi: 10.1029/2005GL23528.

Lal M. 2001. Future climate change: Implications for Indian summer monsoon and its variability, *Current science* 81(9): 1205.

Lal M., K.K. Singh, G. Srinivasan, L.S. Rathore, D. Naidu, C.N. Tripathi,1999. Growth and yield responses of soybean in Madhya Pradesh, India to climate variability and change. *Agricultural and Forest Meteorology* 93 : 53-70.

Ray Deepak K., James S. Gerber, Graham K. MacDonald and Paul C. West (2015). Climate variation explains a third of global crop yield variability. Nature communications.: 1-www.nature.com/naturecommunications

Singh Ram Narayan, Mukherjee Joydeep, Sehgal V. K., Bhatia Arti, Krishnan P., Das Deb Kumar, Kumar Vinod and Harit Ramesh.(2017). Effect of elevated ozone, carbon dioxide and their interaction on growth, biomass and water use efficiency of chickpea (*Cicer arietinum* L.). *Journal of Agrometeorology* 19 (4) : 301-305.

Srivastava, A.K. (2003). Validation and Application of CROPGRO Model for crop management and yield prediction in chickpea crop, grown under agroclimatic conditions of north and central India. *Ph.D. thesis (Unpublished), Dept. of Geophysics, B.H.U., Varanasi*, pp. 1-151.

Srivastava, A. K., Silawat, S., and Agrawal, K. K. (2016). Simulating the impact of climate change on chickpea yield under rainfed and irrigated conditions in Madhya Pradesh. *Journal of Agrometeorology, 18*(1), 100-105.

State knowledge Management Centre on Climate change,(2014). EPCO, Housing and Environment Department, Govt. of M.P., 2014. Madhya Pradesh state action plan on climate change, pp. 1-156.

Vanaja, M.P., Reddy, R.R., Lakshmi, N.J., Yadav, S.K., Reddy, A.N., Maheswari, M. and Venkateswarlu, B. (2011). Yield and harvest index of short and long duration grain legume crops under twice the ambient CO_2 levels. *Indian J. Agric. Sci.,* 81(7): 666-668.

2018, *Climate Risks Management: Sustainable Pulse Production*
Editors: *A K Srivastava and Yogranjan*
Published by: **ASTRAL INTERNATIONAL PVT. LTD., NEW DELHI**   *Pages 97–107*

Chapter 4

Evapotranspiration and Water Use Efficiency of Pulses

Ram Niwas[1] and M.L. Khichar[2]

[1]Professor and Ex Head and [2]Professor,
Department of Agricultural Meteorology,
CCS HAU, Hisar, Haryana

ABSTRACT

Evapo-transpiration (ETc) and water use efficiency (WUE) are the two important parameters which could be maximized for enhancing pulse productivity. Intercropping, higher doses of fertilizers and soil moisture conditions are few the factors directly affecting the evapo-transpiration of a crop. The ETc of crop is more under irrigated conditions than the rainfed conditions. The ETc of pigeonpea, chickpea, mung bean, green gram, blakgram and lentil are examined and reported here. The water requirement for lentil is low as compared other reported pulse crop.

Keywords: Evapo-transpiration, Irrigation, Crop water use, Water use efficiency, Pulse.

In arid and semi-arid regions, water is the main ecological constraint for plant survival, and ecological functioning of soil-vegetation systems (Li, 2011). Therefore, understanding soil, water and plant interactions can aid to know the surface evapo-transpiration (ET) process in dry lands. Soil moisture dynamics are the central component of the hydrological cycle (Legates *et al.*, 2011) and are mainly determined by processes including infiltration, percolation, evaporation and root water uptake.

The actual evapo-transpiration can be determined by the analysis of the concurrent record of rainfall and runoff from a watershed. Transpiration is associated with plant growth and hence evapotranspiration occurs only when the plant is growing, resulting thereby in diurnal and seasonal variations. Transpiration thus superimposes these variations on soil surface evaporation.

Improving water productivity of pulses in dry or hot environments require higher inputs, chiefly high fertilizer doses that need to be considered for the risk trade-offs. Likewise, the low harvest index of pulse crop that contributes to its low water productivity needs to be considered in the context of a trade-off between grain production and crop residues. Trade-offs between water productivity and nutrient use efficiency need to be considered because maximizing water productivity in some farming systems may require additional but costly nitrogen and that to with risks or environmentally unsound impact associated with it. Likewise, trade-offs between yield and water productivity, which are mediated by amountand method of water supply are common. All these trade-offs need to be considered, as the aim of improving water productivity on its own is not necessarily the best pathway to sustainability involving specific production, environmental and social targets.

There are a few available technologies that can increase the productivity and production of pulses. Improved varieties with drought tolerance can provide cost-effective long-term solutions against adverse effects of drought. Supplemental irrigation with a limited amount of water, if applied to rainfed crops during critical stages can result in substantial improvement in yield and water productivity. Application of supplemental irrigation based on water production function will be more efficient towards increasing grain yield/unit water use. Water use is an important component of pulse production being it is a rainfed/limited water or irrigated crop in India. The water use variation in pulses varied due to temperature and relative humidity variation during the growing period, along with wind and soil moisture, all determine the rate of evaporation from the soil and transpiration from the plant (evapo-transpiration or ET_c). Therefore a practical and accurate method for ET estimation and the water use of some important pulse crops have been discussed in this chapter.

Evapo-transpiration

Food and Agricultural Organization (FAO) modified the original Penman equation in 1977, includes a revised wind function, an adjustment factor to account for local conditions and the assumption that the daily average ground heat flux is zero.This method uses mean daily climatic data, with an adjustment for day and night time weather conditions. The modified Penman method (Doorenbos and Pruitt 1977) is expressed as:

$$ET_r = \frac{C}{\rho_w}\left[W\frac{(RN-G)}{\lambda} + (1-W)f(u)(e_a - e_d) \right]$$

where,

ET_r = Reference evapotranspiration, mm/day

W = Weighting factor (temperature and altitude dependent) = $W = \dfrac{\Delta}{\Delta + \gamma}$

(Appendix - I)

Δ = Slope of saturation vapour pressure with temperature curve, $hPa°C^{-1}$

γ = Psychometric constant, $hPa°C^{-1}$

RN = Net solar radiation, $MJm^{-2}d^{-1}$

G = Soil heat flux in $MJm^{-2}d^{-1}$

λ = Latent heat of evaporation, $MJ\ kg^{-1}$

ρ_w = Density of water (1000 kgm^{-3})

f(u) = Wind function, $kg\ hPa^{-1}m^{-2}d^{-1}$ (*Appendix - II*)

f(u) = 0.27(1+0.86 u_2) $kg\ hPa^{-1}m^{-2}d^{-1}$

Wind speed is measured at different heights above the ground surface, can be adjusted to a height of 2m by the equation:

$$u_2 = u_z \left(\frac{2}{z}\right)^{0.2}$$

u_2 and u_z = Wind speed at height of 2 m and z m, in ms^{-1}

z = Height of wind measurement, m.

e_a = saturation vapour pressure at mean air temperature, hPa

e_d = Mean actual vapour pressure of the air, hPa

C = Adjustment factor to account for day and night weather conditions

(*Appendix - III*)

Crop Water Use

In the context of Indian agriculture and daily life, the contribution of pulses has a great importance. Besides, its importance in daily diet, pulse fixes a large amount of atmospheric N in association with *Rhizobium bacteria* (60 – 100 kg N/ha/year), which is a unique feature signifying its great value in agricultural production and maintaining soil fertility through ages. Pigeon pea, Chick pea, green gram, lentil and mung bean are the major pulse crops in India. The crop ET/water use estimated by different workers in respect of pulse crop has been reviewed and presented under sub heading.

Pigeonpea

The water use efficiency of pigeon pea was affected by its varieties. Medium variety showed more (> two times) water use efficiency in comparison with long duration variety. The water use efficiency (WUE) of both variety was improved when pigeon pea intercropped with sorghum (Figure 4.1.). As intercrop the WUE of pigeon pea was higher in case of long duration variety. However, the application of fertilizer increased the water use efficiency of sole as well as intercrops and long duration pigeon pea variety and was more efficient as compared to short duration variety under fertilized condition.

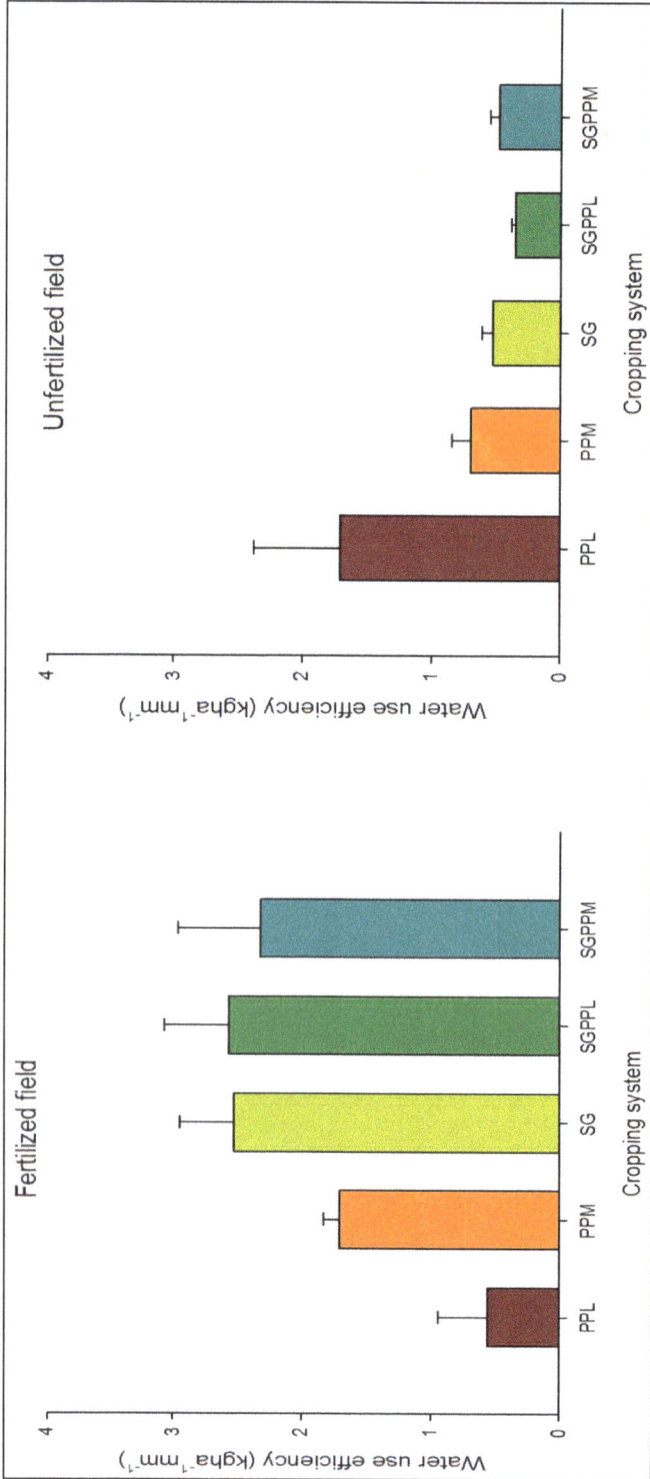

Figure 4.1: Water Use Efficiency of Pigeon Pea under different Cropping Systems.

(PPL: Pigeon pea long, PPM: Pigeon pea medium, SG: Sorghum, SGPPL: Sorghum Pigeon pea long, SGPPM: Sorghum Pigeon pea medium)

(*Source:* **SiegSnapp, Eva Weltzien, and SaakaBuah, Pan-African Grain Legume and World Cowpea Conference, March 2, 2016).**

Chickpea

Gupta *et al.* (1998) collected Lysimeter ET data of chickpea and reported that under rainfed conditions of the years 1994-95 and 1995-96 from Raipur and Srivastva (2003) reported chickpea ET during the year 1990-91 for Delhi were given in Table 4.1.

Table 4.1: Observed ET of Chickpea at Raipur and Delhi under Rainfed Condition

Station	Raipur		Delhi
	1994-95	1995-96	1990-91
Total ET (mm)	131	166	148

Table indicates that the Etc of chickpea under rainfed contions varied between 130 and 170 mm with a mean value of 146 mm. Water use of chickpea varied between 350 and 500 mm of water depending on seasonal conditions (Mace and Haris (2012). Table 4.2, summarizes the irrigation needed to grow chickpeas at three locations Emerald (15 May planting date), Dalby and Narrabri (1 June planting date). Results show a large variation in seasonal crop water demand, and irrigation demand between locations and season types. Zhang *et al.* (2000) found that chickpea water use was 268 mm over 12 seasons. The water use efficiency of chickpea for dry and wet were differing significantly and seed yield production was 8.7 and 3.2 kg/ha.mm respectively.

Table 4.2: Comparison of Average Water Requirements for Chickpeas Planted on the 15 May at Emerald, and the 1 June at Dalby and Narrabri Based on Historical Weather Data

Stations	Narrabari			Dalby			Emerald		
	Season Type			Season Type			Season Type		
	Dry	Avg	Wet	Dry	Avg	Wet	Dry	Avg	Wet
Crop ETc (mm)	424	396	364	449	419	398	413	393	366
crop rainfall (mm)	130	221	367	114	188	279	36	93	199
Irrigation demand (ML/ha)	4.0	2.8	1.7	4.3	3.5	2.7	4.6	3.9	2.9
No. of irrigations	4	3	2	4	3	3	4	4	3

Source: Irrigated Chick pea-Best practice guide (Mace, and Harris, 2012).

Chickpea water use was 426 mm in the fully irrigated crop but only 175 mm in rainfed conditions. There was a significant positive relationship between water use, and seed yield:

$$\text{Seed yield}\left(\frac{g}{m^2}\right) = 80.4 + 1.34 \text{ water use (mm)}$$

$$R^2 = 0.75$$

Maximum potential soil moisture deficit caused a reduction in chickpea seed yield. The critical maximum potential soil moisture deficit was approximately 150 and 90 mm for November and December sowings, respectively (Anwar *et al.*, 1999).

Mungbean

Mungbean crop ETc 331 and 102 mm under no stress condition and stress/rainfed condition respectively in semi-arid environment (Sissay *et al.*, 2014).The yield obtained varied between 1366 kg/ha under 331 mm optimal seasonal irrigation to 492 kg/ha when one-fourth of ETc (102 mm) was uniformly applied though out the growing season. The irrigation water use efficiency (IWUE) of mungbean varied from 0.248 kg/m^3 under optimum irrigation to 0.304 kg/m^3 when 50 per cent deficit was applied. Its water use efficiency is increased under water stress condition and the highest was obtained under 50 per cent uniform deficit irrigation.

The flowering/reproductive stage was observed as the most sensitive growth stage with a 24.9 per cent yield reduction compared to the optimal seasonal irrigation. Irrigation water use efficiency (IWUE) ranged from 0.248 to 0.304 kg/m^3 and concluded that stress at the midseason stage should be avoided and depending on the volume of water available. Deficit irrigation application at flowering/maturity stage resulted in the lowest water use efficiency among stage-wise deficit application treatments (0.25 kg/m^3) which were at par with the control treatment.

Rao and Singh (2003) reported from the experimental results conducted at Jodhpur, Rajasthan that mungbean evapo-transpiration (ET) rate was 310 mm with a water use efficiency of 2.17 kg ha^{-1} mm^{-1}.Grain yield of mungbean is severely affected by soil moisture stress at flowering and pod filling stages ultimately affecting water use efficiency (Thomas *et al.*, 2004).

According to Doorenbos and Kassam (1986) finding, relative yield reduction was less than the relative evapo-transpiration deficit. Irrigation application in pigeon pea at 0.6 IW/CPE ratios produced maximum seed yield (25.72 q/ha) and followed by irrigation at 0.8 and 0.4 IW/CPE ratio values (Table 4.3). Irrigation application in urdbean at 0.4 IW/CPE yielded higher (4.48q/ha) than the irrigation at 0.8 IW/CPE ratio.

Table 4.3: Effect of Irrigation Application Based on IW/CPE Ratio on Seed Yield in Pulse Crop

Sl.No.	IW/CPE	Pigeonpea	Urdbean
1	0.4	2334	448
2	0.6	2572	448
3	0.8	2438	425
	CD at 5 per cent	NS	NS

Transpired water use efficiency of winter pulses (faba bean, chickpea, and lentil) varied between 9 and 20 kg grain/ha/mm. However, ET and water use efficiency was comparatively in lowrange than the transpiration use efficiency. Evapo-transpiration and water use efficiency was ranged from 3 to 8 and 2 to16 kg grain yield/ha/mm under irrigated and rain fed conditions, respectively (Sadras, *et al.*, 2011).

Lentil

Sharma and Prasad (1984) found a seasonal water use of 149 mm indicative of a very dry soil. In these circumstances, irrigation increased consumptive water use to 344 mm and seed yield was increased by 56 per cent over control. ET calculated was maximum (465 mm) in April sown irrigated (Olympic variety), while the October sowing of the same variety had ET value of 318 mm (McKenzie, 1987). Evapo-transpiration of unirrigated lentil crops ranged from 387 mm in the April sown (Olympic lentil) to 253 mm in the October sowing for both cultivars (Table 4.4). All irrigated lentil crop transpirations were more than that of irrigated lentils ranging from 313 mm in the irrigated April sown Olympic to 129 mm in the rainfed (October sown) Titore variety. Zhang *et al.* (2000) found that water use was 259 mm in lentil over 12 seasons. The water use efficiency of lentil and grain yield production was 13.7 and 3.8 kg/ha.mm, respectively.

Table 4.4: Crop Evapotranspiration of Titore and Olympic Lentils with and without Irrigation Sown at Tree Sowing Dates at Canterbury, 1984-85

Sowing date	Titore		Olympic	
	Irrigated	*Unirrigated*	*Irrigated*	*Unirrigated*
April	277	201	313	227
July	270	203	277	210
October	179	129	181	132

The effect of irrigation water on seed yield of lentils is extremely variable. Seed yield increases with irrigation have been reported (Murari and Pandey, 1985) or decreases in seed with irrigation yield have been reported (Kaiser and Horner, 1980; Mckenzie *et al.*, 1985). McKenzie *et al.* (1985) reported a general decrease in seed yield with irrigation during a season in Canterbury that had only 70 per cent of normal rainfall.

Veeranna and Mishra (2017) estimated the lentil evapo-transpiration for deciding the proper sowing time in semi-arid agro-climatic conditions. They reported that the best sowing dates were last week of September to 1st week of October, and ETc is 69.7mm to165 mm and 98.7 mm to 230.8 mm respectively, for these two dates. They concluded that the lentil crop, needs relatively less water requirement and can tolerate the water stress for sowing of mid-September to mid-October.

Black Gram

Kumar and Kumar (2012) reported that the evapotranspiration of black gram under Uttarakhand Tarai conditions is 414.1 mm. The total rainfall received during black gram season was 1037.2 mm.

ETc under Rainfed and Irrigated Conditions in India

Srivastava (2003) Using the daily weather data from 1970 to 1999 for Jabalpur, Hisar, Delhi and 1971-2000 for Raipur, long -term simulations of water balance were performed for rainfed and irrigated treatments using water balance subroutine of CHICKPGRO model. The amount for each irrigation application, one at branching (30mm) and one at pod formation (60mm) stage were used for Raipur and Jabalpur. Two irrigations each of 60mm water, one at branching and one at pod formation stage were used at Delhi. One irrigation of 30mm of amount at pod formation was used at Hisar. From the multiyear simulation, the mean were calculated and presented in Tables 3.5–3.8) for analysis.

Table 3.5: ETc Components under Irrigated and Rainfed Conditions on different Sowing Dates at Raipur

Water Balance Parameters (mm)	Irrigated		Rainfed	
	Early	Normal	Early	Normal
Evapotranspiration	265	249	186	166
Transpiration	180	171	126	117
Soil evaporation	85	78	60	48

Table 3.6: ETc Components under Irrigated and Rainfed Conditions on different Sowing Dates at Jabalpur

Water Balance Parameters (mm)	Irrigated		Rainfed	
	Early	Normal	Early	Normal
Evapotranspiration	226	223	144	151
Transpiration	157	150	99	100
Soil evaporation	69	73	45	51

Table 3.7: ETc Components under Irrigated and Rainfed Conditions on different Sowing Dates at Delhi

Water Balance Parameters (mm)	Irrigated		Rainfed	
	Early	Normal	Early	Normal
Evapotranspiration	220	244	122	128
Transpiration	169	178	92	85
Soil evaporation	51	66	30	43

The Etc of chickpea under rainfed condition in central and north east india varied from 122 to 355 mm whereas under irrigated condition, it varied from 220 to 397 mm. The soil evaporation is very low under irrigated condition as compared to transpiration values.

ET values of cultivar JG 315 under rainfed condition at Raipur varied between 120 to 166mm in three years (1994-1997) of experiments during 14 to 29 November sowing dates (Gupta *et al.*, 1998). Rathore *et al.* (1996) reported during four year of experiment at Raipur that chickpea evaop-transpiration, for last week of October sowing, varied between 160 to 205mm. Chickpea water use varied from 252mm to 324mm at Jabalpur in five years field experiments (Tomar *et al.*, 1996).

Table 3.8: ETc Components under Irrigated and Rainfed Conditions on different Sowing Dates at Hisar

Water Balance Parameters (mm)	Irrigated		Rainfed	
	Early	Normal	Early	Normal
Evapotranspiration	397	385	349	335
Transpiration	323	279	298	265
Soil evaporation	73	106	51	69

Meena (1999) reported a consumptive use of water of 222 and 226mm by cultivar BG 256 in two-year experiments at Delhi environment under rainfed condition for last week of October sowing.

Conclusions

Pulse crop water requirements are relatively low and their water use efficiency is high under limited water conditions and semi-arid environment. Water use efficiency may be increased by incorporation of many cultural practices like use of shorter duration cultivars, normal sowing, inter cropping and use of high dose of fertilizer. High to medium ETc requirement may be maximized the yield of pulse. Among pulses, lentil is most water efficient crop in semi-arid environment under rainfed conditions.

REFERENCES

Anwar, M.R.,McKenzie B.A. and Hill G.D. (1999) Water use efficiency of chickpea (*Cicer arietinum* L.) cultivarsin Canterbury: effect of irrigation and sowing date. *Agronomy NZ.***29**. 1999, 1-8.

Doorenbos, J. and W.O. Pruitt. (1977).Crop water requirements. In: FAO Irrigation and Drainage.

Doorenbos, J., Kassam, A.H. (1986). Yield response to water. In: FAO Irrigation and Drainage Paper No. 33. FAO, Rome, Italy.

Gupta, B.D.; Sastri, ASRAS,; Urkurkar,J.S.; Naidu, D. and Srivastava, A.K. (1998). Water balance studies of a soybean chickpea sequential cropping under rainfed conditions in agroclimatic zone IV: Chhattisgarh region, Final Report, Deptt. of Physics and Agrometeorology, *IGKVV*, Raipur.

Kaiser, W.J. and Horner, G.M. (1980). Root rot of irrigated lentils in Iran. *Canadian Journal of Botany* **58** : 2549-2556.

Kumar N. and Kumar S. (2012). Evaluation of different evapo-transpiration models for wheat (*Triticum aestivum* L.) and black gram (*Vigna mungo* L. hepper) for mollisol of tarai region of uttarakhand by using lysimeter. *Indian J. Agric. Res.*, 46(1): 30-35.

Legates, D. R.; Mahmood, R.; Levia, D. F.; DeLiberty, T. D.; Quiring, S.; Houser, C., and Nelson, F. E. (2011). Soil Moisture: A central and unifying theme in physical geography, *Prog. Phys. Geog.*, **35**: 65–86.

Li, X. Y. (2011).Mechanism of coupling, response and adaptation between soil, vegetation and hydrology in arid and semiarid regions. *Sci. Sin. Terrae.*, **41**: 1721–1730 (in Chinese).

Mace, G. and Harris G. (2012). Irrigated Chickpeas – Best Practice Guide. InL WATER pakage guide for irrigation management in cotton and grain farming systems (Third edition).

McKenzie, B.A.; Sherrell, C.; Gallagher, I.N.; Hill, G.D. (1985). Response of lentils to irrigation and sowing date. *Proceedings of the Agronomy Society of New Zealand* 15: 47-50.

McKenzie,B.A. (1987). The growth, development and water use of lentils (Lensculinaris Medik.). Ph D thesis, in the University of Canterbury New Zealand. pp. 215

Meena, L.R. (1999). Effect of conserved soil moisture on chickpea (*Cicer arietinum* L.) in relation to phosphorus and biofertilizer applicaation under dryland conditions, (Ph.D. thesis) P.G. School, *IARI,* New Delhi.

Murari, K. and Pandey, S.L. (1985). Influence of soil moisture regimes, straw mulching and kaolin spray on yield-attributing characters, and correlation between yield and yield attributes in lentils. *Lens Newsletter* 12 (1): 18-20.

Rao, A. S.; Singh, R. S.(2003) Evapotranspiration, water use efficiency and thermal time requirements of green gram (*Phaseolus radiatus*). *Indian Journal of Agricultural Sciences.* 73(1): 18-22.

Rathore, A.L.; Pal, A.R.; Sahu, R.K. and Chaudhary, J.L. (1996). On-farm rainwater and crop management for improving productivity of rainfed areas. *Agri. Water Manag* 31 : 253-267.

Sadras, V. O.; Grassini, P.andSteduto P. (2011). Status of water use efficiency of main crops. In SOLAW Background Thematic Report - TR07(FAO), pp. 1-41.

Sharma. S.N. and Prasad, R. (1984). Effect of soil moisture regimes on the yield and water use of lentil (*Lens culinaris* Medic.). *Irrigation Science* **5**: 285-293.

Sissay Ambachew, Tena Alamirew, Assefa Melese (2014). Performance of mungbean under deficit irrigation application in the semi-arid highlands of Ethiopia. *Agricultural Water Management* 136: 68–74.

Srivastava, A.K. (2003). Validation and Application of CROPGRO Model for crop management and yield prediction in chickpea crop, grown under agroclimatic conditions of north and central India. *Ph.D. thesis (Unpublished),* Dept. of Geophysics, B.H.U., Varanasi, pp. 1-151.

Thomas, R., Fukai, M.J.S., Peoples, M.B. (2004). The effect of timing and severity ofwater deficit on growth, development, yield accumulation and nitrogen fixationof mungbean. *Field Crops Res.* **86**: 67–80.

Tomar, S.S., Tembe, G.P., Sharma, S.K., Bhadauria, U.P.S. and Tomar, V.S. (1996). Deptt. of Soil Science and Agricultural Chemistry, *JNKVV*, Jabalpur, pp. 40-41.

Veeranna J. and Mishra A. K. (2017). Estimation of evapotranspiration and irrigation scheduling of lentil using CROPWAT 8.0 Model for Anantapur District, Andhra Pradesh, India. *Journal of AgriSearch*, 4(4): 255-258.

Zhang, H., Pala, M., Oweis, T. and Haris, H. (2000). Water use and water use efficiency of chick pea and lentil in Mediterranean environment. *Aust. J. Agric. Res.*, **51** : 295–304.

2018, *Climate Risks Management: Sustainable Pulse Production*
Editors: A K Srivastava and Yogranjan
Published by: **ASTRAL INTERNATIONAL PVT. LTD., NEW DELHI**   *Pages 109–131*

Chapter 5

Chickpea Production under Changing Climate: Option for Yield Maximization

V. Jayalakshmi*, D. Peddaswamy and M. Sudha Rani

Acharya N.G. Ranga Agricultural University
Regional Agricultural Research Station, Nandyal – 518 502
*E-mail: veera.jayalakshmi@gmail.com

ABSTRACT

Chickpea, a crop of temperate region is also being grown in sub-tropical and tropical regions of the world. Climate variables viz., photoperiod, temperature, and precipitation differ in these chickpea growing areas and have significant effect on growth and development of chickpea. Climate change is already affecting the earth's temperature, precipitation and hydrological cycles. Chickpea a nutritious pulse crop with ever increasing demand from increasing population and is no exception from the rest of the crops facing the current scenario of climate change. Increase in temperature will have more adverse effect on cool-season crops like chickpea than the rainy season crops. About 90 per cent of the world's chickpea crop is grown under rainfed conditions on a progressively depleting soil moisture profile and experiences terminal drought, a condition in which grain yield of chickpea is low. Climate change, coupled with increased cultivation of chickpea in the warmer and drier environments in future will further aggravate the detrimental impacts of drought and heat stress on its productivity. Freezing stress is common during vegetative growth in areas of West Asia and North Africa, Europe and Central Asia where chickpea is sown in winter (autumn) or in early spring. Frost at reproductive stage is also a problem in South Asia and Australia. Chickpea productivity, and quality is also affected by several soil borne diseases and insect pests and climate change may affect these plant patho-systems at various levels. Change in climate variables due to climate change might lead to appearance of different races of the pathogens which are not active but might cause even sudden epidemic. Breeding strategies of chickpea must aim to design varieties with greater tolerance to elevated temperatures and drought, improved responsiveness to rising CO_2. Agronomic technologies should be modified for increased *adaption of chickpea to the adverse climate due to climate change. Plant breeders and physiologists have already identified plant traits that impart drought and heat tolerance in chickpea for breeding new varieties that are high yielding with improved drought and heat tolerance. A genomics-led breeding to integrate molecular markers in chickpea breeding programmers to develop cold, drought and heat tolerant varieties coupled with increased tolerance to key pests and diseases will pave the way to cultivate chickpea even under challenging situations and thus stabilizing the yield levels in tune with the demand.*

Keywords: *Chickpea, Climate change, Abiotic stress, Yield maximization.*

Chickpea is a good source of protein (20-22 per cent) and is rich in carbohydrates (around 60 per cent), dietary fiber, minerals and vitamins (William and Singh, 1987). There is a growing international demand for chickpea and the number of chickpea importing countries has increased from about 60 in 1989 to over 140 in 2009. This due to increased awareness about the health benefits of chickpea. Chickpea has several potential health benefits; including beneficial effects on some of the important human diseases such as cardiovascular diseases, type 2 diabetes, digestive diseases and some forms of cancer (Jukanti *et al.*, 2012). In addition to being an important ingredient of human food, chickpea also plays an important role in sustaining soil fertility by fixing the atmospheric nitrogen. Globally, chickpea is cultivated on about 14.8 million ha and produces 14.2 million tonnes with an average productivity of 962 kg/ha (FAOSTAT, 2016).The major chickpea producing countries include India, Australia, Pakistan, Turkey, Myanmar, Ethiopia, Iran, Mexico, Canada, and the United States. India stands first in terms of area (68 per cent) and production (70 per cent). However, its (960 kg ha^{-1}) productivity is slightly less than productivity of countries like Canada (1707 kg/ha), Ethiopia (1404 kg/ha), Myanmar (1335 kg/ha), Australia (1171 kg/ha) and Turkey (1145 kg/ha).

Crop production is now facing the 'perfect storm' of climate change and high population levels, combined with increasing costs (both monetary and environmental) of phosphate and nitrate fertilizer representing a significant challenge to increasing the intensity of food production further, using a single agricultural paradigm (Fischer *et al.*, 2009). This challenge may become overwhelming in the face of climate instability. Climate change will impact food supply unless actions are taken to increase the resilience of crops. Climate change, which is largely a result of burning fossil fuels, is already affecting the earth's temperature, precipitation and hydrological cycles. There is evidence from observations gathered since 1950 of change in some extremes. Many weather and climate extremes are the result of natural climate variability (including phenomena such as El Nino) but, there is evidence that some extremes have changed as a result of anthropogenic influences, including increases in atmospheric concentrations of greenhouse gases. It is established fact that temperature, moisture and greenhouse gases are the major elements of climate change. Current estimates indicate an increase in global mean annual temperatures of 1°C by 2025 and 3°C by the 2100. The carbon dioxide (CO_2) concentration is rising @ of 1.5 to 1.8 ppm/year and is likely to be doubled by the end of 21st century. Variability in rainfall pattern and intensity is expected to be high. Greenhouse gases (CO_2 and O_3) would result in increase in global precipitation of 2 ±

0.5°C per 1°C warming. Crop biodiversity and the distribution of wild crop relatives which is increasingly important genetic resource for the breeding of crops, will be severely affected leading to fragmentation of the distribution and even extinction.

The economic consequences of climate change would be rise in prices for the most important agricultural crops- rice, wheat, maize, and soybeans. This, in turn, leads to higher feed and therefore meat prices. As a result, climate change will reduce the growth in meat consumption slightly and cause a more substantial fall in cereals consumption, leading to greater food insecurity. The net effect of climate change on world agriculture is likely to be negative. Although some regions and crops will benefit, most will not. While increases in atmospheric CO_2 are projected to stimulate growth and improve water use efficiency in some crop species, climate impacts, particularly heat waves, droughts and flooding, will likely dampen yield potential. Indirect climate impacts include increased competition from weeds, expansion of pathogens and insect pest ranges and seasons, and other alterations in crop agro ecosystems.

Chickpea Production under Changing Climate

Chickpea (*Cicer arietimtm* L.) is a cool-season food legume and is grown between 20° N and 40° N in the northern hemisphere and is also cultivated on a small scale between 10° N and 20° N in India and Ethiopia at relatively higher elevations (Berger *et al.*, 2006). In the Southern hemisphere, where chickpea is relatively recent introduction, it is grown between 27° S and 38° S. Although chickpea is a crop of temperate region, its cultivation is gradually spread to sub-tropical and tropical regions of Asia, Africa, North America and Oceania. For example, Africa's share in global chickpea area has increased from 3.8 per cent in 1981-1983 to 4.7 per cent in 2008-2010 (FAOSTAT, 2012). About 73 per cent of chickpea is produced in South Asia. 13 per cent from west Asia and North Africa, 6 per cent from North America, 4 per cent from East Africa and surrounding area and 2 per cent from Australia. These chickpea growing areas differ in photoperiod, temperature, and precipitation, all of which have significant effect on growth and development.

There are two types of chickpea, *desi* (light to dark brown in color) and *kabuli* type (white or beige colored seed). The *desi* type covers about 85 per cent chickpea area and is predominantly grown in South and East Asia, Iran, Ethiopia and Australia, while *kabuli* types are grown in the countries of Mediterranean region, West Asia, North Africa and North America (Gaur *et al.*, 2008). Studies on climate change have already indicated that average surface temperatures are expected to rise by 3-5 °C, posing a major threat to crop production (including legumes) and agricultural systems worldwide, especially in the semiarid tropics (IPCC, 2007 and Hall, 2001). Moreover, any increase in temperature will have more adverse effects especially on cool-season crops like chickpea than the rainy-season crops (Kumar, 2006). Since chickpea is cultivated on large scale in arid and semiarid environment, terminal drought and heat stress, among other abiotic and biotic stresses, are the major constraints of yield in most regions of chickpea producing. About 90 per cent of the world's chickpea is grown under rainfed conditions where the crop grows and matures on a progressively depleting soil moisture profile and experiences terminal drought, a condition in which grain yield of chickpea is low. Flowering and podding in chickpea are known to be very sensitive to the changes in external environment. Exposure to heat stress (35°C) at these stages is known to lead to reduction in seed yield (Summerfield *et al.*, 1984 and Wang *et al.*, 2006). Climate change, coupled with increased cultivation of chickpea in the warmer and drier environments in the future will further aggravate the detrimental impacts of drought and heat stress on its productivity. However, in the cooler environments climate change may have a beneficial impact on the crop in the short term before the optimum temperature thresholds (20-26°C) (Devasirvatham *et al.*, 2012) are exceeded. Crop yields are also expected to increase with the increase in CO_2 concentration in the atmosphere. Free air carbon enrichment (FACE) experiments showed that crop productivity could increase in the range of 15-25 per cent for C_3 crops like wheat, rice and soybean (Tubiello *et al.*, 2007). Temperature increases are likely to support positive effects of enhanced CO_2 until temperature thresholds are reached. Beyond these thresholds, crop yields will decrease despite enhanced CO_2.

There are very limited studies conducted in chickpea to assess the impact of climate change. Gholipoor (2007) assessed the effect of individual versus simultaneous changes in climate variables such as solar radiation, precipitation, and temperature on rainfed- and irrigated-chickpea. He found that the change in biomass, harvest index (HI), and evapotranspiration (ET) was higher for solar radiation, compared to precipitation and temperature. For irrigated-chickpea, the biomass and ET were directly, but HI inversely, affected by solar radiation. Temperature had no impact on HI and ET. When solar radiation, precipitation and temperature were simultaneously changed precipitation and temperature slightly interacted with huge effect of solar radiation. For rainfed chickpea, it was found that nonlinear response to solar radiation for biomass, but linear response for HI. The decreased levels of precipitation positively affected biomass and HI. There was proportionally change in biomass and HI with changing temperature.

Gholipoor and Soltani (2009) conducted a study across six locations, one at ICARDA, Aleppo, Syria, and five in Iran to investigate the effects of two future climate scenarios on chickpea. The scenarios were reduction of 10 per cent historic P (rainfed conditions) + 525 ppm CO_2 + 2 °C warmer temperature (T) by year 2050 and declining of 20 per cent historic R (rainfed conditions) + 700 ppm CO_2 + 4 °C warmer T (year 2100). The results indicated that for both scenarios, the differential grain yield of rainfed chickpea will be positive in all locations. Since the differential HI tended to be mainly negative, the increase in grain yield was not proportional to increase in biomass. Thus, by year 2050, it is expected that the stability of yield to be increased for most locations. However, yield would be less stable for Tabriz, Mashhad (Iran locations) and ICARDA, but more stable for other locations in year 2100. Under irrigated conditions, the differential grain yield appeared to be negative (0–18 per cent) for year 2050; this was also true for year 2100 (6.3–17.1 per cent). These results suggested that under temperatures higher than ceiling temperature, future yield loss could be avoided in irrigated conditions through development of heat tolerant chickpea varieties. The negative effect of rising temperature on chickpea yield under rainfed ecosystem of India state Madhya Pradesh has also been reported (Bhadauria *et al.*, 2009).

Srivastava *et al.* (2009) reported that chickpea will be benefitted by rise in temperature to certain extent and the yield is forecasted to be increased by 45–47 per cent under double level of CO_2. Hajarpoor *et al.* (2014) studied the interactions between different aspects of climate change on chickpea. This study indicated that increasing temperature and CO_2 concentration, simultaneously, would have a positive effect on grain yield and water use efficiency of the chickpea in water-limited conditions.

Singh *et al.* (2014) investigated the impacts of climate change on the productivity of chickpea (*Cicer arietinum* L.) at selected sites in South Asia (Hisar, Indore and Nandyal in India and Zaloke in Myanmar) and East Africa (Debre Zeit in Ethiopia, Kabete in Kenya and Ukiriguru in Tanzania). As compared to the baseline climate, the climate change by 2050 (including CO_2) increased the yield of chickpea by 17 per cent both at Hisar and Indore, 18 per cent at Zaloke, 25 per cent at Debre Zeit and 18 per cent at Kebete. Whereas the yields decreased by 16 per cent at Nandyal and 7 per cent at Ukiriguru. The yield benefit due to increased CO_2 by 2050 ranged from 7 to 20 per cent across sites as compared to the yields under current atmospheric CO_2 concentration. While the changes in temperature and rainfall had either positive or negative impact on yield at the sites. Yield potential traits (maximum leaf photosynthesis rate, partitioning of daily growth to pods and seed filling duration each increased by 10 per cent) increased the yield of virtual cultivars up to 12 per cent. Yield benefit due to drought tolerance across sites was up to 22 per cent under both baseline and climate change scenarios. Heat tolerance increased the yield of chickpea up to 9 per cent at Hisar and Indore under baseline climate and up to 13 per cent at Hisar, Indore, Nandyal and Ukiriguru under climate change. At other sites (Zaloke, Debre Zeit and Kabete) the incorporation of heat tolerance under climate change had no beneficial effect on Yield.

Options for Yield Maximization in Chickpea under Changing Climate Scenario

Because agriculture will not experience the same kind of vulnerability to climate change in all regions, site-specific improved crop varieties and management practices will be needed to match the characteristics of each area. Boote *et al.* (2011) suggested genetic improvement of crops for greater tolerance to elevated temperatures and drought, improved responsiveness to rising CO_2 and the development of new agronomic technologies to adapt crops to the current adverse climate and climate change.

In case of chickpea, the plant breeders and physiologists have already identified plant traits that impact drought and heat tolerance to the crop (Krishnamurthy *et al.*, 2010 and Krishnamurthy *et al.*, 2011). Various sources of drought and heat tolerance traits in the germplasm accessions have been identified for breeding new varieties that are high yielding as well as having improved drought and heat tolerance.

Chickpea Varieties/Germplasm Lines for Tolerance to Abiotic and Biotic Stresses

☆ Warmer climate (heat) : JG 14, JSC 55, JSC 56

☆ Cold climate: ICCV 88506, : ICCV 88503, : Phule G 96006, ICC 8923, PDG 84-10

☆ Salinity : CSG 8962, CSG 88101

☆ Drought : RSG 888, Vijay, Phule G 5, BGD 72, RSG 44 RSG 963, RSG 888, ICCV 10, Vijay, GL 769, BGD 72

☆ Superior lines for drought tolerance identified : ICC 4958, RSG 143-1

☆ Superior lines for drought and heat tolerance combined : ICC 4958

☆ Multiple diseases resistance: GNG 1581, Pusa 362, KWR 108, Bharati

☆ Short duration: ICCV 2, JG 14, Vijay, JG 11, JG 16, JG 9218.

The research efforts made in chickpea and future strategies to cope with changing climatic conditions are briefly presented hereunder.

Genetic Improvement for Heat Tolerance

Heat stress is going to be a demanding concern for chickpea productivity under upcoming climatic conditions. The optimal temperature for chickpea growth ranges between 10 and 30 °C (Maesan, 1972). The reproductive phase (flowering and seed development) of chickpea is particularly sensitive to heat stress. A few days of exposure to high temperatures (35°C or above) during the reproductive phase can cause heavy yield losses through flower and pod abortion.

Breeding efforts exclusively dedicated to developing heat tolerant chickpea cultivars have been limited. There is also a possibility of developing a pollen selection method for heat tolerance, similar to that developed for cold tolerance (Clarke and Siddique, 2004). Pollen selection through heat treatment will further improve the efficiency of chickpea breeding for heat tolerance.

Several breeding lines (*e.g.*, ICCV 07104, ICCV 07105, ICCV 07108, ICCV 07109, ICCV 07110, ICCV 07115, ICCV 07117, ICCV 07118 and ICCV 98902) and cultivars (JC 14, JG 16, JG 130, JAKI 9218, JGK 2, KAK 2, ICCC 37, NBeG 3, Vishal and Vaibhav) developed from the breeding material selected at ICRISAT were found to have good levels of tolerance to heat stress at the reproductive stage (Gaur *et al.*, 2014).

Dua (2001) screened 25 genotypes for heat tolerance and identified two genotypes (ICCV 88512 and ICCV 88513) as heat tolerant. Large genotypic variation for reproductive-stage heat tolerance was observed by delaying the planting by two months compared to normal in the Mediterranean climate (Canci and Toker, 2009). Several heat-tolerant genotypes were identified from screening of 377 germplasm accessions. The *kabuli* types were generally more drought and heat susceptible than the *desi* types. The *desi* chickpea lines ACC 316 and ACC 317 exhibited tolerance to drought and heat (above 40 °C) under field conditions. The seed size was not much affected by adverse climatic conditions and showed the highest heritability. It was suggested that days to first flowering, days to maturity, harvest index, biological yield and pods per plant should be considered ahead of other traits while breeding for heat and drought tolerant genotypes. Canci and Toker (2009) evaluated 68 accessions of eight annual wild *Cicer* species (C. *bijugum*, C. *chorassanicum*, C. *cuneatum*, C. *echinospennum*, C. *judaicum*, *C.pinnatifidum*, C. *reticulatum*, and C. *yamasltitae*) for heat (up to 41.8°C) and drought tolerance and identified large genetic variability for these traits. Based on heat and drought tolerance scores, four accessions of C. *reticulatum* (AWC 605, AWC 616, AWC 620 and AWC 625) and one accession of C. *pinnatijiduin* (AWC 500) were identified as promising.

The reference set consists of 300 genotypes and represents genetic variability present in the chickpea germplasm available at ICRISAT and the ICARDA (Upadhyaya *et al.*, 2008). The reference set (n=280), excluding 20 genotypes (accessions of wild species and very late genotypes), was evaluated under heat stress conditions at Patancheru and Kanpur. 18 accessions (ICC 456, ICC 637, ICC 1205, ICC 3362, ICC 3761, ICC 4495, ICC 4958, ICC 4991, ICC 6279, ICC 6874, ICC 7441, ICC 8950, ICC 11944, ICC 12155, ICC 14402, ICC 14778, ICC 14815 and ICC 15618) were identified as stable tolerant. Some of these genotypes (*e.g.*, ICC 4958 and ICC 14778) were earlier identified as drought tolerant (Krishnamurthy *et al.*, 2010), thus these are good sources for both drought and heat tolerance. Upadhyaya *et al.* (2011) screened 35 early maturing chickpea germplasm accessions for heat tolerance. They identified ICC 14346 to be highly tolerant to heat stress along with nine other tolerant entries (ICC 5597, ICC 5829, ICC 6121, ICC 7410, ICC 11916, ICC 13124, ICC 14284, ICC 14368 and ICC 14653).

Devasirvatham *et al.* (2012) screened 167 chickpea genotypes for heat tolerance over two years at ICRISAT. The genotype ICCV 98902 had a critical temperature of 38°C or above during the pod-filling period and produced the highest grain yield under heat stress. In another study, it was found that the heat tolerant genotypes ICC 1205 and ICC 15614 had greater pod-setting ability compared to the heat sensitive genotypes ICC 4567 and ICC 10685 when exposed to heat stress at the reproductive stage under both field and controlled environmental conditions (Devasirvatham *et al.*, 2015).

A heat-tolerant chickpea breeding line ICCV 92944 developed at ICRISAT has been released for cultivation in Myanmar (as Yezin 6) and India (as JG 14). Owing to its heat tolerance, it was specifically released for late-sown conditions in India. JG14 has emerged as a promising variety for late-sown conditions in India, particularly in rice-fallows where sowing is delayed due to late harvest of rice (Gaur *et al.*, 2014).

Genetic Improvement for Drought Tolerance

Chickpea is predominantly grown as a rainfed crop on residual soil moisture stored during the previous rainy season with very less or no rainfall during the growing season. The soil moisture recedes to deeper soil layers with the advancement in crop growth and the crop experiences increasing soil moisture deficit at the critical stage of pod filling and seed development (called terminal drought). Terminal drought is a major constraint to chickpea production in over 80 per cent of the global chickpea area.

Resistance to drought can be described by various mechanisms, including escape, dehydration avoidance and dehydration tolerance (Blum, 1988; Ludlow and Muchow 1990; Wery *et al.,* 1994). Early flowering and maturity with high yield potential are the components of drought escape in chickpea (Saxena *et al.,* 1993; Silim and Saxena 1993). Progress has been made to develop early maturing genotypes without penalizing yield (Imtiaz and Malhotra, 2009). Dehydration avoidance is related to root attributes (root size, depth, length, density, hydraulic conductance, *etc.*). Genotypic variation for root characters has been reported in chickpea (Nagarajarao *et al.,* 1980; Brown *et al.,* 1989; Kashiwagi *et al.,* 2005; Kashiwagi *et al.,* 2006; Yadav *et al.,* 2006; Ur Rehman, 2009; Canci and Toker, 2009). Some breeding programs have used genotypes with deep and vigorous root system, such as ICC 4958, as one of the parents in crosses, but selection of breeding lines was invariably for seed yield under water-stress conditions rather than on root traits.

Adoption of early maturing varieties has shown high impacts on enhancement of chickpea area and productivity in short-season environments, *e.g.* Myanmar and southern India. Several early maturing high yielding cultivars have been developed, *e.g.*, ICCV 2 (Swetha in India, Wad Hamid in Sudan and Yezin 3 in Myanmar), ICCV 92311 (PKV *Kabuli* 2 or KAK 2 in India) and ICCV 92318 (Chefe in Ethiopia) in *kabuli* type and ICCC 37 (Kranthi in India), ICCV 88202 (Sona in Australia, Yezin 4 in Myanmar and Pratapchana 1 in India), ICCV 93954 (JG 11 in India) and ICCV 93952 (JAKI9218 in India) in *desi* type. Super early breeding lines have been developed (*e.g.* ICCV 96029, ICCV 96030) which further expand opportunities for cultivation of chickpea in areas and cropping systems where the cropping window available for chickpea (Gaur *et al.,* 2014).

Genetic Improvement for Tolerance to Low Temperature

Freezing stress is common during vegetative growth in areas West Asia and North Africa, Europe and Central Asia where chickpea is sown in winter (autumn) or in early spring (Singh *et al.,* 1994). Frost at reproductive stage is also a problem in South Asia and Australia (Srinivasan *et al.,* 1998; Clarke and Siddique, 2004). Night temperature is increasing more than that of maximum day temperature due to climate change (Basu *et al.,* 2009). Screening for cold tolerance has been reported under both field and controlled environments using growth rooms. ICARDA has developed a reliable field screening technique for cold tolerance evaluations where chickpea is planted early in the season (October) to provide sufficient time for the plants to acclimatize to low temperature before the onset of cold.

Breeding at ICARDA has resulted in the expansion of genetic variability for flowering at low temperatures using cultivated x wild *Cicer* crosses. The genes responsible for flowering at low temperature have been transferred from wild to cultivated lines (Chaturvedi *et al.,* 2009). Singh *et al.* (1990) reported sources for tolerance to cold in *Cicer* species while evaluating 137 lines of eight wild species from 1987 to 1989. Subsequently, Singh *et al.* (1995) identified additional sources of tolerance to cold in cultivated and wild *Cicer* species by evaluating 4284 *kabuli* and 2137 *desi* germplasm lines, 857 *kabuli* breeding lines and 59 lines of seven wild annual *Cicer* species. During vegetative stage, chickpea lines which can survive at as low as

-20°C temperature without snow cover and -24°C under snow cover were identified by ICARDA. Some of resistance lines are ILC 8262, ILC 8617, FLIP 03-2C, FLIP 03-3C, FLIP 03-5C, FLIP 03-6C, FLIP 03-9C, FLIP 03-11C, FLIP 03-12C, FLIP 03-13C and FLIP 03-14C. However, the resistance to frost or chilling at reproductive stage is not detected at the high level. The only reported chilling tolerant cultivars are Sonali and Rupali from Australia (Clarke *et al.*, 2004). Using pollen as a selection method, Clarke *et al.* (2004) confirmed the cold tolerance of ICCV 88516 and 88510 and the sensitivity of Amethyst, Dooen, Tyson and FLIP84-15C in Western Australia. Accessions of cultivated and wild *Cicer* sps. were screened for cold tolerance at ICARDA (Singh *et al.*, 1995). These authors reported cold tolerance in the lines ILC 8262, ILC 8617 (a mutant) and a FLIP 97-82C from cultivated *Cicer* along with wild annual chickpea such as *C. bijugum* and *C. reticulatum*. Later, Toker (2005) identified chilling tolerance (<–1.5°C) in annual wild *Cicer* sp. of yamashitae. Heidarvand *et al.* (2011) identified the genotypes Sel 95Th1716 and Sel 96Th11439 as chilling tolerant based on field screening at the vegetative stage where plants were exposed to –11°C to –25°C at the Dryland Agriculture Research Institute (DARI) of Iran.

In India, different sources of resistance to cold tolerance are reported by Chaturvedi *et al.* (2009) and several cold tolerant breeding lines such as ICCVs 88502, 88503, 88506, 88510 and 88516 have been developed that set pods at less than 15°C in India (ICRISAT, 1994). The Indian Agricultural Research Institute (IARI) has also developed a few cold-tolerant genotypes (BGD 112 green, BG 1100, BG 1101, PUSA 1103, BGD 1005, PUSA 1108, DG 5025, DG 5027, DG 5028, DG 5036 and DG 5042) (Gaur *et al.*, 2007).

Both additive and non-additive gene effects govern cold tolerance in chickpea. Cold tolerance was observed to be dominant over susceptibility for at least five sets of genes (Malhotra and Singh, 1990).

Genetic Improvement for Tolerance to Higher Levels of CO_2

Very limited information is available on the response of crops to elevated atmospheric carbon dioxide. The present change in climate is closely linked with the rise in atmospheric carbon dioxide (CO_2) levels from 280 to 387 ppm since the start of the Industrial Revolution. And current levels of CO_2 are expected to double by 2100 (IPCC 2007). Such rise in CO_2 levels affects the biological system of living organisms (Guerenstein and Hildebrand, 2008). Alteration in phytochemistry of plants under the elevated CO_2 concentrations is well documented (Hunter 2001). Irrespective of the biochemical pathway (C_3 and C_4), crop exhibit reduced 'N', increased 'C' and C: N ratio due to rapid photosynthesis and growth (Norby *et al.*, 1999) of the plant.

Because of their indeterminate growth habit and symbiosis relationship, grain legumes have no sink limitation under high CO_2 environment and thus can respond better by utilizing more photo assimilates. A few studies conducted with increased CO_2 levels revealed that doubling CO_2 has increased grain yield of grain legumes by 54 per cent at intermediate temperatures (Kimball, 1983). Under optimal temperatures, similar to other C_3 species, photosynthetic rates of chickpea also increase substantially with increases in CO_2 (Srivastava *et al.*, 2009). Deshmukh *et al.* (2009) studied the effect of elevated CO_2 on yield attributes and Rubisco enzyme of chickpea. They have tested chickpea at 550 ± 35 µLL-1 elevated level of CO_2 and found that plants grown under elevated CO_2 produced more number of pods and grains with 33 per cent increase in pods number. Similar effect was observed for biomass with 12 per cent and 15 per cent increase, respectively. Rogers *et al.* (2009) reported that legumes increase N_2 fixation at elevated CO_2, which is critical in maintaining the C to N ratio in the ecosystem. In chickpea, exposure to elevated CO_2 increased the carbon content and decreased the nitrogen content, resulting in higher C/N ratio similar to those reported in mungbean under high CO_2 (Deshmukh *et al.*, 2009; Srivastava *et al.*, 2001).

Genetic Improvement for Improved Tolerance to Pests and Diseases

Chickpea productivity, and quality, is affected by several foliar (Ascochyta Blight, Botrytis gray mold), soil borne (wilt/root rots and nematodes) diseases, and insect pests (pod borer, leaf miner). Climate change may affect plant pathosystems at various levels *viz.*, from genes to populations, from ecosystem to distributional ranges, from environmental conditions to host vigour/susceptibility and from pathogen virulence to infection rates.

In West Asia and North Africa, pod borer attacks chickpea at very late stage and does not cause much damage; however, for the last few years, the timing and level of infestation is changing, which could be a major threat to chickpea production in West Asia and North Africa like it is in South Asia. Similarly, late rains can cause heavy pod infection by Ascochyta Blight, leading to heavy yield losses and poor grain quality. Also, climate change models suggest a high probability of late season rains and thus increase ascochyta infection at poding stage and more intense extreme events such as rainfall above critical threshold, leading to blight epidemics (Abang and Malhotra 2008). Different temperature regimes were found to affect virulence resistance in *Ascochyta*-chickpea pathosystems. In addition to the existing key biotic constraints to chickpea productivity, new or minor pests are likely to become important under climate change. For example, increased dry conditions can favor dry root rot in some countries, and Fusarium wilt in combination with cyst nematode is already on rise in some countries probably due to soil temperature rise (Barhate and Dake 2009; Asaad *et al.*, 2009).

The Indian sub-continent has witnesses a shift in cropping pattern in pulses last three decades. Chickpea has shifted from highly productive irrigated condition in Northern India to rainfed areas in Central and Southern India. This has made diseases *viz.*, Ascochyta blight and Botrytis grey mould less frequent with wilt and root rots are becoming important in newer niches. Changing climatic conditions and abrupt rise in temperature at flowering and pod filling (March-April) accompanied by rains make the crop vulnerable to botrytis gray mold attack. Rust in chickpea was found to cause in areas of Karnataka where the pathogen could overwinter. Similarly cool and dry weather favoured the higher incidence of powdery mildew at Arnej, Gujart where almost all the genotypes were having heavy infestation at pod filling stage. Change in temperature might lead to appearance of different races of the pathogens which are not active but might cause sudden epidemic (Om gupta, 2014). *Spodoptera exigua,* the beet army worm has recently been reported as one of the emrging pests in India (Sharma, 2014).

Studies on temperature response of chickpea cultivars to races of FOC indicated that use of resistant cultivars and adjustments of sowing dates are important measures of management of *Fusarium* wilt. Greenhouse experiments indicated that the chickpea cultivar Ayala was moderately resistant to *F. oxysporum f.sp. ciceris* when inoculated plants were maintained at a day/night temperature regime of 24/21°C but was highly susceptible at 27/25°C. Field experiments in Israel over three consecutive years indicated that the high level of resistance of Ayala to *Fusarium* wilt when sown in mid to late January different from a moderately susceptible reaction under warmer temperature when sowing was delayed to late February or early March. Experiments in growth chambers showed that a temperature increase of 3°C from 24 to 27°C was sufficient for the resistance reaction of cultivars Ayala and PV-1 to race 1A of the pathogen to shift from moderately or highly resistant at constant 24°C to highly susceptible at 27°C. A similar but less pronounced effect was found when Ayala plants were inoculated with *F. oxysporum f. sp. ciceris* race 6. Conversely, the reaction of cultivar JG-62 to races 1A and 6 was not influenced by temperature, but less disease developed on JG-62 plants inoculated with a variant of race 5 of *F. oxysporum f.sp. ciceris* at 27°C compared with plants inoculated at 24°C. These results indicate the importance of appropriate adjustment of temperature in tests for characterizing the resistance reactions of chickpea cultivars to the pathogen, as well as when determining the races of isolates of *F. oxysporum* f.sp. ciceris (Om gupta, 2014).

Dry root rot was found as potentially emerging constraints to chickpea production than wilt. Increased incidence is directly correlated with the climate change variables *viz.,* temperature and moisture on the disease development. Prolonged moisture may create a new scenario of potential diseases as collar rot, wet root rot, anthracnose and Alternaria blight, and rusts. Other pathogens such as powdery mildew tends to strive in conditions with lower moisture (Pande *et al.,* 2010).

Crops with C_3 photosynthesis respond markedly to increasing CO_2 concentrations by inhibiting photorespiration, making photosynthesis more efficient, however, leaf nitrogen and protein concentrations ultimately decrease by more than 12 per cent (Ainsworth and Long 2005). Such a loss of nitrogen and protein significantly diminishes the nutritional value of plant affecting growth and development of insect herbivores either directly or indirectly. In contrast, plants with C_4 photosynthesis will respond little to rising atmospheric CO_2 due to saturation of photosynthesis (Leon Hartwell and Vara Prasad, 2004). Influence of increased CO_2 concentrations (550 and 700 ppm) on chickpea, and Gram caterpillar was studied in relation to ambient CO_2 (380 ppm) concentration under laboratory conditions. The foliar chemistry of chickpea under elevated CO_2 revealed low nitrogen and high carbon content with increased C: N ratio but no change in phenol content. This alteration in food quality significantly affected the growth parameters of Gram caterpillar in the form of increased food consumption, gain in larval weight and more fecal matter production (Khadar *et al.*, 2014). Studies on elevated CO_2 (350, 550 and 750 ppm) on host plant defense response in chickpea against *Helicoverpa armigera by* Sharma *et al.*, 2016 indicated that the activities of defensive enzymes [peroxidase (POD), polyphenol oxidase (PPO), phenylalanine ammonia lyase (PAL) and tyrosine ammonia lyase (TAL)] and amounts of total phenols and condensed tannins increased with an increase in CO_2 concentration in chickpea. The nitrogen balance index was greater in plants kept at 350 ppm CO_2 than in plants kept under ambient conditions. The *H. armigera*-infested plants had higher H_2O_2 content; amounts of oxalic and malic acids were greater at 750 ppm CO_2 than at 350 ppm CO_2. Plant damage was greater at 350 ppm than at 550 and 750 ppm CO_2.

Most of the available data clearly suggests that increased CO_2 would affect the physiology, morphology and biomass of crops (Challinor *et al.*, 2009). Elevated CO_2 and associated climate change have the potential to accelerate plant pathogen evolution, which may, in turn, affect virulence. Pathogens fecundity increased due to altered canopy environment and was attributed to the enhanced canopy growth that resulted in conducive microclimate for pathogen's multiplication (Pangga *et al.*, 2004). Foliar diseases like Ascochyta blights, Stemphylium blights and Botrytis gray mold can become a serious threat in pulses under the higher canopy density. Increased CO2 will lead to less decomposition of crop residues and as a result soil borne pathogens would multiply faster on the crop residues. The greenhouse gases *viz.*, ozone, nitrous oxide, carbon monoxide are secondary pollutants and current climate change scenarios predict an increase of these greenhouse gases which will interfere with photosysthesis and growth process of plants. Ozone can make plants sensitive for attack particularly for root rot fungi *etc.*

Use of Genomics to Develop Climate Resilient Chickpea

A genomics-led breeding strategy for new cultivars for the development of new cultivars that are "climate change ready" (Varshney *et al.*, 2005) commences by defining the stress(es) that will likely affect crop production and productivity under certain climate change scenarios. Data from multi-environment testing provide an opportunity for modeling "stress-impacts" on crops and target populations of environments. Plant breeders and genebank curators will search for morphological and physiological traits in available germplasm that could enhance crop adaptation under such climate variability. In this regard, crop physiology may help define the ideotypes to be pursued for enhancing such adaptation. In case of chickpea efforts at national and international levels have led to the development of large-scale genetic and genomic resources (Varshney *et al.*, 2013). These resources have been used to understand the existing genetic diversity and exploit it in breeding programs. In chickpea, several intra- and inter-specific genetic maps have been developed (Gaur *et al.*, 2011; Gujaria *et al.*, 2011; Thudi *et al.*, 2011; Hiremath *et al.*, 2012) and genomic regions responsible for different biotic stresses (Anbessa *et al.*, 2009; Kottapalli *et al.*, 2009; Anuradha *et al.*, 2011), abiotic stress (Rehman *et al.*, 2011; Vadez *et al.*, 2012) and agronomic traits (Cobos *et al.*, 2009; Rehman *et al.*, 2011; Bajaj *et al.*, 2014, 2015; Das *et al.*, 2015; Kujur *et al.*, 2015a,b) have been reported. The draft genome sequence of both *kabuli* (http://www.icrisat.org/gt-bt/ICGGC/GenomeSequencing.htm) and *desi* (http://www.nipgr.res.in/CGWR/home.php) chickpeas have recently been published (Jain *et al.*, 2013; Varshney *et al.*, 2013a). These sequence data of chickpea will assist in enhancing their crop productivity and lead to conserving food security in arid and semi-arid environments.

Efforts are already being made to integrate molecular markers in chickpea breeding programmes to develop drought tolerant chickpeas. Identification of molecular markers linked to major genes controlling root traits can facilitate marker assisted breeding (MAB) for root traits (Varshney *et al.*, 2013). There are numerous molecular marker analyses of stress tolerance in chickpea have been published to tag traits controlled by several genes (Dita *et al.*, 2006; Millan *et al.*, 2006, Chandra *et al.*, 2004, Imtiaz and Malhotra, 2010, Ur Rehman, 2009). For the future, efforts are going on to identify perfect markers for heat and cold related traits for use in marker-assisted selection (MAS).

Many studies targeted Ascochyta Blight, *Fusarium* wilt diseases and reported a number of QTL studies (Millan *et al.*, 2006; Rajesh *et al.*, 2007). Anbessa *et al.* (2009) studied four population putatively derived from different sources of resistance and found that 5 QTL distributed across linkage groups (LG) 2, 3, 4, 6, and 8, in different cultivars studied. They concluded that linked markers have the potential for gene pyramiding to build durable resistance to safeguard the crop against existing or potential new strains. Similarly, other researchers reported linked markers for AB (Santra *et al.*, 2000; Tekeoglu *et al.*, 2002; Udupa and Baum 2003). Genes for *Fusarium* wilt resistance have also been reported for various races (Millan *et al.*, 2006; Sharma *et al.*, 2004). However, very recently, Gowda *et al.* (2009) reported tightly linked molecular markers and validated those in the commercial cultivars. Advances are being made to develop and validate linked molecular markers in breeding for pest and disease resistance in chickpea (Ali *et al.*, 2011; Allahverdipoor *et al.*, 2011; Ahmad *et al.*, 2014).

Efforts are also under way to develop chickpeas tolerant to drought (Bhatnagar Mathur *et al.*, 2009; Yu X *et al.*, 2016) and *Helicoverpa* pod borer (Ganguly *et al.*, 2014; Chakraborty, 2016) through transgenics.

Conclusion

Change in climate variables in climate change scenario are likely to aggravate the impact of stress (es) *viz.*, drought, heat, low temperature and elevated CO_2 levels on chickpea production and productivity. Tackling drought and heat through crop improvement approaches has given successful results in chickpea while limited progress was reported in breeding for tolerance to low temperature and elevated CO_2 levels. The advent of new generation sequencing, chickpea genome sequencing and availability of large-scale genetic and genomic resources should move along to further unravel the new genes that could be exploited in breeding programs expected at coping with climate change.

REFERENCES

Abang, M and Malhotra, R. S. (2008). Chickpea and climate change. *ICARDA Caravan; Special Issue on Climate Change* 25, 9–12.

Ahmed, Z., Mumtaz, A. S., Ghafoor, A., Ali, A. and Nisar, M. (2014). Marker Assisted Selection (MAS) for chickpea *Fusarium oxysporum* wilt resistant genotypes using PCR based molecular markers. *Molecular Biology Reports* 41 (10), 6755-6762.

Ainsworth, E. A., Beier, C. and Calfapietra, C. (2008). Next generation of elevated (CO_2) experiments with crops, A critical investment for feeding the future world. *Plant, Cell and Environment* 31, 1317–1324.

Ainsworth, E. A, Long, S. P. (2005). What have we learned from 15 years of free-air CO2 enrichment (FACE)? A meta-analytic review of the responses of photosynthesis, canopy. *New Phytology* 165, 351–71.

Ali, Q., Ashan, M., Tahir, M. H. N., Farooq, J., Waseem, M., Anwar, M. and Ahmed, W. (2011). Molecular markers and QTLs for Ascochyta rabiei resistance in chickpea. *International Journal for Agro Veterinary and Medical Sciences* 5 (2), 249-270.

Allahverdipoor, K. H., Bahramnejad, B. and Amini. J. (2011). Selection of molecular markers associated with resistance to *Fusarium* wilt disease in chickpea (*Cicer arietinum* L.) using multivariate statistical techniques. *Austalian Journal of Crop Sciences* 5(13), 1801-1809.

Anbessa, Y., Taran, B., Warkentin, T. D., Tullu, A. and Vandenberg, A. (2009). Genetic analyses and conservation of QTL for Ascochyta blight resistance in chickpea (*Cicer arietinum* L.) *Theory of Applied Genetics* 119, 757–765.

Anuradha, C., Gaur, P. M., Pande, S., Gali, K. K., Ganesh, M., Kumar, J. (2011). Mapping QTL for resistance to botrytis grey mould in chickpea. *Euphytica* 182, 1–9.

Asaad, S., Malhotra, R. S. and Imtiaz, M. (2009). Management of chickpea wilt (*Fusarium oxysporium* f.sp. ciceri) through fungicide seed treatment. *International Conference on Grain legumes-Quality Improvement, Value Addition and Trade (ICGL)*, Indian Institute of Pulses Research, Kanpur, India, 263.

Bajaj, D., Saxena, M. S., Kujur, A., Das, S., Badoni, S., Tripathi, S. (2014). Genome-wide conserved non-coding micro satellite (CNMS) marker- based integrative genetical genomics for quantitative dissection of seed weight in chickpea. *Journal of Experimental Botany* 66, 1271–1290.

Bajaj, D., Upadhyaya, H. D., Khan, Y., Das, S., Badoni, S., Shree, T. (2015). A combinatorial approach of comprehensive QTL- based comparative genome mapping and transcript profiling identified a seed weight- regulating candidate gene in chickpea. *Scientific Reports* 5, 9264.

Barhate, B. G. and Dake, G. N. (2009). Screening of chickpea genotypes to Fusarium wilt. *International Conference on Grain legumes—Quality Improvement, Value Addition and Trade (ICGL)*, Indian Institute of Pulses Research, Kanpur, India, 261.

Basu, P. S., Ah, M. and Chaturvedi, S. K. (2009). Terminal heat stress adversely affects chickpea productivity in Northern India -strategies to improve thermo tolerance in the crop under climate change. *ISPRS Arch.,* XXXVIII-8/W 3 (*Workshop Proceedings, Impact of Climate Change on Agriculture*) 189-193.

Berger, J. D., Ali, M., Basu, P. S., Chaudhary, B. D., Chaturvedi, S. K., Deshmukh, P. S., Dharmaraj, P. S., Dwivedi, S. K., Gangadhar, G. C., Gaur, P. M., Kumar, J., Pannu, R. K., Siddique, K. H. M., Singh, D. N., Singh, D. P., Singh, S. J., Turner, N. C., Yadava, H. S. and Yadav, S. S. (2006). Genotype by environment studies demonstrate the critical role of phenology in adaptation of chickpea (*Cicer arietinum* L.) to high and low yielding environments of India. *Field Crops Research* 98, 230–244.

Bhadauria, U. P.S., Jain, S. and Agrawal, K. K. (2009). Climate change and chickpea (*Cicer arietinum* L.) productivity in central part of Madhya Pradesh. *International Conference on Grain legumes-Quality Improvement, Value Addition and Trade (ICGL), Indian Institute of Pulses Research, Kanpur, India, February,* 14–16.

Bhatnagar-Mathur, P., Vadez, V., Jyostna Devi, M., Lavanya, M., Vani, G. and Sharma, K. K. (2009). Genetic engineering of chickpea (*Cicer arietinum* L.) with the P5CSF129A gene for osmoregulation with implications on drought tolerance. *Molecular Breeding* 23, 591–606.

Blum, A. (1988). Plant Breeding for Stress Environments. CRC Press, Boca Raton, FL.Boote, K. J., Amir, M. H. I., Lafitte, R., Mc Culley, R., Messina, C., Murray, S. C., Specht, J. E., Taylor, S., Westgate, M. E., Glasener, K., Bijl, C. G. and Giese, J. H. (2011). Position statement on crop adaptation to climate change. *Crop Science* 51, 2337-2343.

Brown, S. C., Gregory, P. J., Cooper, P. J. M. and Keatinge, J. D. H. (1989). Root and shoot growth and water use of chickpea (*Cicer arietinum*) grown in dryland conditions, Effect of sowing date and genotype. *Journal of Agricultural Science (Cambridge)* 113, 41–49.

Canci, H. and Toker, C. (2009). Evaluation of yield criteria for drought and heat resistance in Chickpea. *Journal of Agronomy and Crop Sciences* 195, 47-54.

Canci, H. and Toker., C. (2009). Evaluation of annual wild Cicer species for drought and heat resistance under field conditions. *Genetic Resource and Crop Evolution* 56, 1-6.

Chakraborty, J., Sen, S., Ghouse, P., Senugupta, A., Basu, D. and Das, S. (2000) Homologous promoter derived constitutive and chloroplast targeted expression of synthetic *cry1Ac*in transgenic chickpea confers resistance against *Helicoverpa armigera*. *Plant cell, tissue and organ culture* 125 (3), 521-535.

Challinor, A. J., Ewert, F., Arnold, S., Simelton, E. and Fraser, E. 2009. Crops and climate change: Progress, trends and changes in simulating impacts of informing adaptation. *Journal of Experimental Botany* 60, 2775-2789

Chaturvedi, S. K., Mishra, D. K., Vyas, P. and Mishra, N. (2009). Breeding for cold tolerance in chickpea. *Trends in Biosciences* 2, 1–6.

Clarke, H. J., Khan, T. N. and Siddique, K. H. M. (2004). Pollen selection for chilling tolerance at hybridization leads to improved chickpea cultivars. *Euphytica* 139, 65-74.

Clarke, H. J. and Siddique, K. M. M. (2004). Response of chickpea genotypes to low temperature stress during reproductive development. *Field Crop Research* 90, 323-334.

Cobos, M. J., Winter, P., Kharrat, M., Cubero, J. I., Gil, J., Milian, T. (2009). Genetic analysis of agronomic traits in a wide cross of chickpea. *Field Crops Research* 111, 130–136.

Croser, J. S., Clarke, H. J., Siddique, K. H. M. and Khan, T. N. (2003). Low temperature stress, implications for chickpea (*Cicer arietinum* L.) improvement. *Critical Review on Plant Sciences* 22, 185–219.

Das, S., Upadhyaya, H. D., Bajaj, D., Kujur, A. and Badoni, S., Laxmi. (2015). Deploying QTL-seqquence for rapid delineation of a potential candidate gene under lying major trait- associated QTL in chickpea. *DNA Research* 22, 193–203.

Deshmukh, P. S., Uprety, D. C., Dwivedi, N., Kumar, A. R, Bhagat, K, Talwar, S, Kushwaha, S.R. and Singh, T. P. (2009). Impact of climate change on productivity of pulses. In, Ali *et al.*(eds) *Legumes for Ecological Sustainability*, pp. 175–194. *Indian Society of Pulses Research and Development*, Indian Institute of Pulses Research, Kanpur, India.

Devasirvatham, V., Gaur, P. M., Mallikarjuna, N., Raju, T. N., Trethowan, R. M. and Tan, D. K.Y. (2012). Effect of high temperature on the reproductive development of chickpea genotypes under controlled environments. *Functional Plant Biology* 39, 1009-1018.

Devasirvatham, V., Tan, D. K.Y., Gaur, P. M., Raju, T. N. and Trethowan, R. M. (2012). Effects of high temperature at different developmental stages on the yield of chickpea, presented at capturing opportunities and overcoming obstacles in Australian Agronomy. *16th Australian Agronomy Conference, Armidale.*

Devasirvatham, V., Tan, D. K. Y., Gaur, P. M., Raju, T. N. and Trelhowan, R. M. (2012b). High temperature tolerance in chickpea and its implications for plant improvement. *Crop Pasture Science* 63, 419-428.

Dita, M. A., Rispail, N., Prats, E., Rubiales, D. and Singh, K. B. (2006). Biotechnology approaches to overcome biotic and abiotic stress constraints in legumes. *Euphytica*, 147, 1–24.

Dua, R.P. (2001). Genotypic variations for low and high temperature tolerance in gram (*Cicer arietinum*) *Indian Journal of Agricultural Science* 71, 561-566.

FAOSTAT, 2016. Available in http,//faostat.fao.org/site.

Fischer, R. A., Byerlee, D. and Edmeades, G. O. (2009). Can technology deliver on the yield challenge to 2050? Rome, *Food and Agriculture Organization of the United Nations.*

Ganguly, M., Molla, K.A., Karmakar, S., Datta, K., Datta, S.K. (2014). Development of pod borer-resistant transgenic chickpea using a pod-specific and a constitutive promoter-driven fused cry1Ab/Ac gene. Theoretical and Applied Genetics 127(12): 2555-65.

Gaur, P.M., Gowda, C. L. L., Knights, E. J., Warkentin, T. D., Acikgoz, N., Yadav, S. S. And Kumar, J. (2007). Breeding achievements, in Chickpea Breeding and Management (eds S. S. Yudav, B. Redden, W. Chen, and B. Sharma), *CABI, Wallingford*, pp. 391-416.

Gaur, P. M., Srinivasan, S. Gowda, C. L. L. and Rao, B.V. (2007). Rapid generation advancement in chickpea. *Journal of Semi Arid Tropical Agriculture Research* 3, 1-3.

Gaur, P. M., Krishnamurthy, L. and Kashiwagi, J. (2008). Improving drought avoidance root traits in chickpea (*Cicer arietinum* L.), current status of research at ICRISAT. *Plant Production Science* 11, 3-11.

Gaur, P. M., Kumar, J., Gowda, C. L. L., Pande, S., Siddique, K. H. M., Khan, T. N., Warkentin, T. D., Chaturvedi, S. K., Than, A. M. and Ketema, D. (2008). Breeding chickpea for early phenology, perspectives, progress and prospects, in Food legumes for Nutritional Security and Sustainable Agriculture. (ed. M.C. Kharkwal), *Ind ian Society of Genetics and Plant Breeding, New Delhi* 2, 39-48.

Gaur, P. M., Chaturvedi, S. K., Tripathi, S., Gowda, C. L.L., Krishnamurthy, L., Vadez, V.,Mallikarjuna, N. and Varshney, R.K. (2010). Improving heat tolerance in chickpea to increase its resilience to climate change. Abstract presented at the *5th International Food Legumes Research Conference (IFLRCV) and 7th European Conference on Grain Legumes (AEP VII), Antalya.*

Gaur, R., Sethy, N. K., Choudhary, S., Shokeen, B., Gupta, V. and Bhatia, S. (2011). Advancing the stms genomic resources for defining new locations on the intra specific genetic linkage map of chickpea (*Cicer arietinum* L.). *BMC Genomics* 12, 117.

Gaur, P. M., Jukanti, A. K., Srinivasan, S. and Gowda. C. L. L. (2012). Chickpea (*Cicer arietinum* L.), in *Breeding of Field Crops* (ed. D.N. Bharadwaj), *Agrobios (India), Jodhpur,* 165 -194.

Gaur, P. M., Jukanli, A. K. and Varslmey., R. K. (2012). Impact of genomic technologies on chickpea breeding strategies. *Agronomy* 2, 199-221.

Gaur, P. M., Jukanti, A. K. and Srinivasan, S. (2014). Climate change and heat stress tolerance in chickpea. In, Tuteja N, Gill SS (eds), *Climate Change and Plant Abiotic Stress Tolerance. Wiley-VCH, Weinheim, Germany* 839–855.

Gholipoor, M. (2007). Potential effects of individual versus simultaneous climate change factors on growth and water use in chickpea. *International Journal of Plant Production* 1(2),September, 2007.

Gholipoor, M., Soltani, A. (2009). Future climate impacts on chickpea in Iran and ICARDA. *Research Journal of Environmental Sciences* 3, 16–28.

Gowda, S. J. M., Radhika, P., Kadoo, N. Y., Mhase, L. B. and Gupt, V. S. (2009). Molecular mapping of wilt resistance genes in chickpea. *Molecular Breeding* 24, 177–183.

Guerenstein, P. G. and Hildebrand, J. G. (2008). Roles and effects of environmental carbon dioxide in insect life. *Annual Review of Entomology* 53, 161-178.

Gujaria, N., Kumar, A., Dauthal, P., Dubey, A., Hiremath, P. and Prakash, A. B. (2011). Development and use of genic molecular markers (GMMS) for construction of a transcript map of chickpea (*Cicer arietinum* L.). *Theory of Applied Genetics.* 122, 1577–1589.

Hall, A.E. (2001). Crop Responses to Environment. C R C Press, Boca Raton, FL.

Hajarpoor, A., Soltani, A., Zeinali, E. and Sayyedi, F. (2014). Potential benefits from adaptation to climate change in chickpea. *Agriculture Science Developments* 3(7), 230-236.

Haware, M. P. and Nene, Y. L. (1982). Races of *Fusarium oxysporum* f. sp. *ciceris*. *Plant Disease* 66, 809–810.

Heidarvand, L., Amri, R. M., Naghavi, M. R., Farayedi, Y., Sadeghzadeh, B. and Alizadeh, K.H. (2011). Physiological and morphological characteristics of chickpea accessions under low temperature. *Russian Journal of Plant Physiology* 58, 157–163.

Hiremath, P. J., Kumar, A., Penmetsa, R. V., Farmer, A., Schlueter, J. A., Chamarthi, S. K. (2012). Large-scale development of cost-effective snp marker assays for diversity assessment and genetic mapping in chickpea and comparative mapping in legumes. *Plant Biotechnology Journal* 10,716–732.

Hunter, M. D. (2001). Effect of atmospheric carbon dioxide on insect-plant interactions. *Agriculture and Forest Entomology* 3, 153-159.

Imtiaz, M. and Malhotra, R. S. (2009). Reduce stress, Breed for drought tolerance. *ICARDA Caravan, Science for Food Security* 26, 34–36.

Imtiaz, M., Malhotra, R. S. (2010). Breeding chickpea for drought tolerance, Conventional and molecular approaches. *5th International Food Legumes Research Conference and 7th European Conference on Grain Legumes on Grain legumes, Antalya, Turkey*, 192.

IPCC (2007). Climate change 2007. The physical science basis, in Contribution of Working Group I to the Fourth Assessment Report of the Intergovernmental Panel on Climate Change (eds S. Solomon, D. Qin, M. Manning, Z. Chen, M. Marquis, K.B. Averyt, M. Tignor and H.L. Miller), *IPCC, Geneva.*

Jain, M., Misra,G., Patel, R. K., Priya, P., Jhanwar, S., Khan, A. W. (2013). A draft genome sequence of the pulse crop chickpea (*Cicer arietinum* L.). *Plant Journal* 74, 715–729.

Jim´enez-D´yaz, R. M., Alcal´a-Jim´enez, A. R., Herv´as, A and Trapero-Casas, J. L. (1993). Pathogenic variability and host resistance in the *Fusarium oxysporum f. sp. ciceris/C.arietinum* pathosystem. *Proceedings of the 3rd European Seminar on*

Fusarium Mycotoxins, Taxonomy, Pathogenicity and Host Resistance, Plant Breeding and Acclimatization, Inst, Radzik´ov, Poland, pp. 87–94.

Jimenez-Gasco, M. M., Milgroom, M. G. and Jimenez-Diaz, R. M. (2004). Stepwise evolution of races in *Fusarium oxysporum* f. sp. *ciceris* inferred from fingerprinting with repetitive DNA sequences. *Phytopathology* 94, 228–235.

Jukanti, A. K., Gaur, P. M., Gowda, C. L. L. and Chibbar. R. N. (2012). Chickpea, nutritional properties and its benefits. *British Journal of Nutrition* 108, S 11-S 16.

Kashiwagi, J., Krishnamurthy, L., Upadhyaya1, H. D., Krishna, H. and Chandra, S. (2005). Genetic variability of droughtavoidance root traits in the mini core germplasm collection of chickpea (*Cicer arietinum* L.) *Euphytica*.146,213–222.

Kashiwagi, J., Krishnamurthy, L., Crouch, J. H. and Serraj, R. (2006). Variability of root length density and its contributions to seed yield in chickpea (*Cicer arietinum* L.) under terminal drought stress. *Field Crops Research*. 95, 171–181.

Khadar, A. B., Prabhuraj, A., Srinivasa Rao, M., Sreenivas, A. G. and Naganagoud, A. (2014). Influence of elevated CO_2 associated with chickpea on growth performance of gram caterpillar, helicoverpa armigera. *Applied Ecology and Environmental Research* 12 (2), 345-353.

Kimball, B. A. (1983). Carbondioxide and agricultural yield, An assemblage and analysis of 430 Prior observations. *Agronomy Journal*, 75, 779–788.

Kottapalli, P., Gaur, P. M., Katiyar, S. K., Crouch, J. H., Buhariwalla, H. K., Pande, S. (2009). Mapping and validation of QTLs for resistance to an indian isolate of ascochyta blight pathogen in chickpea. *Euphytica* 165, 79–88.

Krishnamurthy, L. Kashiwagi, J., Gaur, P. M., Upadhyaya, H. D. and Vadez, V. (2010). Sources of tolerance to terminal drought in the chickpea (*Cicer arietinum* L.) minicore germplasm. *Field Crop Research* 119, 322-330.

Krishnamurthy, L., Gaur, P. M., Basu, P. S., Chaturvedi, S. K., Tripathi, S., Vadez, V., Rathore, A., Varshney, R. K. and Gowda, C. L. L. (2011). Large genetic variation for heat tolerance in the reference collection of chickpea (*Cicer arietinum* L.) germplasm. *Plant Genetic Resource* 9, 59-69.

Kujur, A., Bajaj, D., Upadhyaya, H. D., Das, S., Ranjan, R., Shree, T. (2015b). A genome-wide SNP scan accelerates trait-regulatory genomic loci identification in chickpea. *Scientific Reports*. 5, 11166.

Kujur, A., Upadhyaya, H. D., Shree, T., Bajaj, D., Das, S., Saxena,M. (2015a). Ultra-high density intra-specific genetic linkage maps accelerate identification of functionally relevant molecular tags governing important agronomic traits in chickpea. *Scientific Reports* 5, 9468.

Kumar, S. (2006). Climate change and crop breeding objectives in the twenty first century. *Current Science* 90, 1053-1054.

Leon Hartwell and Vara Prasad. (2004). Crop response to elevated Carbon dioxide. *Encyclopedia of Plant and Crop Science*. DOI: 10.1081/E-EPCS 120005566.

Levitt, J. (1980). Responses of plants to environmental stresses. Vol. II. Water, Radiation, Salt and other Stresses. 2nd edn., *Academic Press, New York, NY.*

Ludlow, M. M. and Muchow, R. C. (1990). A critical evaluation of traits for improving crop yields in water-limited environments. *Advances in Agronomy* 43, 107–153.

Maesen, V. L. J. G. (1972). *Cicer* L., A Monograph of the Genus, with special references to the Chickpea (*Cicer arietinum L.*) its ecology and cultivation. *Mendelingen Land bouwhogeschool, Wageningen.*

Malhotra, R. S. and Singh, K. B. (1990). The inheritance of cold tolerance in chickpea. *Journal on Genetics and Plant Breeding,* 44, 227–230.

Malhotra, R. S. and Singh, K. B. (1991). Gene action for cold tolerance in chickpea. *Theoretical and Applied Genetics,* 82, 598–601.

Millan, T., Clarke, H. J., Siddique, K. H. M., Buhariwalla, H.K., Pooran, M. G., Kumar, J., Gill, J., Kahl, G. and Winter, P. (2006). Chickpea molecular breeding, New tools and concepts. *Euphytica* 147, 81–103.

Nagarajarao, Y., Mallick, S. and Singh, G. C. (1980). Moisture depletion and root growth of different varieties of chickpea under rainfed conditions. *Indian Journal of Agronomy* 25, 289-293.

Norby, R. J., Willschleger, S. D., Gunderson, C. A., Johnson, D. W. and Ceulemans, R. (1999). Tree responses to rising CO_2 in field experiments: implications for the future forest. *Plant cell and Environment* 22, 683-714.

Pande, S., Sharma, M., Rao, S. K. and Sharma, R. N. (2010). *Annual Progress report 2009-10.* Collaborative work on "Enhancing chickpea production in rainfed rice fallow lands (RRFL) of Chhattisgarh and Madhya Pradesh states of India following Improved Pulse Production and Protection Technologies (IPPPT)". International Crops Research Institute for the Semi-Arid Tropics (ICRISAT), Indira Gandhi Krishi Vishwavidyalaya (IGKV), Raipur, CG and Jawharlal Nehru Krishi Vishwa Vidyalaya (JNKVV), Jabalpur, MP.

Pangga, I. B., Chakaraborthy, S. and Yates, D. (2004). Canopy size and induced resistance *Stylosanthes scabra* determine anthracnose severity at high CO2. *Phytopathology* **94**, 221-227.

Rajesh, P. N., Mc Phee, K. E., Ford, R., Pittock, C., Kumar, J. and Muehlbauer, F. J. (2007). Ciceromics, Advancement in genomics and recent molecular techniques In, SS Yadav *et al.* (eds) *Chickpea Breeding and Management*, pp. 445–457. *CAB International*, Oxford shire.

Rehman, A. U., Malhotra, R. S., Bett, K., Tar'an, B., Bueckert, R., and Warkentin, T. D. (2011). Mapping QTL associated with traits affecting grain yield in chickpea (*Cicer arietinum* L.) under terminal drought stress. *CropScience* 51, 450–463.

Rogers, A., Ainsworth, E. A. and ADB Leakey. (2009). Will elevated carbon dioxide concentration amplify the benefits of nitrogen fixation in legumes? *Plant Physiology* 151, 1009–1016.

Santra, D. K., Tekeoglu, M., Ratnaparkhe, M., Kaiser, W. J. and Muehlbauer, F. J. (2000). Identification and mapping of QTL conferring resistance to Ascochyta blight in chickpea. *Crop Science,* 40, 1606–1612.

Saxena, N. P., Johansen, C., Saxena, M. C. and Silim, S. N. (1993). Selection for drought and salinity, A case study with chickpea. In, KB Singh, MC Saxena (eds) *Breeding for Stress Resistance in Cool-Season Food Legumes,* p. 11. JohnWiley and Sons, New York, NY.

Sharma, H. C. (2014). Climate change effects on pest spectrum and population dynamics in grain legumes. National conference on pulses: challenges and opportunities under changing climate scenario, 29th September- 1st October, 2014, pp.79.

Sharma, K. D., Winter, P., Kahl, G. and Muehlbauer, F. J. (2004). Molecular mapping of *Fusarium oxysporum* f. sp. *ciceris* race 3 resistance gene in chickpea. *Theoretical and Applied Genetics* 108, 1243–1248.

Silim, S. N. and Saxena, M. C. (1993). Adaptation of spring-sown chickpea to the mediterranean basin. II. Factors influencing yield under drought. *Field Crops Research* 34, 137–146.

Singh, K. B., Malholra, R. S., Halila, M. H., Knights, E. J., and Verma, M. M. (1994). Current status and future strategy in breeding chickpea for resistance to biotic and abiotic stresses, in Expending the production and Use of Cool Season Food Legumes. (*eds F.J. Muehlbauer and W.J. Kaiser*), *Kluvver, Dordrecht,* pp. 572-591.

Singh, F. and Diwakar, B. (1995). Chickpea Botany and Production Practices, Skill Development Series. 16, *International Crops Research Institute for the Semi-Arid Tropics, Patancheru.*

Singh, K. B., Malhotra, R. S. and Saxena, M. C. (1990). Sources for tolerance to cold in *Cicer* species. *Crop Science* 30, 1136–1138.

Singh, K. B., Malhotra, R. S. and Saxena, M. C. (1995). Additional sources of tolerance to cold in cultivated and wild *Cicer* species. *Crop Science* 35, 1491–1497.

Singh, P., Nedumaran, S., Traore, P., Boote, K., Rattunde, H., Prasad, P., Singh, N., Srinivas, K.and Bantilan, M., 2014. Quantifying potential benefits of drought and heat tolerance in rainy season sorghum for adapting to climate change. *Agricultural and Forest Meteorology* 185, 37–48.

Solomon, S., Qin, D. and Manning, R. B. (2007). Technical summary. In, Solomon, S., Qin, D., Manning, M., Chen, Z., Marquis, M., Averyt, K. B., Tignor, M. and Miller, H. L. (eds) *Climate Change 2007, The Physical Science Basis. Contribution of Working Group I to the Fourth Annual Assessment Report of the Intergovernmental Panel on Climate Change,* p. 996. Cambridge University Press, Cambridge, UK, and New York, NY.

Srinivasan, A., Johansen, C. and Saxena, N. P. (1998). Cold tolerance during early reproductive growth of chickpea (*Cicer arietinum* L.) characterization of stress and genetic variation in pod set. *Field Crop Research* 57, 181-193.

Srivastava, A. C., Sengupta, U. K. and Pal, M. (2001). Growth, CO_2 exchange rate and dry matter partitioning in mungbean grown under elevated CO2. *Indian Journal of Experiemntal Biology* 39, 572–577.

Srivastava, A. K., Gupta, B. R. D. and Panda, B. C. (2009). Impact of climate change on Chickpea yield in central and Northwest India. *Challenges and Opportunities in Agrometeorology*, New Delhi, February 23–25.

Summerfield, R. J., Hadley, P., Roberts, E. H., Min chin, F. R. and Rawsthrone, S. (1984). Sensitivity o f chickpea (*Cicer arietinum* L.) to hot temperatures during the reproductive period. *Experimental Agriculture* 20,77-93.

Thudi, M., Bohra, A., Nayak, S. N., Varghese, N., Shah, T. M., Penmetsa, R. V. (2011). Novel SSR markers from bac-end sequences, dart arrays and a comprehensive genetic map with 1,291 marker loci for chickpea (Cicer arietinum L.). *PLOS ONE* 6, e27275. doi: 10.1371/journal.pone.0027275

Toker, C. (2005). Preliminary screening and selection for cold tolerance in annual wild *Cicer* species. *Genetic Resource and Crop Evolution* 52, 1–5.

Tekeoglu, M., Rajesh, P. N. and Muehlbauer, F. J. (2002). Integration of sequence tagged micro satellites to the chickpea genetic map. *Theoretical and Applied Genetics* 105, 847–854.

Tubiello, F. N., Soussana, J. F., Howden, M. and Easterling, W. 2007. Crop and pasture response to climate change; fundamental processes. *Proceedings of National Academy of Sciences* 104, 19686-19690.

Udupa, S. M. and Baum, M. (2003). Genetic dissection of pathotype-specific resistance to ascochyta blight disease in chickpea (*Cicer arietinum* L.) using microsatellite markers. *Theoretical and Applied Genetics*, 106,1196-1202.

Upadhyaya, H. D., Dwivedi, S. L., Baum, M., Varshney, R. K., Udupa, S. M., Gowda, C. L. L., Hoisington, D. and Singh, S. (2008). Genetic structure, diversity, and allelic rich ness in composite collection and reference set in chickpea (*Cicer arietinum* L.) *BMC Plant Biology* 8, 106.

Upadhyaya, H. D., Dronavalli, N., Gowda, C. L. L. and Singh, S. (2011). Identification and evaluation of chickpea germplasm for tolerance to heat stress. *Crop Science* 51, 2079- 2094.

Uprety, D. C., Dwivedi, N. and Mohan, R. (1997). Interactive effect of rising atmospheric carbon dioxide and drought on nutrient and constitutes of *Brassica juncea* seeds. *Science and Culture* 63, 291–292.

Ur Rehman, A. (2009). *Characterization and Molecular Mapping of Drought Tolerance in Kabuli Chickpea* (*Cicer arietinum* L.) A PhD thesis Submitted to the College of Graduate Studies and Research, Department of Plant Sciences University of Saskatchewan, Saskatoon, Canada.

Vadez, V., Krishnamurthy, L., Thudi, M., Anuradha, C., Colmer, T. D., Turner, N. C. (2012). Assessment of ICCV2 × JG62 chickpea progenies shows sensitivity of reproduction to salt stress and reveals QTL for seed yield and yield components. *Molecular Breeding* 30, 9–21.

Varshney R. K., Graner A. and Sorrells M. E. (2005) Genomics-assisted breeding for crop improvement. *Trends in Plant Sciences* 10, 621–630.

Varshney R. K., Mohan S. M., Gaur P. M., Gangarao N. V., Pandey M. K., Bohra A., Shrikant, L., Sawargaonkara, Annapurna, C., Paul, K., Janilaa, P., Saxenaa, K.B., Fikrei, A., Sharmaa, M., Rathorea, A., Pratap, A., Tripathi, S., Datta, S., Chaturvedi, S.K., Mallikarjuna, N., Anuradhak, G., Babbar, A., Choudhary, A. K., Mhase, M.B., Bharadwaj, Ch., Mannur, D.M., Harern, P.N., and Guo, B. (2013a). Achievements and prospects of genomics-assisted breeding in three legume crops of the semi-arid tropics. *Biotechnology Advances* 31, 1120–1134.

Varshney R. K., Song, C., Saxena, R. K., Azam, S., Yu, S., Sharpe, A. G., Cannon, S. B., Baek, J., Tar'an, B., Millan, T., Zhang, X., Rosen, B., Ramsay, L. D., Iwata, A., Wang, Y., Nelson, W., Farmer, A. D., Gaur, P. M., Soderlund, C., Penmetsa, R. V., Xu, C., Bharti, A.K., He, W., Winter, P., Zhao, S., Hane, J. K., Carrasquilla-Garcia, N., Condie, J. A.,Upadhyaya, H. D., Luo, M., Singh, N. P., Lichtenzveig, J., Gali, K. K., Rubio, J.,Nadarajan, N., Thudi, M., Dolezel, J., Bansal, K. C., Xu, X., Edwards, D., Zhang, G.,Kahl, G., Gil, J., Singh, K. B., Datta, S. K., Jackson, S. A., Wang, J. and Cook, D. (2013). Draft genome sequence of kabuli chickpea (*cicer arietinum*): Genetic structure and Breeding constraints for crop improvement. *Nature Biotechnology* 31, 240–246.

Wang, J., Gan, Y.T., Clarke, F. and McDonald, C. L. (2006). Response of chickpea yield to high temperature stress during reproductive development. *Crop Science* 46, 2171–2178.

Wery, J, Silim, S. N., Knights, E. J., Malhotra, R. S. and Cousin, R. (1994). Screening techniques and sources of tolerance to extremes of moisture and air temperature in cool season food legumes. *Euphytica* 73, 73–83.

William, P. C. and Singh, U. (1987). The chickpea - nutritional quality and the evaluation of quality in breeding programs, in the chickpea (eds M.C. Saxena, K.B. Singh), *ICARDA, Aleppo*, 329 -356.

Yadav, S. S., Kumar, J., Yadav, S. K., Singh, S., Yadav, V. S., Turner, N. C. and Redden, R. (2006). Evaluation of Helicoverpa and drought resistance in desi and kabuli chickpea. *Plant Genetic Resources* 4(3), 198–203.

Yu, X., Liu, Y., Wang, S., Tao, Y., Wang, Z., Mijiti, A., Wang, Z., Zhang, H. and Ma, H. (2016). A chickpea stress-responsive NAC transcription factor, *CarNAC5*, confers enhanced tolerance to drought stress in transgenic *Arabidopsis*. 79 (2), 187-197.

Section C

Nutritional Heritage of Pulses

2018, *Climate Risks Management: Sustainable Pulse Production*
Editors: A K Srivastava and Yogranjan
Published by: **ASTRAL INTERNATIONAL PVT. LTD., NEW DELHI**   *Pages 133–146*

Chapter 6

Underutilized Pulses: The Keys to Food and Nutritional Security

Kavita Bisht[1], Preeti Bora[2] and Pratibha Joshi[3]

[1, 3]ICAR–Indian Agricultural Research Institute, New Delhi
[2]Uttarakhand Open Universtiy, Haldwani, Uttarakhand

ABSTRACT

Pulses are considered as most economical source of protein, and are enriched with high fibre content ample quantity of vitamins and minerals. The pulse protein is of low quality as they are deficient in methionine. But being rich in lysine, which is lacking in cereals, they can supplement proteins of cereals. Pulses also contain a number of bioactive substances that cannot be considered as nutrients but which exert metabolic effect on the humans or animals that may be regarded as positive, negative or both. Antinutrients like phytates, tannins, trypsin inhibitors, haemagglutinins, goitrogens etc. present in pulses can significantly reduce their nutritional value. These anti-nutrients can be reduced or degraded completely with simple cooking methods. The present article highlights the nutritional composition and culinary uses of underutilized pulses along with their non-nutrient components and culinary uses of pulses.

Keywords: Underutilized pulses, Nutrient, Non-nutrient components, Culinary uses.

The contribution of pulses in global concern highlights the position of India as largest producer and consumer in the world. India contributes to 34 per cent of cultivable area and 25 per cent production of world food legumes. It is the top pulse producing country with 26.17 million hectare area under pulses cultivation and the production of pulses for the year 2014-15 was around 17.15 million tons. In India, major pulse crops grown are chickpea (chana), pigeonpea (arhar), mungbean (mung), urdbean, lentil (masoor)), lathyrus (khesri dal), cowpea (lobhiya), moth bean, horse gram (kulthi) *etc.* It has been documented that the major pulse producing states in the country are Madhya Pradesh, Uttar Pradesh, Maharashtra, Rajasthan, Karnataka, Gujarat, Tamilnadu, Jharkhand, Bihar and Andhra Pradesh, which together contribute for 75 per cent of the total pulse production in the country. Among a diverse array of pulses being grown in India, a number of pulses are still underutilized and mainly cultivated in marginal land with low inputs; hence, their productivity is considerably low. However, these underutilized pulses have capacity to fit it many cropping systems and also ability to survive adverse climatic risks.

Nutritional Composition and Culinary Uses of Pulses

The protein content of pulses ranges from 18 to 25 per cent. In a vegetarian diet, they are an important source of protein. The pulse protein is of low quality as they are deficient in methionine. But being rich in lysine, these pulses can supplement proteins of cereals. Pulses with their higher content of the total fat contribute along with cereals to meet the essential fatty acid needs of an adult. They are good source of thiamine, riboflavin and phosphorus and fair source of niacin, calcium and iron. They also being good source of complex carbohydrate are categorized as low glycemic index food and recommended for people suffering from diabetes.

Lentil

Lentil or *Masoor* (*Lens esculenta*) is an important pulse crop grown globally in many countries like India, Canada, Turkey, Bangladesh, Iran, China, Nepal and Syria. India ranks first in the world with respect to its production. Cultivated in an area of around 1.42 million hectare lentil has a productivity of 1.13 million tones with a yield of 787kg/hectare. Lentil derives its name *Lens* from its lens shaped seeds. The protein content of lentil is about 25-28 per cent and it is also an excellent source of vitamins like niacin, minerals like calcium and iron; and dietary fibers (Tables 6.1–6.4) that contribute to its several health benefits. Lentil is mainly used as split *dhal*. The whole grain is also used in some culinary preparations. Young pods are eaten as vegetables. Lentil flour is used for thickening of soups. It is also mixed with wheat flour in bread and cake preparations.

Table 6.1: Proximate Composition of Underutilized Pulses (per 100 g)

Pulse	Moisture (g)	Protein (g)	Fat (g)	Minerals (g)	Crude Fiber (g)	Carbo-hydrates (g)	Energy (kcal)
Lentil	12.4	25.1	0.7	2.1	0.7	59.0	343
Moth Bean	10.8	23.6	1.1	3.5	4.5	56.5	330
Cow pea	13.4	24.1	1.0	3.2	3.8	54.5	323
Horse gram (whole)	11.8	22.0	0.5	3.2	5.3	57.2	321
Kidney Bean	12.0	22.9	1.3	3.2	4.8	60.6	346
Lathyrus	10.0	28.2	0.6	2.3	2.3	56.6	345

Source: Nutritive Value of Indian Foods by C. Gopalan, B.V. Rama Sastri and S. C. Balasubramanian.

Moth Bean

Moth bean or *Matki* Bean (*Phaseolus aconitifolius*) is also known as Terapy bean. It is an important kharif pulse crop native to India owing to widespread cultivated and wild forms. India is a major producer of Moth bean cultivating it in an area of 1.37 million hectare with a production of 0.47 million tones leading to a yield of 346 kg/hectare. The yield of the grain is low compared to other pulses. Moth bean has been identified as one of the potential protein food source. It is rich in protein and calcium, micronutrients (Tables 6.1–6.4) which makes it an excellent supplement to cereal diet. The mature dry grain can be used either whole or split *dhal*. Besides this its seeds can be used in other popular Indian preparations like *papad, kheech, bhajis* and *dalmoth*. The green pods are used as vegetables too. Moth bean is the most drought tolerant crop among the pulses and it enriches the soil by its biological nitrogen fixation.

Table 6.2: Vitamin Contents of Underutilized Pulses (per 100 g)

Pulse	Carotene (µg)	Thiamine (mg)	Riboflavin (mg)	Niacin (mg)	Total Folic Acid (µg)
Lentil	270	0.45	0.20	2.6	36
Moth Bean	9	0.45	0.09	1.5	-
Cow pea	12	0.51	0.20	1.3	133
Horse gram (whole)	71	0.42	0.20	1.5	-
Kidney Bean	-	-	-	-	-
Lathyrus	120	0.39	0.17	2.9	-

Source: Nutritive Value of Indian Foods by C. Gopalan, B.V. Rama Sastri and S. C. Balasubramanian.

Cow Pea

Cow pea or *lobia* (*Vigna unguiculata*) is an important and commonly used food legume in India. In global context being drought tolerant and ability to grow in poor soil. It forms major staple food in many parts of Africa. Although not extensively cultivated it is one of the pulse used from primitive agriculture in India. Area under cow pea cultivation in India is around 3.9 million hectares with a production of 2.2 million tons having a yield of around 683 kg/hectare. Many cultivars of cowpea are commercially grown for its long green pods as vegetable, for its dry seeds as pulse and for its foliage as fodder of animals. Cow pea is a good source of protein, minerals, micronutrients and carbohydrates (Tables 6.1–6.4). In India cowpea is used as whole or as split *dhal*. It is dehusked and also used for making ground flour for various culinary preparations. The seeds may also be consumed after germination. However cowpea is considered to be 'difficult to digest' and hence less popular as compared to other popular pulses like black gram for consumption.

Horse Gram

Horse gram or Kulthi (*Dolichos biflorus*) is a crop of dry tropics and mainly cultivated in southern parts of India. Being an important pulse crop of India, it is

Table 6.3: Mineral and Trace Elements of Underutilized Pulses (mg per 100 g of edible portion)

Pulse	Ca	P	Fe	Mg	Na	K	Cu	Mn	Mo	Zn	Cr	S	Cl
Lentil	69	293	7.58	80	40.1	629	1.87	1.04	0.171	2.8	0.024	104	19
Moth Bean	202	230	9.5	225	29.5	1096	0.85	-	-	-	-	180	9
Cow pea	77	414	8.6	210	23.2	1131	0.87	1.34	1.890	4.6	0.029	165	10
Horse gram (whole)	287	311	6.77	156	11.5	762	1.81	1.57	0.749	2.8	0.024	181	8
Kidney Bean	260	410	5.1	184	-	-	1.45	1.60	-	4.5	0.029	-	-
Lathyrus	90	317	6.3	92	37.7	644	0.77	-	-	-	-	144	36

Source: Nutritive Value of Indian Foods by C. Gopalan, B.V. Rama Sastri and S. C. Balasubramanian.

Table 6.4: Essential Amino Acid Composition of Underutilized Pulses (mg per g N)

Pulse	Total N (g/100g)	Arginine	Histidine	Lysine	Tryptophan	Phenyl-alanine	Tyrosine	Methionine	Cystine	Threonine	Leucine	Isoleucine	Valine
Lentil	4.02	540	160	440	60	270	200	50	70	220	470	270	310
Moth Bean	3.78	-	210	340	40	280	-	60	30	-	420	310	200
Cow pea	3.86	420	200	430	70	320	230	90	80	230	480	270	310
Horse gram (whole)	3.52	530	190	520	70	380	-	70	130	230	540	370	390
Kidney Bean	3.66	370	180	460	60	340	100	60	40	270	470	300	330
Lathyrus	4.51	490	160	470	50	260	-	30	70	140	410	410	250

Source: Nutritive Value of Indian Foods by C. Gopalan, B.V. Rama Sastri and S. C. Balasubramanian.

grown in an area covering 0.45 million hectares with a productivity of 0.21 million tons. The yield of horse gram is approximately 464 kg/hectare. The seeds of horse gram vary from light red, brown to grey, black or mottled which is affected by growth and moisture conditions during seed ripening period. It is a good source of protein (18 to 29 per cent), minerals like calcium and iron and vitamins like thiamine, riboflavin, niacin and vitamin C (Tables 6.1–6.4). The pulse is extensively used for cattle and horse feed and with regard to human use it is considered to be a poor man's pulse. The whole grains of this pulse are eaten both in boiled and fried form while the other uses include as sprouts and in other culinary preparation in the form of flour.

Kidney Bean

Kidney bean or *Rajmah* (*Phaseolus vulgaris*) is known by many other names like haricot bean, common bean, snap bean or navy bean. It is globally cultivated and Brazil is the leading producer of Kidney beans. In India it is grown in the states of Maharashtra, Jammu and Kashmir, Himachal Pradesh and Uttar Pradesh hills, Uttarakhand, Nilgiri hills, Kerala hills and Darjeeling (West Bengal) hills. Kidney bean has an average yield of 1000-1200 kg/hectare. It is named for its visual resemblance in shape and colour to a kidney. Kidney beans are a very good source of dietary fiber. It is also a good source of protein, folate, thiamin, vitamin K and minerals like manganese, phosphorus, iron, cobalt, magnesium and potassium (Tables 6.1–6.4). In addition to this it is also a good source of molybdenum, a trace mineral that play an essential role in detoxification. As it is grown globally it varies in size, shape and colour of pods and seeds. The beans are used as vegetable when the seeds are immature and green. French bean is the most important variety being used as immature pods and less fiber. Dry bean is used as *dhal*. Red kidney beans called as *Rajmah* are an integral part of cuisine in northern region of India.

Lathyrus

Lathyrus or *khesari dhal* (*Lathyrus sativus*) is commonly grown for human consumption and livestock food in Asia and East Africa. It is a pulse that grows in adverse conditions like drought and famine. It is commonly known as grass pea which has an immense potential as a food, feed, fodder as well as green forage manure. In India, edible seeds of plants are cultivated in some states *i.e.* Bihar, Chattisgarh, West Bengal, Orissa and Andhra Pradesh for local cuisine. Although being a good source of dietary protein, carbohydrates and minerals (Tables 6.1–6.4), consumption of *Khesari dhal* leads to the pathological condition of Lathyrism amongst people who consume it in large quantities, commonly seen in India. This condition results in muscular rigidity, weakness and paralysis of leg muscle. The dehusked seed of *Khesari dhal* resembles Bengal gram *dhal* and red gram *dhal*, therefore sometimes also used as adulterant in other pulses. The neurotoxin responsible for lathyrism in *Khesari dhal* can be removed by steeping or parboiling.

Non-nutrient Bioactive Compositions

Pulses are sound reservoir of bioactive substances that cannot be considered as nutrients but which exert metabolic effect on the humans or animals that may be regarded as positive, negative or both. Most bioactive substances have been classified as "antinutritional factors" and known by different names as toxic constituents, anti-nutrients or antinutritional factors (or compounds), bioactive substances, nutritive factors, associated substances, micronutrients and phytochemicals *etc.* (Champ, 2002). Anti-nutritional factors, although are nontoxic but generate adverse physiological responses and interfere with the utilization of nutrients. The important anti- nutritional compositions are trypsin inhibitors, phytates, oxalates, tannins, lectins and goitrogens. Some of the compounds like lathyrogens may be toxic while others like fibers are beneficial. These anti-nutritional factors, toxins and dietary fibers are important for assessing the overall food value of a pulse.

Phytates

Phytates is hexa phosphate of inositol which is located in the endosperm portion of legume seeds. It occurs as a mineral complex and is insoluble at the physiological pH of the intestine. Many human studies demonstrate that phytates inhibit absorption of iron, zinc and calcium (Brune *et al.*, 1992; Hurell *et al.*, 1992; Champ, 2002; Sandberg, 2002). Phytic acid is very much reactive with positively charged ions, such as minerals (especially Zn, Ca and Fe), thereby forming insoluble complexes that are less available for digestion and absorption in the small intestine. Alternatively, the ability of phytic acid to chelate minerals may have protective effects, such as decreasing the risk of iron-mediated colon cancer and lowering serum cholesterol and triglycerides in experimental animals (Cheryan, 1980; Sandberg, 2002). Oral administration of phytic acid inhibited colon carcinogenesis in rodents during the initiation and post initiation stages. Phytic acid seems to act mainly as an antioxidant, reducing the rate of cell proliferation and augmenting the immune response by enhancing the activity of natural killer cells (Reddy, 1999). The possible mechanism of phytic acid as a lipid lowering agent include its ability to bind Zn and thus lower the plasma Zn:Cu ratio to reduce plasma glucose and insulin concentrations, which may lead to a reduced stimulus for hepatic lipid synthesis (Klevery, 1977; Wolever, 1990). The underutilized pulses which are rich in phytates include horse gram, kidney bean, lentil, cow pea *etc.* However, simple processing like germination, cooking, roasting, soaking and fermentation cause significant reduction in phytic acid content in these legumes (Khamgaonkar *et al.*, 2013).

Tannins

Tannins are one of several antinutritional factors which are located mainly in the seed coat or testa of dry beans. The tannin content of dry beans ranges from 0 to 2.0 per cent depending on the bean species and color of the seed coat. Many high tannin bean varieties are of lower nutritional quality than low tannin varieties of beans. Researches have demonstrated that bean tannins decrease protein digestibility, either by inactivating digestive enzymes or by reducing the susceptibility of the substrate proteins after forming complexes with tannins and absorbed ionizable iron (Reddy *et al.*, 1985; Broughton *et al.*, 2003). Mainly cowpea,horse gram are the rich source of tannins. The presence of tannins enable these underutilized pulses to minimize the effect of draught.

Trypsin Inhibitor

Among the underutilized pulses, trypsin inhibitors are present in cowpea, double beans, and *lathryus sativus*. Trypsin inhibitors are proteins that inhibit the activity of trypsin in the gut and interfere with digestibility of dietary proteins thereby reducing their utilization. Pancreas enlargement and growth retardation occur in animals that consume diet containing trypsin inhibitors. The release of essential amino acids, particularly, methionine is hampered by the presence of inhibitors.

Haemagglutinins

These are proteinous in nature and also referred as phytoagglutinins or lectins. These occur widely in leguminous seeds and have been isolated from field bean, white bean, double bean and horse gram. Haemagglutinins combine with the cells lining the intestinal wall, in almost the same way as it combines with red blood cells thus causing an impairment with the absorption of amino acids. The lectins protect the plants from fungal attack and insect infestation.

Goitrogens

These substances interfere with iodine uptake by thyroid gland. Thiocyanate, isothiocyanates and their derivatives are present in lentil. Excessive intake of these foods and marginal intake of iodine from foods and water may result in goiter. Goitrogens are primarily found in lentils and cowpea.

Lathyrogens

Lathyrism is a nervous disease caused due to excessive consumption of the pulse *Lathyrus sativus* (Khesari dal). The neurotoxin responsible for lathyrsim is β-N-Oxalyl-L-α-β diamino propionic acid which interferes with the formation of normal collagen fibers in the connective tissues. *Lathyrus sativus* is mainly grown in dry districts of Madhya Pradesh, Uttar Pradesh, Bihar, Bengal, Maharashtra, Mysore and Andhra Pradesh. The dehusked seed resemble red gram dal and sometimes, is used as an adulterant. When eaten in small quantities, lathyrus seeds are valuable as food since it contains 28 per cent protein. But if they are the main source of energy providing more than 50 per cent, a severe disease of spinal cord may result. Lathyrism progress is characterized by a typical manner of walking with short steps and jerky movements and a kind of scissors or crossed gait. The patient may remain in this condition for the rest of his life or may progress to the next stage, where the muscular stiffness is increased and this makes it necessary to perform all walking by tilting the pelvis to such a degree that a stick is necessary to maintain balance. At the severity, patient needs two sticks for support and there is marked tilting of the pelvis sideways to maintain balance and patient is unable to walk upright on account of considerable bending of the knees and extreme stiffness of the lower limbs.

Dietary Fibers

The dietary fiber content of legumes depends on the species, the variety and processing of legume seeds. In most grain legumes the dietary fiber content ranges from 8 to 27.5 per cent (Guillon and Champ, 2002). High dietary fiber content of pulses contribute to their role in prevention of diseases like cardiovascular diseases, type II diabetes, obesity and colon cancer. Pulses have low glycemic index, which makes them valuable foods for diabetics (Ofuya and Akhidue, 2005).

Processing of Pulses and Effect on Nutritive Value

Milling or Decortications

The seed coat tightly envelops the cotyledons probably through a layer of gum and lignin. Legumes with seed coat require considerable longer time to improve their nutritional properties and to make them palatable. A large amount of abrasive force is applied to separate husk from grains that result in high losses in the form of broken and powdered grains. Dehusking reduces the fiber content causing significant nutritional losses particularly vitamins B and minerals that exists mainly in the husk. The extent of losses varies according to the degree of milling and the distribution pattern of nutrients in the grain (Rao and Belavady, 1978). Dehusking of whole grain pulses bring about a significant reduction in total dietary fiber content (34.9 to 59.6 per cent) as well as insoluble dietary fiber content (38.9 to 65.7 per cent). As insoluble dietary fiber constitutes 85 to 89 per cent of total dietary fiber content in whole pulses, so the decrease in total dietary fiber content of pulse over those of whole grain pulse is mainly due to the decrease in their insoluble dietary fiber content. This indicates that fiber present in the husk of whole grain pulses is mainly insoluble fiber (Rao and Ramulu, 1998). The soluble dietary fiber content of pulses is low (2 to 3.2 per cent) and dehusking of pulses results in a marginal decrease in their soluble dietary fiber content (Ramulu and Rao, 1997).

Soaking

A number of pulses having hard outer covering need soaking prior to cooking. Singh *et al.* (1988) observed that soaking reduced cooking time for lentil and chickpea. Soaking makes the pulse tender and hastens the process of cooking. During soaking, water enters through the hilum where the bean is attached to pod. From there, it sees around the periphery of the bean and causes the seed coat to wrinkle. These wrinkles are eliminated when the cotyledons swell and fill the seed coat. However, time required to soften the pulse grains increased if hard water is used for soaking and cooking. Cooking time varies with calcium content of the variety due to the fact that calcium with pectin in the seed coat prevents the absorption of water and softening of grains on cooking (Stanley and Cline, 1950). Soaking in salt solution is preferred to loosen the seed coat and enhance water absorption. Addition of soda to water reduces cooking time significantly but it also results in loss of thiamine as a result of alkaline environment. Wolf (1975) suggested that by hydrating the beans in a solution of NaCl, sodium carbonate which is followed by soaking, draining, rinsing and drying, cooking time could be decreased from 35 to 75 minutes to 25 to 35 minutes.

Boiling/Pressure Cooking

It has been observed that pressure cooking of pulses result in significant increase in their dietary fiber contents (Rao and Ramulu, 1998). An increase in soluble dietary fiber content of some beans during cooking was reported (Rao and Ramulu, 1998), which might be attributed to formation of resistant starch. In traditionally cooked pulses, resistant starch increased to 1.6 to 9 fold whereas pressure cooked pulses showed a 2.1 to 8 fold increase over that of uncooked pulses (Mahadevamma and Tharanathan, 2004). However, a decrease in total dietary fiber content of chickpea, pigeon pea, green gram and black gram was observed after germination and fermentation.

Germination/sprouting

Sprouting is a process by which the dormant embryo in the seed begins to grow into a seedling. In this process, the grains are wrapped in a wet paper towel and laced in a seed germinator set at 30°C and 85 per cent relative humidity for 24 to72 hours. The sprouting process may vary depending on the pulse, season, temperature and humidity. During sprouting, dormant enzymes get activated and digestibility and availability of nutrients is improved. Germination of seeds causes hydrolyses of macromolecules which facilitates digestion.

Sprouting reduces trypsin inhibiting factors due to release of enzymes. The enzymes such as cytases and pectinases are released during sprouting. Phytic acid content is reduced resulting in increased availability of proteins and minerals like iron, calcium and zinc. Vitamin C is synthesized during germination process resulting in an increase in vitamin C content. Sprouting also result in an increase in riboflavin, niacin, folic acid, cholin and biotin content. The oligosaccharides present in pulses gets metabolized and hence germinated pulses do not produce gas or flatulence.

Fermentation

Fermented foods may be defined as the foods that have been subjected to the action of micro-organisms or enzymes to bring out the desirable biochemical changes so as to cause significant modification to the food. The fermentation process results in food with more nutritious (increased vitamin B and C), good digestibility and have better flavor (Raghuvanshi and Singh, 2009). The soybean meal mixed with chickpea fermented for 18 hours at 30?C resulted in a decrease in trypsin inhibitor activity (Kanekar, 1992) and increase in protein quality (Khader, 1983). The examples of fermented products of pulses are- *idli* and *dosa, dhokla* and *bari etc.*

Parching

Parching *i.e.* roasting without the presence of fat is a traditional Indian practice. The process involves sprinkling the grains with a little water containing salt. The pulse is then mixed with four times of its own volume of preheated sand or salt of about 240 to 335°C. The pulse is then subsequently roasted by rapid mixing in the pan using a ladle. During the process the pulse temperature increases from 26°C to 132°C in a short interval of 2 to 3 minutes. The roasted material is separated from the sand or salt by sieving. This process results in a light, porous texture in the pulse. The parching of grains results in a decrease in phytic acid content (Khetarpaul and

Goyal, 2008) which might be attributed to the formation of insoluble complexes between phytic acid and other components (Kumar *et al.*, 1978; Chitra *et al.*, 1996).

Frying

Frying may involve deep-oil frying, shallow frying or sautéing. In deep-oil frying, the food is totally immersed in hot oil. Cooking is rapidly completed as the temperature is 180-220?C and results in an increased calorific value of the food.

Extrusion

Extrusion is the process by which moistened, starchy or proteinaceous materials are plasticized by a combination of high pressure and mechanical shear (Hauck, 1980). Extrusion cooking reportedly reduces the moisture content of products and hence prolongs shelf life (Osundahunsi, 2006).

Conclusion

Pulses are a low fat source of protein with a high fiber content and low glycemic index. Pulses provide important vitamins and minerals like folate, thiamine, niacin, iron, potassium, magnesium and zinc. Pulses contain about double the amount of protein found in whole grain cereals and in most developing countries constitute the main source of protein for most population specially for the vegetarians. Antinutrients like phytates, tannins, trypsin inhibitors, haemagglutinins, goitrogens *etc.* are present in pulses that can significantly reduce their nutritional value. However, the action of these antinutrients can be reduced or degraded completely with simple cooking methods such as boiling, soaking, sprouting, fermentation or a combination of these.

REFERENCES

Broughton, W.J., Hernandez, G., Blair, M., Beebe, S., Gepts, P. and Vonderleyden, J. (2003). Beans (*Phaseolus* spp.)- model food legumes. *Plant and Soil*, 252: 55-128.

Brune, M., Rossander, H.L., Hallberg, L., Gleerup, A. and Sandberg, A.S. (1992). Human iron absorption from bread: Inhibiting effects of cereal fiber, phytate and inositol phosphates with different numbers of phosphate groups. *J. Nutr.*, 122: 442–449.

Champ, M.J. (2002). Non-nutrient bioactive substances of pulses. *Br. J. Nutr.*, 88 (Suppl.3): S307-S319.

Cheryan, M. (1980). Phytic acid interactions in food systems. *CRC Critical Review of Food Science*, 13: 297–335.

Chitra, U., Singh, U. and Rao, P.V. (1996). Phytic acid, *in vitro* protein digestibility, dietary fiber and minerals of pulses as influenced by processing methods. *Plant Foods for Human Nutrition* 49, 307–316.

Guillon, F. and Champ, M.J. (2002). Carbohydrate fractions of legumes: Use in human nutrition and potential for health. *Br. J. Nutr.*, 88 (Suppl.3): S293-306.

Hauck, K. (1980). Marketing opportunities for extrusion cooked products. *American Association of Cereal Chemists*. 25: 594-595.

Hurrell, R.F., Juillerat, M.A., Reddy, M.B., Lynch, S.R., Dassenko, S.A. and Cook, J.D. (1992). Soy protein, phytate, and iron absorption in humans. *Am. J. Clin. Nutr.*, 56: 573–578.

Jenkins, D.J.A., Wolever, T.M. and Taylor, R.M. (1980). Rate of digestion of foods and postpradial glycemia in normal and diabetic subjects. *Br. Med. J.*, 281: 14-17.

Johnson, K.W. and Snyder, H.E. (1978). Soymilk – a comparison of processing methods on yield and composition. *Journal of Food Science* 43, 349–353.

Kanekar, P., Joshi, N., Sarnaik, S. and Kelkar, A. (1992). Effect of fermentation by lactic acid bacteria from soybean seeds on trypsin inhibitor (TI) activity. *Food Microbiology*, 9: 245–249.

Khader, V. (1983). Nutritional studies on fermented, germinated and baked soy bean (*Glycine max*) preparations. *Journal of Plant Foods* 5, 31–37.

Khamgaonkar, S.G., A. Singh, K. Chand, N.C. Shahi, and. U.C. Lohani (2013). Processing technologies of Uttarakhand for lesser known crops: An overview. *J. Acad. Indus. Res.* 1(8): 447-452.

Klevay, L.M. (1977). Hypocholesterolemia due to sodium phytate. *Nutrition Reports International*, 15: 587–593.

Kumar, K.G., Venkataraman, L.V., Jaya, T.V. and Krishnamurthy, K.S. (1978). Cooking characteristics of some germinated legumes: Changes in phytins, Ca++, Mg++ and pectins. *Journal of Food Science* 43, 85–93.

Mahadevamma, S. and Tharanathan, R.N. (2004). Processing of legumes: resistant starch and dietary fiber contents. *Journal of Food Quality*, 27(4): 289-303.

Nelson, A.I., Steinberg, M.P. and Wie, L. S. (1976). Illinois process for preparation of soymilk. *Journal of Food Science* 41, 57–61.

Ofuya, Z.M. and Akhidue, V. (2005). The role of pulses in human nutrition: A review. *Journal of Applied Sciences and Environmental Management*, 9 (3): 99-104.

Osundahunsi, O.F. (2006). Functional properties of extruded soybean with plantain flour blends. *Journal of Food, Agriculture and Environment.* 4: 57-60.

Raghuvanshi, R. and Singh D.P. (2009). Preparation and use. In: Erskine, W., Muehlbauer, F. J., Sarker, A. and Sharma, B. *The Lentil: Botany, Production and Uses.* CAB International, Wallingford, UK, pp. 408-424.

Ramulu, P. and Rao, P.U. (1997). Effect of processing on dietary fiber content of cereals and pulses. *Pl. Fd. Hum. Nutr.*, 50: 249-157.

Rao, P.U. and Belavady, B. (1978). "Oligosaccharides in pulses: Varietal differences and effects of cooking and germination." *Journal of Agriculture and Food Chemistry*, 26: 316-319.

Rao, P.U. and Ramulu, P. (1998). Dietary fiber-Research highlights. *Nutrition News.* 19 (2).

Reddy, B.S. (1999). Possible mechanisms by which pro- and prebiotics influence colon carcinogenesis and tumour growth. *J. Nutr.*, 129: 1478S–1482S.

Reddy, N.R., Pierson, M.D., Sathe, S.K. and Solunkhe, D.K. (1985). Dry bean tannins: An overview of nutritional implications. *Journal of American Oil Chemists' Society*, 62(3): 541-549.

Sandberg, A.N. (2002). Bioavailability of minerals in legumes. *Br. J. Nutr.*, 88 (Suppl.3): S281-285.

Singh, K.B., Erskine, W., Robertson, L.D., Nakkoul, H. and Williams, P.C. (1988). Influence of pre-treatment on cooking quality parameters of dry food legumes. *J. Sci. Food. Agric.* 42(2): 135-142.

Singh, S. (1978). Potential of soyprotein in improving Indian diet. In: *Proceedings of the International Protein Food Conference*, Singapore, 25–27 January 1978, pp. 70–73.

Stanley, L. and Cline, J.A. (1950). *Food: Their selection and preparation.* reved. Ginn and Company USA: 262p.

Wolever, T.M. (1990). The glycemic index. *World Rev. Nutr. Diet.*, 62: 120-185.

Wolf, W.J. (1975). Effect of refining operations on legumes. In: Nutritional Evaluation of Food Processing. 2nd ed. Harris, R.S. and Karmas, E. The Avi Publishing Company. Inc. West Port. Conneticut. pp. 158-187.

2018, *Climate Risks Management: Sustainable Pulse Production*
Editors: A K Srivastava and Yogranjan
Published by: **ASTRAL INTERNATIONAL PVT. LTD., NEW DELHI**   *Pages 147–168*

Chapter 7

Pulse Processing and Nutrition for Better Health Assurance

Aparana Sharma

Assistant Professor (Food Science)
College of Agriculture, JNKVV, Ganj Basoda, M.P.

ABSTRACT

Pulses come from the family leguminosae, which is mainly dominated by variety of species, cultivated all over the world. Botanically, pulses are either the edible fruits or plants with edible seeds. Pulses are an important ingredient of many traditional medicines and have various health's promoting effects. The consumption of pulses should be encouraged in both adults as well as children and among the malnourished population because of their high protein and poly unsaturated fatty acid contents. The use of pulses as components of weaning foods in combination with cereals is also recommended, as this would give cheaper cereals with more complete protein. The present review is an attempt to document health benefits, utilization and nutritional importance of pulses and also signifies pulses for a better health assurance.

Keywords: Pulses, Processing techniques, Nutrition, Food value.

Introduction

Agriculture and agro products are the most important sectors of the Indian economy and Indian agriculture is beginning to appear globally competitive. Both cereals and pulses play a vital role for improving human nutrition. Pulses come from the family leguminosae (COPR, 1981), which is mainly dominated by variety of species, cultivated all over the world (Rubatzky and Yamaguchi, 1997). Most of the pulses are cultivated for their green pods, green seeds as well as dried seeds. Pulses form major source of cheap protein for human diet, especially when there is a lot of chaos on animal food for being expensive. In addition, supplementing the diet with legume seeds helps to alleviate protein deficiency (Sathe and Salunkhe, 1981) in human beings. Botanically speaking, pulses are the fruits, that are edible or plants that bear pods with edible seeds. Additionally, pulses/legumes are rich source of lysine and hence in some researches as well as traditional cuisines, a combination of cereal protein and legume protein is recommended, which is helpful in providing ideal dietary proteins for human consumption. In the developing countries like India, the nutritional complementary of cereals and pulses is extremely important. Moreover most of the pulses are rich sources of edible and digestible proteins, carbohydrates, all the B-group vitamins, Vitamin E, iron, some trace minerals and fiber.

Pulses are important for their food proteins, yet pulse production in most of the countries is stable or is declining. The value of pulses is contributing to increased food productivity, nutritional sustainability of any production system. Pulses form an important part of human diet and also that of animals in the form of supplements. However, comparing pulses with meat, or the vegetarian source with non-vegetarian source of protein, these edible seeds are found to be deficient in sulfur-containing amino-acids, which are categorized as essential for human growth. Pulses are a potential dietary source of fifteen essential minerals required by man (Gurusak, 2002). However, the concentrations of iron, zinc and calcium; the other essential minerals are low as compared to animal based protein sources.

Despite being nutritionally rich, most of the legumes are not normally included in regular diet because of their hard seed coat which requires long cooking time and expensive, scarce fuels. Furthermore, these pulses may reduce protein digestibility due to the presence of some anti nutritional factors that adversely affect the nutritive value of these pulses (Sharma and Sehgal, 1992; Oboh *et al.,* 2000).

Nutritive Values of Pulses

Pulses are important both nutritionally and metabolically. These can be a valuable source of energy (Ofuya and Akhidue, 2005), which are needed for carrying out various metabolic activities. In addition, pulses are also rich source of various kinds of amino acids and minerals (Jambunathan and Singh, 1981).

Majority of proteins in pulses consist of storage proteins in the form of salt soluble globulins, which are synthesized during the process of seed development. These globulins are stored in protein bodies and during germination; it is further hydrolyzed to yield nitrogen and carbon skeletons helpful in the development of plant seedlings. The other storage protein is albumin which covers many 'housekeeping' proteins, lectins and lipoxygenases. On the basis of their sedimentation co-effecinets, as given by Utsumi *et al.* (2002), the pulse globulins comprise of two classes which are commonly termed as 7/8S and 11/12S. Amongst these, the 11/12S are hexameric and are generally known as legumins (or glycinin in soybean). The other one *i.e.* the 7/8S are called as trimmers and is named variously as vicilin, convicilin, beta-conglucinin, phaseolin, canavalin and other trivial names. These names also reflect their species of origin or the pulse they belong to. Both in the form of Cys and Met are nutritionally deficient, also the 7/8S type is more deficient than the 11/12S.

However, both types of proteins possess certain physiochemical and/or "functional" properties that affect their further utilization in food and food products.

Pulses are so versatile in their use and the edible product varies with the cuisine and variety of cooking techniques involved. The product list starts from simple boiling, to roasting, steaming to simple dhal, salads, breads, entrees and desserts and many more. Flatulence is a common problem encountered by most of the people consuming pulses. This is basically due to the presence of biochemical compound Raffinose oligosaccharides, one of the anti-nutritional factor (*e.g.*, stachyose, raffinose, verbascose) that dominates pulses (Oboh, *et al.*, 2000). They produce flatulence both in man and animals as their digestive systems lack enzyme α-galactosidase required for the hydrolysis of α-1,6-galactosidic linkage present in oligosaccharides, especially in the lower intestine. During the process of digestion, these sugars undergo fermentation (in the absence of oxygen) by anaerobic bacteria that ultimately produce carbon-di-oxide, hydrogen and small amount of methane gas causing flatulence. It is characterized by abdominal rumbling, cramps, diarrohea and nausea. Legumes are deficient in metheonine and cystine whereas cereals are richer in the same (Dogra *et al.*, 2004). Food legumes are rich in sulfur containing amino acids and therefore a combination of proteins both from cereal source and legume source helps in providing an ideal amount of dietary proteins as suggested for human consumption. However to achieve optimum amino acid balance, cereal and legumes must be consumed in the ratio of 63:35 of cereal and legumes respectively. A combination of 67 per cent wheat and 33 per cent chickpea was found to be ideal for supplying ideal amount of amino acids as per the WHO recommendations. Also the regular household wheat flour can be conveniently supplemented with chickpea flours upto 20 and 40 per cent for the preparation of bread and cookies respectively (Dhaliwal and Kalia, 1996).

Pulses are also utilized as flour after milling, individually or in combination. These flours are further used in the preparation of variety of products for human consumption. Flours and flakes can be used as starting material for different types of products. Flours can be prepared using different methods/techniques *viz.*, soaking, germination, roasting, drying and others (Figures 7.1 and 7.2). These flours in the form of dough and batters are utilized for preparation of variety of edible products including leavened and unleavened breads, roasted and plated breads, extruded products and snacks, noodles and many types of fried and steamed food products. Flours and graded milled products of varying thickness and quality are also used as porridge type of products and their consumption is very common in traditional backdrop societies.

Weaning foods are another type of product which is made by combining or mixing of cereal and pulses and is provided either as a mixture of flours and/or flakes. Such mixtures are reconstituted with the help of milk or water and is widely accepted for feeding young children. These mixtures are made nutritionally balanced so as to provide required amount of energy, digestible protein and other nutrients which helps to maintain good health and growth of a child. These types of food stuff are vital for a growing child as it is in this transition period when a child's diet is shifted from breast milk to family pot (grain based diet) that the child is exposed

Legumes

↓

Cleaning

↓

Dehulling ◄──────────────── Soaking
(Stone *Chakki*)   (6 hrs)

↓   ↓

Milling   Alkali treatment
(1.0per cent Ca(OH) for 15 min. at 100°C)

↓   ↓

Flour-I   Dehulling
(by hand)

↓

Drying
(at 55°C for 4hrs)

↓

Milling

↓

Flour-II

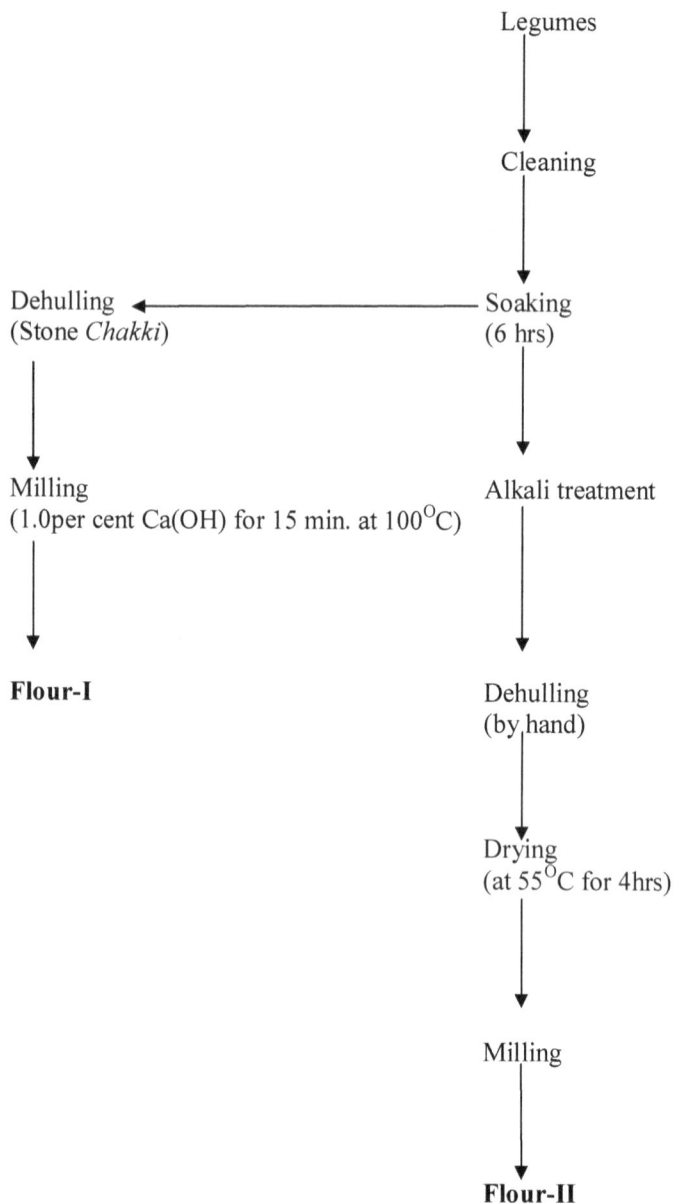

Figure 7.1: Preparation of Flours from Legume(s).
Source: Dhaliwal and Attri (2004).

Legumes

Cleaning ———————→ Dehulling ————————→ Milling

Soaking (6 hrs)

Flour-I

Germination
(for 36hrs at 35°C)

Germination
(for 36hrs at 35°C)

Drying
(at room temperature)
70$\pm$5°C)

Drying
(at 70$\pm$5°C)

Drying
(at room temperature)

Drying
(at

Dehulling

Dehulling

Dehulling

Dehulling

Milling

Milling

Milling

Milling

Flour-II

Flour-III

Flour-IV

Flour-V

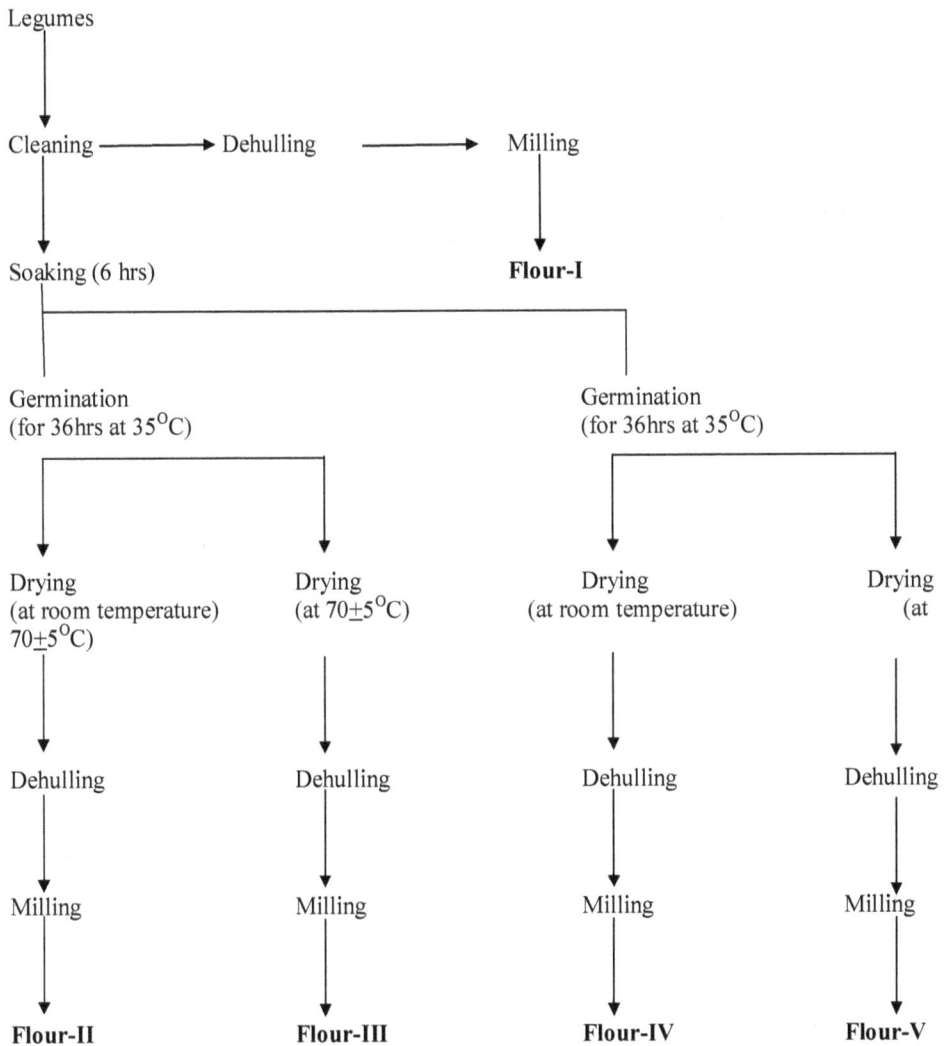

**Figure 7.2: Preparation of different Kinds of Flours Using
different Cooking/Processing Methods.**
Source: **Dhaliwal and Attri (2004).**

to the most common health problem *viz.*, protein-energy malnutrition and other nutritional deficiency diseases, which may or may not be present earlier.

The nutritive values of some of the pulses of human use have been indicated in Table 7.1. Most of the pulses are important sources of protein, carbohydrate, dietary fibre and minerals Extensive investigations have been made for their fat and fatty acid components of various types of oilseeds, also used as pulses. Amongst these Soy beans and peanuts are found to be rich in oil content. The average oil content of these pulse based oilseeds is found to be between 18.0-22.0 per cent in soybeans and 40- 50 per cent in peanuts.

Table 7.1: Nutritive Value of Pulses of Human Use

Nutrient (per 100g)/ Pulse(s)*	Energy (kcals)	Mois-ture (g)	Protein (g)	Fat (g)	Mineral (g)	Carbo-hydrate (g)	Fibre (g)	Calcium (mg)	Phos-phorus (mg)	Iron (mg)
Bengal gram (whole)	360	10	17	5	3	61	4	202	312	5
Bengal gram (dhal)	372	10	21	6	3	60	1	56	331	5
Bengal gram (roasted)	369	11	22	5	2	58	1	58	340	9
Black gram (dhal)	347	11	24	1	3	60	1	154	385	4
Cow pea	323	13	24	1	3	54	3	77	414	9
Field bean (dry)	347	10	25	1	3	60	1	60	433	3
Green gram (whole)	334	10	24	1	3	57	4	124	326	4
Green gram (dhal)	348	10	24	1	3	60	1	75	405	4
Horse gram (whole)	321	12	22	0	3	57	5	287	311	7
Khesari dhal	345	10	28	1	2	57	2	90	317	6
Lentil	343	12	25	1	2	59	1	69	293	7
Moth beans	330	11	24	1	3	56	4	202	230	9
Peas green	93	73	7	0	1	16	4	20	139	1
Peas dry	315	16	20	1	2	56	4	75	298	7
Peas roasted	340	10	23	1	2	57	4	81	345	6
Rajmah	346	12	23	1	3	61	5	260	410	5
Red gram (dhal)	335	13	22	2	3	58	1	73	304	2
Red gram (tender)	116	65	10	1	1	17	6	57	164	1
Soybean	432	8	43	20	5	21	4	240	690	10

*Adopted from Gopalan *et al.*, 2004, Tharanathan and Mahadevamma, 2003, Yoshida *et al.*, 2005.

Nutrient Composition, Processing, Health Benefits and Food Uses of Pulses

Mungbean (*Phaseolus aureus*)

Nutritional Composition

The pulse is also called as green gram and is scientifically named as *Phaseolus aureus*. The pulse is a excellent source of protein and approximately contains 27 per cent, where the composition of essential amino-acid compares favourably with that of soybean, kidney bean and FAO/WHO reference protein (El-Adawy, 1996). However the bean is low in sulfur containing amino-acids and has low protein digestibility (Mubarak, 2005). Table 7.2 indicates the bio-chemical composition of mungbean (raw) on dry weight basis, vis-a-vis as affected by traditional domestic processes. The mung bean is admired by nutritionists for its high nutritional value and easy digestibility. As per various researches, approximately 100 gram of mung bean gives about 334 Kcal of energy. It contains good quantity of carbohydrates (56.7g/100g) and is a very rich in some minerals like potassium (843mg/100g), magnesium (127mg/100g), calcium (124mg/100g), phosphorous (326mg/100g) and iron (4.4mg/100g). Mungbean also contains appreciable amounts of healthy vitamins *viz.*, carotene, thiamine, niacin, riboflavin, ascorbic acid and folic acid. One of the cheapest and popular sources of protein, the mungbean is richer in a number of essential amino acids which includes arginine, histidine, lysine, tryptophan, phenylalanine, leucine, isoleucine, tyrosine, valine, threonine, cystine and metheonine. Mung, hence is considered to be a substansive source of dietary proteins and carbohydrates. Mungbean also provides significant amounts of dietary iron to plant based diets, especially in developing countries, where it is consumed (Nair, 2013). It also contains some chemical components like flavanoids (including flavones, isoflavones and isoflavonoids), phenolic acids (namely gallic, vanillic, caffeic, cinnamic, protocatechuic *etc.*). Some of the organic acids are also isolated from mung bean and it has been found to support general health and growth and this adds to various health promoting claims made for the bean (Kavya *et al.*, 2014).

Table 7.2: Effect of Traditional Domestic Processes on the Chemical Composition of Mung Bean Seeds (g/100g dry weight basis)

Components	Raw	Dehulling	Soaking	Germi-nation	Boiling	Auto-claving	Microwave Cooking
Crude protein	27.5	27.6	27.0	30.0	26.80	26.6	26.7
Crude fat	1.85	1.82	1.83	1.45	1.82	1.82	1.81
Crude fibre	4.63	4.10	4.45	4.40	4.50	4.61	4.64
Ash	3.7	3.60	3.32	3.55	3.55	3.54	3.65
Total carbohydrate	62.3	62.9	63.4	61.7	63.3	63.4	63.2
Moisture	9.75	10.10	10.50	11.10	10.13	10.15	10.10

Effect of Cooking/Processing

Soaking, germination and pressure cooking are some of the common processing/cooking methods used traditionally to make the pulses edible and these methods have proved to be effective home-scale strategy that helps in reducing the polyphenol and tannin levels in pulse based foods. This process, in turn enhances the bioavailability of protein from the pulses. Processing reduces the concentration of polyphenols by 19-59 per cent and of tannins by 22-59 per cent (Khandelwal *et al.,* 2010). Various cooking methods, however, was also responsible for significant losses in soluble solids, especially minerals (Barampama and Simard, 1995). The losses are due to leaching of minerals from the pulse -seeds into the water used for cooking water that too in different proportions. Dehulling and soaking were proved to be the most deteriorating methods with respect to mineral retention (Table 9.3). This loss in minerals was attributed to their binding to protein and also to the formation of a phytate-cation complex. Germination of mung beans revealed a slight increase in minerals *viz.,* K, Ca, P, Mg, Fe and Mn.

Table 9.3: Effect of Traditional Domestic Processes on Mineral Composition of Mung Bean Seeds (g/100g dry weight basis)

Components	Raw	Dehulling	Soaking	Germi-nation	Boiling	Auto-claving	Microwave Cooking
Na	12.00	10.20	9.60	11.60	8.20	8.95	8.10
K	3.62	2.90	2.35	3.95	2.90	2.88	2.80
Ca	84.00	80.00	81.00	88.50	75.00	80.00	78.00
P	391	385	38	406	368	370	365
Mg	55.60	54.30	49.90	56.6	44.00	48.00	47.80
Fe	9.70	8.60	8.40	9.65	7.90	8.10	8.00
Mn	1.70	1.50	1.40	1.70	1.30	1.55	1.40

Health Benefits

Mungbean is one of the considerate pulse, which has a unique quality or nutritional composition that initiates various physiological effects in human system. It is considered as a drug in *Ayurveda,* where it is believed to have distinctive properties like easily digestible, and also has a catabolic post-digestive effect and thus helps in nourishing the digestive system. It is recognized as an efficient builder and nourisher for tissues. It is admired as one of the most popular and compatible dietary source which is suitable for the individuals of all age groups (Kavya *et al.,* 2014). It has an astringent like property, which helps in wound healing and is a good absorbent as well as a good cleaning agent, especially for both digestive as well as metabolic pathways. It is considered to be a vivifying, nourishing, strength promoting and growth promoting food (Agnivesha, 2011). Mung bean is known to possess antioxidant, anti-ageing and prolonging life attributes. These properties make the pulse a highly admirable option for inclusion in daily diet, especially in the present scenario, wherein various kinds of lifestyle disorders are affecting people worldwide. Cardio vascular diseases and allied conditions like diabetes, obesity, dyslipidemis are leading threats to the mankind globally.

It has been reported that pulses from mungbeans are found promising in inhibiting LDL oxidation due to their scavenging activity which binds the free-radical (Chung *et al.*, 2011). The pulse has also been found to be effective in playing an important role in controlling high blood pressure which is another factor responsible for increasing the risk of CHD- cardiovascular disease. In a study conducted on rats (hypertensive), their feed was supplemented with the extract of sprouts from mung dhal for one month. It was interfered that the systolic blood pressure of the hypertensive rats reduced to a significant level (Hsu *et al.*, 2011). This antihypertensive effect exhibited by feeding mungbean sprouts was attributed to its high peptide concentration, which initiate the reducing activity of angiotensin converting enzyme (ACE) that otherwise constricts blood vessels and causes a rise in blood pressure.

Food Uses

Mungbean is considered to be a easily digestible pulse and is generally recommended during recoveries from illness and is as one of the best option for light-wholesome food. Whole mungbean is eaten raw after germinating, as salad or a breakfast cereal. Simply soaked mungbean is tossed in a little oil, mixed with fresh veggies like cucumber, tomatoes, chilles and sprinkle of salt and spices make it a snack food. It is taken as a protein rich supplement by the people interested to loose weight, thereby building body muscles. Dried, split and deep fried or roasted mung dal is an all time favourite snack for all age groups. Besides, it is served as cooked boiled dhal with regular meals and is easy and fast to cook and serve. Flour of mungbean is a suitable option for crunchy fritters and is relished in many north Indian cusines as '*vada*' or 'bhajiya'.

Chickpea (*Cicer arietinum*)

Nutritional Composition

Chickpeas are wide and varied in diifernt regions of the world and are also known by different names like, garbanzo beans, Bengal gram, ceci beans and chana beans. The differences in appearance and chemical composition of chickpea varieties are dependent on the growing region and the conditions (Table 9.4), which affect the length of the plant growing season. Both types of pulses *viz.*, *Desi*. and *Kabuli*, differ primarily in their protein, fiber, polyphenols, and carbohydrates contents. The energy value of *Desi* variety grains is 327 kcal/100 g, while for *Kabuli* variety it equals 365 kcal/100g (Maheri-Sis *et al.*, 2008). The pulse is relished for its nutty taste and health-enhancing attributes due to the presence of high levels of fiber, zinc, copper, beta-sitosterol, and iron. The nutritional profile of the chickpea pulse has attracted many workers as well as consumers. Chickpea is an important crop of semi-arid tropics and hence is a important component of their diets. It is also affordable and hence is a good option for those who cannot go for animal food for the sake of nutrition or by choice. Chickpea is a good source of carbohydrates and protein; collectively both of these nutrients contribute towards approximately 80 per cent of the total dry seed mass (Chibbar *et al.*, 2010; Geervani, 1991) in comparison to other pulses. The pulse is cholesterol free and is a good source of dietary fibre, vitamins and minerals (Agriculture, 2006; wood and Gurusak, 2007). Chickpea beans are a rich source of polyphenols and flavonoids, which have high antioxidant properties. Most of their content, 95 per cent, is in the pile of the bean. The darker the chickpea bean color, the greater the content of polyphenols, flavonoids, and higher antioxidant properties (Segev *et al.*, 2010).

Table 9.4: Chemical Composition of *Desi* and *Kabuli* Type Chickpea

Composition	Desi Dry Basis (per cent)	Kabuli Dry Basis (per cent)
Crude protein	22.76	24.63
Crude fiber	9.94	6.49
Total tannin	0.12	0.09
Total phenolic compounds	0.26	0.27
Non fibrous carbohydrate	46.81	49.13
Starch	38.48	39.12
Soluble sugars	7.53	8.43

Source: Adaopted from Meheri_sis *et al.*, 2008.

Effect of Cooking

Chickpea, like other legumes is consumed after proper cooking (Kaur, *et al.*, 2005), frying (Bhat and Bhattacharya, 2001), baking (Gomeza *et al.*, 2008) or roasting (Kaur *et al.*, 2005; Acar *et al.*, 2009). These procedures of cooking/processing helps to improve the flavor and palatability of the food product. It enhances the bioavailability of various nutrients by inactivating various anti-nutritional factors (Chau *et al.*, 1997; Hira and Chopra, 1995). However, they may also affect the activity of its bioactive compounds and their anti-oxidant activity (Han and Baik. 2008; Xu and Chang, 2008). Cooking methods like baking, frying and roasting significantly increased the Total Polyphenol Content (TPC), Total Flavonoid Content (TFC) and Ferreic-Reducing Ability of Plasma Anti-oxidant Activity (FRAP-AA). Also as compared to soaking and traditional cooking method (boiling), where approximately entire quantity of TPC, TFC and FRAP-AA is leached out in the water using the methods *viz.*, frying, baking and roasting of pulses to retain these contents (Aharon *et al.*, 2012).

Health Benefits

Healthwise, chickpea with its potential health benefits is considered as a functional food for human beings. Fibers, both soluble and insoluble found in chickpeas are considered cardio protective. Most of the studies based on the epidemiological and clinical effects reveal that the fiber content of the chickpeas, specifically the soluble fiber, helps to reduce the 'bad cholesterol' (Low Density Lipoproteins) levels, which ultimately to a greater extent fights the risk of coronary heart disease. Also, the high homo-cysteine levels in the blood that triggers the risk of heart disease is controlled by the presence of folate and magnesium in chickpeas.

Additionally, magnesium in the pulse works as a magical cardio protective agent by relaxing the arteries and thereby smoothen the blood flow.

Presence of two important micro-minerals *viz.*, zinc and copper in abundance in chickpea makes it a health boosting pulse, as both these minerals help to boost the immune system and helps to make it defensive against fatal diseases. In addition, presence of fibres in the pulse makes it a suitable food for the persons interested in losing weight or those suffering from obesity. The presence of fibres and proteins helps it to control elevated sugar levels in diabetic people. Consumption of chickpeas may also be beneficial for those who suffer from various types of gastrointestinal problems. The coloured variety of chickpea is found to have high anti-oxidant activity, which might contribute significantly towards the management and/or prevention of free radical damage causing various degenerative diseases. Asia is the major grower of chickpea and is widely been utilized for various kinds of applications *viz.*, food uses, medicine and therapeutic. Astringent properties of chickpea leaves are used for repairing displacements of bones and dislocations after cooking. Whereas, the extract of chickpea leaves is used for treating the gastric conditions like diarrhea or indigestion. Egyptians use the grains of chickpea pulse to improve or increase the body weight, to cure head and throat aches and for the treatment of cold and cough. The fine powder from chickpea grains are used for preparation of facial masks and scrubs. It is also used in the preparation of antidandruff products like soaps and shampoos. Chickpea flour is high in dietary fiber, protein, potassium, calcium, iron and has low glycemic index (GI) compared to rice and wheat flour (Table 9.5). Low glycemic index foods reduce blood sugar level thus preventing diabetes mellitus.

Table 9.5: Comparison of Nutritional Contents of Chickpea Flour verses Wheat and Rice Flours

Nutrient Composition (per 100 g flour)	Chickpea Flour	Wheat Flour	Rice Flour
Total fat (g)	10.4	3.6	0.3
Total Carbohydrate (g)	69.4	75.2	81.3
Dietary fibres (g)	13.7	0.4	2.1
Proteins (g)	19.2	11.0	7.6
Iron (mg)	12.0	4.1	0.0
Calcium (mg)	129.0	122.0	15.0
Potassium (mg)	1.100	137.0	65.0
Glycemic Index	6 (low)	60 (medium)	60 (medium)

Source: Adopted from Pelin B. Belino (2015).

Food Uses

Chickpeas, depending upon their variations in size, shape and colour (Singh, *et al.*, 1991) are classed in two major categories: *kabuli* and *desi* (Nizakat, *et al.*, 2007). All over the world, chickpea is mostly consumed as a food in different forms and different preparations. The product and preparation is influenced by a number of ethnic and regional factors (Muehlbauer and Tullu, 1997; Ibrikei *et al.*, 2003). The cotyledons of chickpea pulse are especially split for consumption as traditional preparation *'dhal'* in the Indian subcontinent. It is also ground finely in the form of flour called *'besan'*, which is further utilized in the preparation of a variety of snacks (Chavan *et al.*, 1986; Hulse, 1991). Other food applications of chickpea in other parts of the world, especially in Asia and Africa includes as stews, soups/salads, roasted seeds, boiled pulses, salted and in fermented food forms. These variety of food products provide consumers with numerous options to consume a valuable pulse with potential nutritional and health benefits. Human beings have been consuming chickpea pulse from ancient times for its nutritional benefits. Immature grains of chickpea pulses are so tender that it can be eaten raw and the ripe seeds, which are slightly harder to chew can be dried and ground into flour and used as animal feed or as a substitute for coffee and/or other flour replacers. The seeds of these pulses after soaking or steam cooking can be used as a crunchy addition to healthy salads. These seeds can be dipped in cooked sugar or jaggery syrup to be consumed as a sweet meat. The dried, roasted seeds of pulses can simply be tossed with a pinch of spices to be eaten as a snack food. Flour from chickpea grains can easily replace regular flours used in the preparation of pasta, soups, and bread (Sekara, 2005). A small replacement of regular flour with chickpea flour is found to significantly lower the carbohydrates and fat content in the final food product. Also, it increases the amount of protein, fiber, and essential mineral substances in food products. In the baked food products including, wheat based cookies, chickpea and its protein may limit the formation of acrylamide, which is otherwise not beneficial for human consumption especially for children. However, it is always recommended to use or add chickpea in the diet only after proper treatment like heating, boiling, soaking, roasting *etc.* so that the antinutritional compounds in the raw pulses are removed and the pulse is more nutritious and safe to consume (Danuta Rachwa-Rosiak *et al.*, 2015).

Faba Bean (*Vicia faba*, L. major)

Nutritional Composition

The faba bean (*Vicia faba*, L. major) ranks sixth in the in the world pulse production list which includes soybean, peanut, beans, peas and chickpeas. It is one of the oldest crops grown in the world (Milner, 1972). Faba bean pulse/seeds contain a appreciable amounts of proteins, carbohydrates, B-group vitamins and minerals. Depending upon the variety of faba bean the nutritional content of the pulse varies and the protein content of faba beans ranges between 20-41 per cent (Chavan, *et al.*, 1989). Faba bean seeds contain 51 per cent to 68 per cent of carbohydrate in total which includes starch in major which ranges between 41–53 per cent (Cerning *et al.*, 1975). Chavan *et al.*,(1986) reported a crude fibre content in faba beans ranging from 5.0 per cent to 8.5 per cent. Hove *et al.* (1978) found dietary fibre values of 15–30 per cent, which seem to depend on the seed variety. Amongst the fiber, hemicellulose forms the major portion in faba bean and it is approximately 60 per cent. The sugars found in faba bean contributes towards the antinutritional factor of the pulse and it includes α-galactosides, including raffinose, stachyose and verbascose. Faba bean seeds are particularly rich in verbascose and stachyose (Sosulski and Cadden, 1982). It is now established fact that the sugars namely, oligosaccharides are the major cause of flatulence in human beings and their abundance presence in faba bean pulse is one of the major drawback in the process of its consumption and utilization as food and food products (Price, *et al.*, 1988). Faba beans are a good source of dietary minerals, such as phosphorus, potassium, calcium, sulphur and iron. Calcium content in faba bean seeds ranges from 120 to 260 g/100 g dry mass (Chavan, *et al.*, 1989).

Effect of Cooking

The processing methods of faba bean include simple house hold cooking procedures like soaking, boiling, pressure cooking *etc.* Soaking of pulses in water is a common process used prior to cooking and consumption. It helps to soften the texture of the pulse and also reduces the cooking time (Luo *et al.*, 2009). Sometimes $NaHCO_3$ added to the cooking water especially to reduce the time required for cooking and also to soften the pulse deep inside. Similarly, heating is a process that results in significantly reducing polyphenols content of the grains, reducing enzyme inhibitors, phytic acid and also some minerals and vitamins are also decreased. However, heating increases protein digestibility especially in faba beans (Alonso *et al.*, 2000; Luo and Xie, 2013). Besides, dry roasting of faba bean pulses helps to increase the tanin content when treated at 149°C for 20 min and 177°C for 18 min, but slightly higher temperature of roasting *viz.*, 204°C for 14 min and 232°C for12 min reduces the tannin content of faba beans (Anderson *et al.*, 1994). Similarly total polyphenols and total flavinoids contents of faba beans and azuki beans are adversely affected by the addition of $NaHCO_3$ in soaking/cooking water (Yuwei Luo *et al.*, 2014).

Pigeonpea (*Cajanus cajan* L.)

Nutritional Composition

Pigeonpea is one of the important tropical grain legume grown across India and is more commonly known by the name of red gram, arhar, tur dhal. The pulse belongs to the family Leguminosae (Ghadge *et al.*, 2008). Pigeon pea pulse is a richer source of starch, protein, crude fiber, fat, major minerals like calcium and manganese and other trace elements (Table 7.6). In addition to its high nutritive value, pigeon pea is also a popular ingredient of traditional folk medicine in India, China, Philippines and some other nations. Nutritionally speaking, pigeon pea is an important source of protein in diets of people especially the economically poor communities of many tropical and sub-tropical regions of the world (Singh *et al.*, 1984). The pulse is found to contain approximately 20-22 per cent protein, 1.2 per cent fat, 65 per cent carbohydrate and 3.8 per cent ash (FAO, 1982). Lipid content of the pulse is less and is free of cholesterol. Additionally, it is a good source of a number of minerals and vitamins. The carbohydrates are present in complex form in these pulses and hence are beneficial for human consumption. Protein in these pulses contains appreciable quantities of sulphur containing amino acids *viz.*, cysteine and metheonine (Saxena *et al.*, 2002). It is a important source of crude fibre, iron, calcium, potassium, manganese and water soluble vitamins especially, thiamine, riboflavin, niacin (Saxena *et al.*, 2010) in the diets.

Table 7.6: Dietary Nutrients of Pigeon Pea

Constituent(s)	Green Seed	Mature Seed	Dhal
Protein (per cent)	21.0	18.8	24.6
Protein digestibility (per cent)	66.8	58.5	60.5
Trypsin inhibitor (units/mg)	2.8	9.9	13.5
Starch (per cent)	48.4	53.0	57.6
Starch digestibility (per cent)	53.0	36.2	-
Amylase inhibitor (per mg)	17.3	26.9	-
Soluble sugar (per cent)	5.1	3.1	5.2
Crude fibre (per cent)	8.2	6.6	1.2
Fat (per cent)	2.3	1.9	1.6
Flatulence factors (g/100g soluble sugar)	10.3	53.5	-
Minerals and Trace elements (mg/100g dry matter)			
Calcium	94.6	120.8	16.3
Magnesium	113.7	122.0	78.9
Copper	1.4	1.3	1.3
Iron	4.6	3.9	2.9
Zinc	2.5	2.3	3.0
Vitamins (mg/100g fresh weight of edible portion)			
Carotene (Vit A/100g)	469.0		
Thiamin	0.3		
Riboflavin	0.3		
Niacin	3.0		
Ascorbic acid	25.0		

Source: Adopted from Faris *et al.*, 1987.

Effect of Cooking

Soaking, sprouting, germination and cooking are some of the simplest and inexpensive domestic and house hold techniques of food processing, especially employed for pulses before consumption (Prodanor *et al.*, 2004). These common processes, soaking, cooking and retorting or pressure cooking, helps to lower the phytate levels and increase the availability of minerals (Tabekhia and Luh, 1980). Soaking reduces anti-nutritional factors particularly, oligosaccharides and raffinose, also the extent of reduction increases with increase in soaking time. This reduction in phytate is mainly due to leaching out of these compounds in water under the influence of 'concentration gradient' (Cheryan, 1982). Blanching is another crucial processing operation that helps in inactivation of enzymes that ultimately helps in retaining nutritive value and aesthetic qualities *viz.*, colour, flavor or texture of the pulse during processing and storage as well. It also helps to remove unpleasant flavours produced by tannins and phytic acid (Erdman and Pneros- Schneier, 1994; Yeum and Russel, 2002; Yadav and Sehgal, 2002)

Health Effects

Occurrence of a number of secondary metabolites including polyphenols and flavonoides makes pigeonpea more therapeutic than other pulses. Different parts of the pulse are used in the management of disorders like ulcers, diahhorea, joint pain, cough, sores, dysentery, hepatitis, measles, as a febrifugr and to stabilize menstrual cycles (Amalraj and Ignacimuthu, 1998; Grover *et al.*, 2002). The dietary fibre present in pigeon pea provide potential health benefits by preventing the risks of various chronic diseases and thus is considered as a functional food (Trinidad *et al.*, 2010). Cooked diet of pigeonpea has revealed significant hypoglycemic effect on healthy human (Panlasigui *et al.*, 1995).

Food Uses

Pigeonpea is a versatile crop mainly grown as a vegetable crop in Carribean and South America and is considered as a multi-use grain crop or *dhal* in India. The whole dry seeds of pulse are cooked alone or together with some other vegetable to be consumed as human food. The immature seeds of pigeon pea can also be taken as vegetable, which has comparatively higher nutritive value than the dry seeds of the pulse. Sometimes very young pods (with underdeveloped seeds) are harvested to be cooked like bean curries. Pigeon pea can extensively be utilized in the preparation of a variety of food products including fresh sprouts, tempeh, ketchup, noodles, snacks and various extruded food products (Saxena *et al.*, 2002; Sharma *et al.*, 2011). Flour from dried pigeon pea is a wonderful component to be used in the snack industry and its incorporation as an ingredient is found to appreciably enhance the nutritive value of food products like pasta, noodles *etc.* without affecting its sensory properties (Torres *et al.*, 2007).

Conclusion

The changing food habits (carbohydrate rich) of a large mass of population has crept many health risks. To minimize the same, pulses having low glycemic index is a preferred option. Although pulses are rich reservoir of not only proteins but also other nutrients such as vitamins, fibres and minerals *etc.*, which may mitigate the global issue of malnutrition. But at the same time, pulses too possess some other bioactive compounds like polyphenols, flavonoids, tannins and phytates *etc.* Ill effects of many of these bioactive compounds can easily be remediated by adopting simple cooking methods. Therefore, the pulses can serve as healthy food for large masses of Indian population.

REFERENCES

Abbiw, D.K. (1990). *Useful Plants of Ghana*. Kew, London, UK: Richmond Intermediate Technology Publications and Royal Botanic Gardens.

Acar, OC., Gokmen, V., Pellegrini, N and Fogliano, V. (2009). Direct evaluation of the total antioxidant capacity of raw and roasted pulses, nuts and seeds. *European Food Research and Technology*, 229(6): 961-969.

Agnivesha, Charaka Samhita, Sutra sthana 27/23, refined and annoted by Charaka, redacted by Dridhabala with Ayurveda Deepika commentary by Chakrapanidatta; edited by Yadavji Trikamji Acharya; Varansi; Chaukhambha Press. 2011 pp. 155.

Agriculture and Agri-Food Canada. (2006). Chickpeas: Situation and outlook. *Bi-weekly Bull*. **19**: 1–14.

Aharon Segev, Hana Badani, Liel Galili, Ran Hovav, Yoram Kapulnik and Han Shomer (2012). Effects of baking, roasting and frying on total polyphenols and antioxidant activity in coloured chickpea seeds. *Food And Nutrition Sciences*. 3: 369-376.

Akpinar, N., Akpinar, MA and Turkoglu, S. (2001). Total lipid content and fatty acid composition of the seeds of some *Vicia* L., species. *Food Chemistry*. 74: 449-453.

Alonso, R., Aguirre, A. and Marzo, F. (2000). Effects of extrusion and traditional processing methods on antinutrients and *in vitro* digestibility of protein and starch in faba and kidney beans. *Food Chemistry* 68: 159-165.

Amalraj, T. and Ignacimuthu, S. (1998). Evaluation of the hypoglycaemic effect of Cajanus cajan (seeds) in mice. *Indian J. Exp. Biol.*, 36: 1032-1033.

Anderson, J. C., Idowu, A. O. and Singh, U. (1994). Physicochemical characteristics of flours of faba bean as influenced by processing methods. *Plant Foods for Human Nutrition* 45: 371-379.

Apata, D.F. and Ologhabo, A.D. (1994). Biochemical evaluation of some Nigerian legume seeds. *Food Chemistry*. 49: 333-338.

Barampama, Z. and Simard, R.E. (1995). Effect of soaking, cooking and fermentation on composition, in vitro starch digestibility and nutritive value of common beans. *Plant Food for Human Nutrition*. 48: 349-365.

Bhat, K.K. and Bhattacharya, S. (2001). Deep fat frying characteristics of chickpea flour suspensions. *International Journal of Food Science and Technology*. 36: 499-507.

Calloway, D.H., Hickey, C.A. and Murphy, E.L. (1975). *J. Food Sci.*, 36: 251–255.

Cerning, J., Saposnik, A. and Guilbot, A. (1975). *Cereal Chem.*, 52: 125–138.

Chau, C.F., Cheung, P.C. and Wong, V.S. (1997). Effect of cooking on content of amino acids and ant- nutrients in three Chinese Indigenous Legume seeds. *Journal of the Science of Food and Agriculture*. 75: 447-452.

Chavan J.K., Kadam, S.S. and Salunkhe, D.K. (1986). Biochemistry and technology of chickpea (*Cicer arietinum* L.) seeds. *Crit. Rev. Food Sci. Nutr.*, **25:** 107-157.

Chavan, J.K., Kute, L.S. and Kadam, S.S. (1989). Broad bean. In: *Handbook of World Food Legumes: Nutritional, Processing, Technology and Utilization, Vol. I*. CRC Press, Boca Raton, Fl, pp. 223–245.

Cheryan, M., (1982). Phytic acid interactions in food systems. *Critical Reviews in Food Science and Nutrition*, 13: 296-335.

Chibbar, R.N., Ambigaipalan, P. and Hoover, R. (2010). Molecular diversity in pulse seed starch and complex carbohydrates and its role in human nutrition and health. *Cereal Chem.*, **87:** 342-352.

Chung, I.M., Yeo, M.A., Kim, S.J., Moon, H.I. (2011). Protective effects of organic solvent fractions from the seeds of Vigna radiate L. wilczek against antioxidant mechanisms. *Hum. Exp. Toxicol.*, 30(8): 904-909.

COPR (Centre for Overseas Pest Research): Pest control in tropical grain legumes. 1981. Crown Publisher. 1st Edition.

Cristofaro, E., Mottu, F. and Wuhrmann, J.J. (1972). *Am. Assoc. Cereal Chem.*, 2: 102–106.

Danuta Rachwa-Rosiak, Ewa Nebesny and Gra¿yna Budryn (2015). Chickpeas: Composition, nutritional value, health benefits, application to bread and snacks– A review. *Critical Reviews in Food Science and Nutrition*, 55(8): 1137-1145.

Dhaliwal, Y.S. and Kalia, M. (1996). Bread and cookie making properties of wheat and chickpea flour blends. *Himachal Journal of Agricultural Research*, 22(1 and 2): 101-108.

Dogra, J., Dhaliwal, Y.S. and Aparana Sharma (2004). Physico-chemical and nutritional characteristics of soybean and its utilization in the preparation of bread and biscuits. *Beverage and Food World*, 31(9): 68-72.

Duke, J.A., Vasquez, R. (1994). *Amazonian Ethnobotanical Dictionary*. Boca Raton, FL, USA: CRC Press Duke.

El-Adawy, T.A. (1996). Chemical, nutritional and functional properties of mung bean protein isolate and concentrate. *Menufiya Journal of Agricultural Research*. 21(3): 657-672.

Elegbede, J.A. (1998). Legumes; *In: Nutritional quality of Plant food*. Osagie, AL., Eka OU (*Eds.*) Post Harvest Research Unit. Department of Biochemistry, University of Benin, Nigeria pp. 53-93.

Erdman, J.W. and Pneros-Schneier, A.G. (1994). Factors affecting nutritive value in processed foods. In: *Modern Nutrition in Health and Disease*, Shils ME, Olson JA, Shile M (eds.). Philadelphia:

Fan, T.Y. and Sosulski, F.W. (1974). Dispersibility and isolation of protein from legume flours. *Canadaian Institute of Food Science and Technology Journal*. 7: 256-261.

FAO (1982). *Legumes in human nutrition*. Food and Agriculture Organization of the United Nations. Food and Nutrition Series, No. 20 Rome.

Faris, D.G., Saxena, K.B., Mazumdar, S. and Singh, U. (1987). Vegetable pigeon pea: A promising crop for India: International Crop Research Institute for the Semi Arid Tropics, Patancheru, AP.

FNRI-DOST (1997). *The Philippine Food Composition Table*. FNRI-DOST, Bicutan, Taguig, Metro Manila.

FNRI-DOST (2009). *Glyceamic index of common foods*. FNRI-DOST, Taguig, Metro Manila

Geervani, P. (1991). Utilization of chickpea in India and scope for novel and alternative uses. In: *Proceedings of a Consultants Meeting*, pp. 47-54. AP, India: ICRISAT.

Ghadge, P.N., Shewalkar, S.V. and Wankhede, D.B. (2008). Effect of processing methods on qualities of instat whole legumes: Pigeon pea (*Cajanus cajan* L.). *Agricultural Engineering International: The CIGR E Journal*. 08(X): 04.

Gomeza, M., Olietea, B., Rosellb, C.M., Pandoe. V and Fernandez, E. (2008). Studies on cake quality made of wheat-chickpea flour blends. *Food Science and Technology*, 41(9): 1701-1709.

Gopalan, C., Rama Sastri, B.V.R. and Balasubramanian, S.C. (2004). *Nutritive Value of Indian Foods*. National Institute of Nutrition, ICMR, Hyderabad.

Gopalan, C., Savitri, B.V.R. and Balasubramanian, S.C. (2007). *Nutritive Value of Indian Foods*. National Institute of Nutrition, Hyderabad: ICMR pp: 161.

Grover, J.K., Yadav, S. and Vats, V.J. (2002(. Medicinal plants of India with antidiabetic potential. *J. Ethnopharmacol.*, 81: 81-100.

Gurusak, M.A. (2002). Enhancing mineral content in plant food products. *Journal of American Clinical Nutrition*. 21: 178-183.

Hafidh, R.R., Abdulamir, A.S. and Bakar, F.A., *et al.* (2012). Novel molecular, cytoxical, and immunological study on promising and selective anticancer activity of Mung bean sprouts. *BMC Complimentary and Alternative Medicine*. 12: 208.

Han, H. and Baik, B.K. (2008). Antioxidant activity and phenolic content of lentils (*Lens culinaris*), chickpeas (*Cicer arietinum* L.), peas (*Pisum sativum*) and soybean (*Glycine max*) and their quantitative changes during processing. *International Journal of Food Science and Technology*, 43(11): 1971-1978.

Hira, C.K, and Chopra, N. (1995). Effects of roasting on protein quality of Chickpea (*Cicer arietinum*) and peanut (*Arachis hypogea*). *Journal of Food Science and Technology*, 32: 501-503.

Hove, E.L., King, S. and Hill, G.D. 1978. *N.Z. J. Agric. Res.*, 21: 457–462.

Hsu, G.S.W., Lu, Y.F., Chang, S.H. and Hsu, S.Y. (2011). Antihypertensive effect of mung bean sprout extracts in spontaneously hypertensive rats. *J Food Biochem.* 35(1): 278-288.

Hulse, J.H. (1991)). Nature, composition and utilization of pulses. In: *Uses of Tropical Grain Legumes, Proceedings of a Consultants Meeting*, pp. 11-27. AP, India: ICRISAT.

Ibrikci, H., Knewtson, S.J.B. and Grusak, M.A. (2003). Chickpea leaves as a vegetable green for humans: Evaluation of mineral composition. *J. Sci. Food Agric.*, **83:** 945-950.

Jambunathan, R. and Singh, U. (1981). Grain quality of pigeon pea. In: *Proceedings of International workshop on Pigeon pea*. Vol. 1. ICRISAT, Hyderabad, AP (India). Dec: 15-19, 1980.

Julie Garden-Robinson (2014). *Pulses: The perfect food. In* A report- North Dakota State University Extension Service. www.northernpulse.com

Kaur, M., Singh, N., and Singh Sodhi, N. (2005). Physico-chemical cooking, textural and roasting characteristics of Chickpea (*Cicer arietinum* L.) cultivars. *Journal of Food Engineering*, 69(4): 511-517.

Kavya, N., Kavya, B., Ramarao, V., Kishore Kumar, R. and Venkateshwarlu, G. (2014). Nutritional and therapeutic uses of Mudga (*Vigna radiata* L.): A potential interventional dietary component. *International Journal of Research in Ayurvedic Pharmaceutical*, 5(2): 238-241.

Khandelwal, Shweta, Udipi, Shobha A. and Ghugee, Padmini (2010). Polyphenols and tannins in Indian Pulses: Effect of soaking, germination and pressure cooking. *Food Research International*, 43(2): 526-530.

Luo, Y.W. and Xie, W.H. (2013). Effect of different processing methods on certain antinutritional factors and protein digestibility in green and white faba bean (*Vicia faba* L.). *CyTA -Journal of Food*, 11: 43-49.

Luo, Y.W., Xie, W.H., Xie, C.Y., Li, Y. and Gu, Z.X. (2009). Impact of soaking and phytase treatments on phytic acid, calcium, iron and zinc in faba bean fractions. *International Journal of Food Science and Technology* 44: 2590-2597.

Maheri-Sis, N., Chamani, M., Sadeghi, A-A., Mirza-Aghazadeh, A. and Aghajanzadeh-Golshani, A. (2008). Nutritional evaluation of kabuli and desi type chickpeas (*Cicer arietinum* L.) for ruminants using *in vitro* gas production technique. *Afr. J. Biotechnol.* 7: 2946–2951.

Meisinger, C., Baumert, J., Khuseyinova, N., Loewel, H., Koenig, W. (2005). Plasma oxidized low-density lipoprotein, a strong predictor for acute coronary heart disease events in apparently healthy, middle-aged men from the general population. *Circulation.* 112(5): 651-657.

Milner, M. (1972). Nutritional improvement of food legumes by breeding. *Proc Symp Protein Advisory Group*, Food and Agriculture Organization, Rome

Mubarak, A.E. (2005). Nutritional composition and antinutritional factors of mung bean seeds (*Phaseolus aureus*) as affected by some home traditional processes. *Food Chemistry* 89: 489-495.

Muehlbauer, F.J. and Tullu, A. (1997) *Cicer arietinum* L. In: *New CROP FactSHEET*, pp. 6. Seattle, WA: Washington State University, USDA-ARS.

Nair, R.M. (2013). Biofortification of mungbean (*Vigna radiata*) as a whole food to enhance human health. *Journal of Science of food and Agriculture.* 93(8): 1805-1813.

Nizakat, B., Amal, B.K., Gul, S.S.K., Zahid, M. and Ihsanullah, I. (2007). Quality and consumer acceptability studies and their inter-relationship of newly evolved Desi type Chickpea genotypes (*Cicer arietinum* L.): Quality evolution of new chickpea genotypes. *International Journal of Food Science and Technology*, 42(5): 528-534.

Oboh, H.A., Muzquiz, M., Burbano, C., Cuadrado, C., Pedrosa, M.M., Ayet, G. and Osagie, A.U. (2000). Effect of soaking, cooking and germination on the oligosaccharide content of selected Nigerian legume seeds. *Plant Foods for Human Nutrition*, 55(2): 97-110.

Ofuya, Z.M., and Akhidue, V. (2005). The role of pulses in Human Nutrition: A Review. *Journal of Applied Sciences and Environmental Management*, 9(3): 99-104.

Panlasigui, L.N., Panlilio, L.M. and Madrid, J.C. (1995). Glycemic response in normal subjects to five different legumes commonly used in the Philippines, *International Journal of Food Science and Nutrition*, 46(2): 155-160.

Pelin, B., Belino Esther T., Botangen, Ines C. Gonzales, Fernando R. Gonzales, and Hilda L. Quindara (2015). Development of chickpea (*Cicer arietinum* L.) food products and its benefits to human nutrition. *International Journal of Chemical, Environmental and Biological Sciences (IJCEBS)*, 3(1): 1-4.

Price, K.R., Lewis, J., Wyatt, G.M. and Fenwick, G.R. (1988). *Nahrung* 32: 609–626.

Pritchard, P.J., Dryburgh, E.A. and Wilson, B.J. (1973). *J. Sci. Food Agric.*, 24 : 663–668.

Prodanor, M., Sierra, I and Vidal-Valverde, C. (2004). Influence of soaking and cooking on the thiamin, riboflavin and niacin contents of legumes. *Food Chemistry*. 84: 271-277.

Reddy, N.R., Pierson, M.D. and Salunkhe, D.K. (1986). *In: Legume Based Fermented Foods*. CRC Press, Inc-Boca Raton, Florida, pp. 11-21

Rubatzky, M.Y. and Yamaguchi, M. (1997). *In: World Vegetables: Principles, Production and Nutritive Value*. Chapman Hall (ITP), New York.

Sathe, S.K. and Salunkhe, D.K. (1981). Preparation and utilization of protein concentrate and isolates for nutritional and functional improvement of foods. *Journal of Food Quality*, 4: 145-233.

Saxena, K.B., Kumar, R.V. and Gowda, C.L.L. (2010). Vegetable pigeon pea: A review. *Journal of Food Legumes*, 23(2): 91-98.

Saxena, K.B., Kumar, R.V. and Rao, P.V. (2002). Pigeon pea nutrition and its improvement. In: *Quality Improvements in Field Crops*. Basra AS and Randhawa IS (*Eds.*) Food Products Press, pp. 227-260.

Saxena, K.B., Kumar, R.V. and Sultana, R. (2010). Quality nutrition through pigeon pea: A review. *Health* doi: 10.4236/health.2010.211199.

Segev, A., Badani, H., Kapulnik, Y., Shomer, I., Oren-Shamir, M. and Galili, S. (2010). Determination of polyphenols, flavonoids, and antioxidant capacity in colored chickpea (*Cicer arietinum* L.). *J. Food Sci.*, 75(2): 115–119.

Sekara, A. (2005). Chickpea milkvetch. *Dzia³kowiec*. 12: 1–2.

Sharma, A. and Sehgal, S. (1992). Effect of processing and cooking on anti-nutritional factors of fababean (*Vicia faba*). *Food Chemistry*. 43: 383-385.

Sheel Sharma, Nidhi Agarwal and Preeti Verma (2011). Pigeon pea (*Cajanus Cajan* L.): A hidden treasure of regime Nutrition. *Journal of Functional and Environmental Botany*. 1(2): 91-101.

Singh, U., Jain, K.C., Jamnunathan, R. and Faris, D.G. (1984). Nutritional quality of vegetable pigeon peas [*Cajanus cajan* (L.) Millsp.]: Mineral and trace elements, *Journal of Food Science*, 49: 645- 646.

Singh, U., Jain, R.C, Jambunathan, R.J. and Faris, P.G. (1984). Nutritive value of vegetable pigeon pea, mineral trace elements. *Journal of Food Science*, 49: 645.

Singh, U., Jambunathan, R., Saxena, K. and Subrahmanyam, N. (1990). Nutrition quality evaluation of newly developed high-protein genotypes of pigeon pea (*Cajanus cajan*). *Journal of the Science of Food and Agriculture*, 50: 201-209

Singh, U., Subrahmanyam, N. and Kumar, J. (1991). Cooking quality and nutritional attributes of some newly developed cultivars of Chickpea (*Cicer arietinum*). *Journal of the Science of Food and Agriculture*, 55: 37-46.

Sinha, S.K. (1977). Food legumes: Distribution, Adaptability and biology of yield. *In: FAO plant production and protection*. FAO, Rome.

Sosulski, F.W. and Cadden, A.M. (1982). *J. Food Sci.*, 47: 1472–1477.

Tabekhin, M.M. and Luh, B.S. (1980). Effect of germination, cooking and canning on phosphorous and phytate retention in dry beans. *Journal of Food Science*. 45: 406-408.

Tharanathan, R.N. and Mahadevamma, S. (2003). Grain legumes: A boon to human nutrition. *Trends in Food Science and Technology*. 14: 501-518.

Thompson, I.U., Hung, L., Wang, N., Rapser, V and Gade, H. (1976). Preparation of mung bean flour and its application in bread making. *Canadaian Institute of Food Science and Technology Journal*. 9: 1-7

Torres, A., Frias, J., Granito, M., Vidal-Valverde, C. (2007). Germinated Cajanus cajan seeds as ingredients in pasta products: Chemical, biological and sensory evaluation. *Food Chemistry*, 101: 202–211

Trinidad, T.P., Mallillin, A.C., Loyola, A.S., Sagum, R.S. and Encabo, R.R. (2010). The potential health benefits of legumes as a good source of dietary fibre. *British Journal of Nutrition*, 103: 569-574.

Utsumi, S., Maruyama, N., Satoh, R. and Adachi, M. (2002). Structure-function relationships of soybean proteins revealed by using recombinant systems. *Enzyme Microbiology and Technology*, 30: 284-288.

Uzogara, S.G. and Ofuya, Z.M. (1992). Processing and utilization of cowpeas in developing countries: A review. *Journal of Food Processing and Preservation*, 16(3): 105-147.

Vijayakumari, P., Siddhuraju, P., Pugalenthi, M. and Janardhanan, K. (1998). Effect of soaking and heat processing on the levels of antinutrients and digestible proteins in seeds of *Vigna sinensis*. *Food Chemistry*, 63: 259-264.

Wang, N., Lewis, M.J., Brennan, J.G. and Westby, A. (1997). Effect of processing methods on nutrients and antinutritional factors in cowpea. *Food Chemistry*, 58: 59-68.

Wood, J.A. and Grusak, M.A. (2007). Nutritional value of chickpea. In: *Chickpea Breeding and Management*. pp. 101-142 [S.S. Yadav, R. Redden, W. Chen and B. Sharma, editors]. Wallingford, UK: CAB International.

Xu, B. and Chang, S.K.C. (2008). Effect of soaking, boiling and setaming on total phenolic content and antioxidant activities of cool season food legumes. *Food Chemistry*. 110(1): 1-13.

Yadav, S.K and Sehgal, S. (2002). Effect of domestic processing and cooking methods on total, HCl extractable iron and *in vitro* availability of iron in spinach and amaranth leaves. *Nutr. Health*, 16: 113–120.

Yeum, K.J. and Russell, R.M. (2002). Carotenoid bioavailability and bioconversion. *Ann. Rev. Nutr.*, 22: 483–504.

Yoshida, H., Hirakawa, Y., Murakami, C., Mizushina, Y and Yamade, T. (2003). Variation in the content of tocopherols and distribution of fatty acids within soybean seeds (*Glycine max* L.). *Journal of Food Composition and Analysis*, 16: 429-440.

Yoshida, H., Hirakawa, Y., Tomiyama, Y., Nagamizu, T and Mizushina, Y. (2004). Fatty acid distributions of triacylglycerols and phospholipids in peanut seeds (*Arachis hypogeal* L.) following microwave treatment. *Journal of Food Composition and Analysis*. 18: 3-14.

Yuwei Luo, Weihua Xie, Zhenping Hao, Xiaoxiao Jin and Qian Wang (2014). The impact of processing on in vitro bioactive compounds bioavailability and antioxidant activities in faba bean (*Vicia faba* L.) and azuki bean (*Vigna angularis* L.). *International Food Research Journal*, 21(3): 1031-1037.

2018, *Climate Risks Management: Sustainable Pulse Production*
Editors: A K Srivastava and Yogranjan
Published by: **ASTRAL INTERNATIONAL PVT. LTD., NEW DELHI**   *Pages 169–180*

Chapter 8

Status of Major Pulses Crop and its Importance in Nutritional Security in North Western Himalaya

Anirban Mukherjee[1], Kushagra Joshi[1], Pratibha Joshi[2],
Shubha[3], Manik Lal Roy[1], Renu Jethi[1] and Nirmal Chandra[4]

[1]*Scientist,* [4]*Principal Scientist,*
ICAR-VPKAS, Almora – 236 601, Uttarakhand
[2]*Scientist, ICAR-IARI, CATAT, New Delhi – 110 012*
[3]*Scientist, ICAR-NBPGR, New Delhi – 110 012*
E-mail: anirban.extn@gmail.com

ABSTRACT

In pulse production neither North Western Himalaya nor India is in self sufficient. There is deficit in production and requirement of total pulse which is highest in Jammu and Kashmir followed by Uttarakhand and Himachal Pradesh. To eradicate appearance of malnutrition, the policy needs not only to promote diverse diets but agricultural diversification. The pulses is such a class of food that has the power to eradicate malnutrition. In hills of North Western Himalaya Rajma, black gram, horse gram, rice bean, moth bean, pegion pea etc. are major pulse crops. In this book chaper outlook on pulse production scenario in North Western Himalaya has been depictd. Emphasis has been given on Pulses from food and nutritional security concern in hills of NWH and a strategy for enhancing productivity of pulse crops in hills has also been dicussed.

Keywords: *Hill agriculture, Public Private Partnership (PPP), Pulses of hills, Traditional pulses, Nutrition.*

Introduction

India has ubiquitous position as the leading producer, the foremost consumer and the largest importer of pulses. The pulse crop is playing an important role and is major source of proteins for the whole vegetarian community in India. It complement the staple cereals in the diets with essential amino acids, proteins, vitamins and minerals. The pulses contain 22-24 per cent protein, which is almost 2 times the protein that of wheat and 3 times that rice contains. Besides the pulses provide significant nutritional and wellness benefits, and are recognised as to reduce several serious diseases such as cardiovascular diseases and colon cancer (Yude *et al.*, 1993; Jukanti *et al.*, 2012).

In India the major pulses grown include pigeonpea or red gram (*Cajanus cajan*), chickpea or bengal gram (*Cicer arietinum*), lentil (*Lens culinaris*), urdbean or black gram (*Vigna mungo*), lablab bean (*Lablab purpureus*), mungbean or green gram (*Vigna radiata*), horse gram (*Dolichos uniflorus*), moth bean (*Vigna aconitifolia*), pea (*Pisum sativum var. arvense*), cowpea (*Vigna unguiculata*), grass pea or khesari (*Lathyrus sativus*), and broad bean or faba bean (*Vicia faba*). More popular among these are chickpea, mungbean, pigeonpea, lentil, urdbean, cowpea and field pea.

Pulses can be produced along a scope of land and climatic conditions and play important role in crop rotation, mixed and inter-cropping, maintaining soil fertility through nitrogen fixation, release of soil-bound phosphorus, and thus contribute importantly to the sustainability of the farming systems.

Status of Pulse Crop in India

India has achieved self-sufficiency in food after green revolution. But in pulse crop still the revolution is awaited. Pulse in India recorded less than 3 per cent annual growth rate in production in the past 63 years while its per capita availability declined from 60 grams a day in the 1950s to 35 grams a day in the 2000s. Now it has been increased up to 43 grams a day in 2013. It has been witnessed a sluggish annual growth rate (AGR) in the total pulse area (0.51 per cent) and productivity (1.16 per cent) over the last 63 years. Currently, even as production has stabilized at 19.27 million tonnes (Table 8.1), our consumption is hovering at 23 million tonnes, which necessitates yearly pulse imports of around 3.5-4 million tonnes. India is losing precious forex (nearly $2.3 billion) every year on importing pulses from players such as Canada and Australia. Price fluctuation is common in the largely unorganized pulses market in the country, and often exacerbated by the lack of assuring procurement. Due to lack of production and distribution efficiencies price of pulse has gone beyond the reach of common man. The inflation rate in pulse WPI (Wholesale Price Index) has increased more than 6 per cent annually. Estimates suggest that India needs an annual growth of 4.2 per cent to ensure projected demand of 30 million tonnes by 2030. To meet this benchmark, constraints to production must be analyzed and effective steps must be taken.

One of the important constraints in any food production system is availability of quality seeds. The growth rate of major pulse quality seed has increased many fold over the last thirteen years (Figure 8.1), there is a steep growth of gram seed

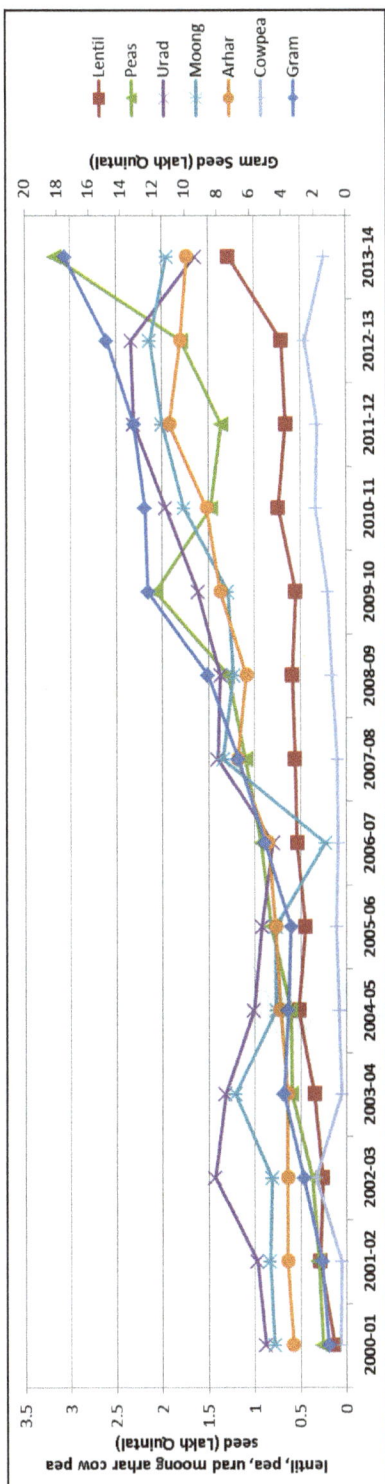

Figure 8.1: Quality Seed Availability Trend of Major Pulses in India.

availability (more than 15 times) followed by field peas (more than 11 times) and Lentil (more than 8 times). Overall quality seed growth rate has increased in a tune of six times over last thirteen years *i.e.* 47 per cent annual growth rate.

Table 8.1: Area, Production and Productivity of Total Pulse Crop in India

Year	Area (mha)	Production (mt)	Yield (kg/ha)	Per cent Area under Irrigation
1950-51	19.09	8.41	441	9.4
1960-61	23.56 (2.37)	12.7 (4.69)	539 (2.25)	8 (-1.78)
1970-71	22.54 (-0.49)	11.82 (-0.79)	524 (-0.31)	8.8 (1.06)
1980-81	22.46 (-0.04)	10.63 (-1.17)	473 (-1.13)	9 (0.25)
1990-91	24.66 (0.04)	14.26 (3.32)	578 (2.25)	10.5 (1.73)
2000-01	20.35 (1.04)	11.08 (-2.76)	544 (-0.67)	12.5 (1.96)
2010-11	26.4 (2.93)	18.24 (5.69)	691 (2.69)	14.8 (1.89)
2013-14	25.23	19.27	764	16.1

Figures in parentheses indicates decadal growth rate.

Source: Agricultural Statistics at a Glance 2014.

At present 27.8 lakh quintal quality seed is available and out of that 17.48 lakh quintals is of Gram (63 per cent), followed by Pea (11 per cent), Moong (7 per cent), Arhar and Urad (6 per cent) (Figure 8.2). Although the amount of quality seed is too less to achieve desired levels of productivity but it is a good progress towards achieving the desired goal.

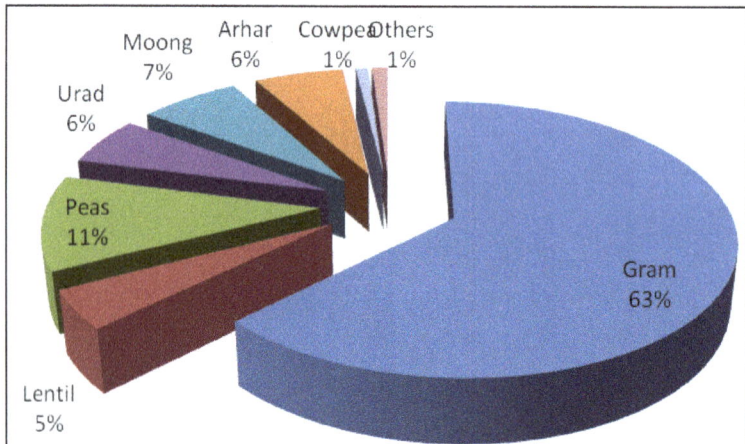

Figure 8.2: Distribution of Pulse as per Availability of Quality Seed (2013-14).

Status of Crops in NWH States

The North West Himalaya (NWH), comprised of three hilly states *viz.*, Jammu and Kashmir (J&K), Uttarakhand (UK) and Himachal Pradesh (HP) plays a crucial role in ecology and economy of Indian agriculture. Geographically, it is hilly region transacted by a number of mountain ranges, and the source of numerous rivers and rivulets. Its physiography is highly undulating with slopes, erodible and fertile soils, sub-tropical intermediate hills, temperate valleys, high hills and cold arids. Agriculture has been a key economic source of this region as lion share of States's GDP comes from agriculture. There are five major farming system prevalent in this region, namely: (i) Cereal-based production system (rice, wheat, millets, maize) (ii) Horticulture or agri-horti-based production system (iii) Vegetables, floriculture and mushroom-based production system (iv) Livestock-based production system and (v) Agri-horti- silvi-pastoral-based production system.

The crop production systems in North Western Himalayan region are based on mainly agriculture (*i.e.* field crops), olericulture (*i.e.* vegetable), horticulture (*i.e.* fruits) or agri-horticulture and agroforestry/agri-horti-silvi-pastoral system. Rice, wheat, maize, finger millet and barnyard millet are the major cereals and lentil, black gram, horse gram, ricebean, cow pea and *bhatt* (black soybean, which is used as pulse in hills of uttarakhand and has better quality proteins and fats which makes it tastier and more digestible than the common soybean, (primarily used as an oilseed) are the major pulse crops. Among the vegetables, cole crops, capsicum, tomato, radish, bean, pea, potato cucurbits, and onions are the major vegetables of this region.

Status of Pulse Crop in NWH States

In hills of NWH mainly Rajma, black gram, horse gram, rice bean, moth bean, pegion pea *etc.* are major pulse crops. But as per productivity is concern Himachal Pradesh is better than UK and J&K. Among all the hilly states, area under total pulse is near about double in UK than HP and J&K (Figure 8.3) due to massive pulse programme organized by the state government during last fifteen years. Himachal Pradesh has shown a considerable increase in pulse productivity but observed more fluctuating trend during last 15 years.

In pulse production neither NWH nor India is in surplus (Table 8.2). There is deficit in production and requirement of total pulse which is highest in J&K (170.3 thousand tonnes), followed by UK (118.7thousand tonnes) and HP (91.3 thousand tonnes).

Table 8.2: Requirement, Production, Productivity and Surplus/Deficit of Pulse Crops in North Western Himalaya

State	Requirement ('000 t)	Production ('000 t)	Productivity (Kg/ha)	Surplus/Deficit ('000 t)
Uttarakhand	154.7	36.0	590	-118.7
Himachal Pradesh	110.9	19.0	713	-91.3
Jammu and Kashmir	183.8	13.5	504	-170.3
NW Himalaya	449.5	69.1	599.3	-379.9
India	18,304	13384.4	598	-4920

Source: FAI (2006-07).

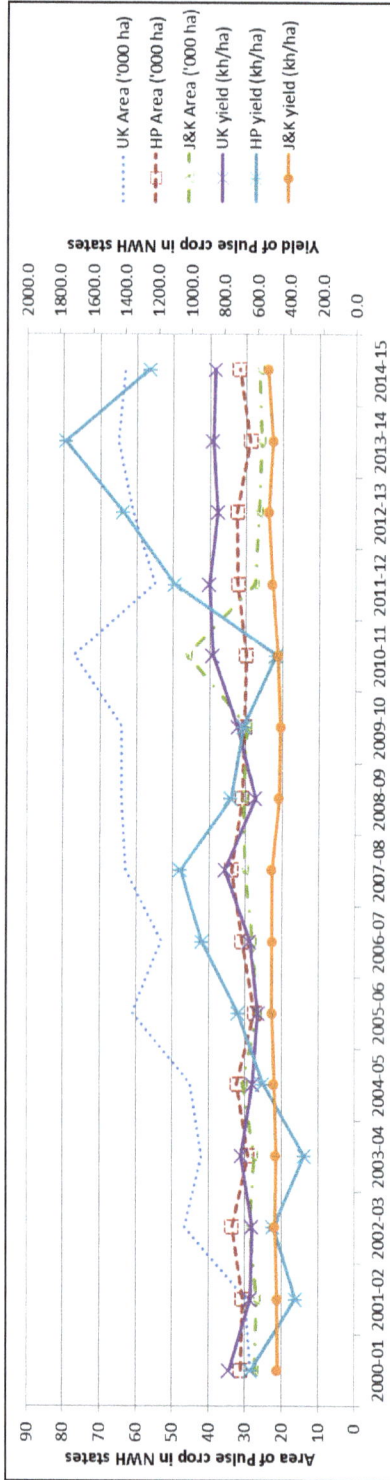

Figure 8.3: Area and Yield Trends of Total Pulse Crop in NWH States.

Pulses from Food and Nutritional Security Concern in Hills

The dietary nutritional status of a population depends fundamentally on the interaction between genetic, environmental, and socio-political factors (Infant and Cordeo 1997). It must be acknowledged that the price of eco-degradation is being paid most heavily by poor hill farmers owning small and fragmented plots of land. The degraded environment is directly linked to reduced productivity and depletion of other natural resources that have been the mainstay of the hill population for centuries. This decrease of indigenous wealth automatically lowers the standard of living of the local people and their nutritional status is the first casualty (Dutta 1998). It therefore confirming that nutritional assessment studies conducted in the Himalayas present a dismal picture.

In hills of Uttarakhand prevalence of Chronic Energy Deficiency (CED) was found high among the farm women (Jethi and Chandra 2013). A study by Dutta *et al.* (2009) on nutritional status of children in Garhwal region of Uttrakhand reveals that malnutrition rate among the children in the Garhwal was very high. The study revealed that a wide children population were severely stunted and wasted, indicating a high prevalence of both acute and chronic malnutrition. In such a condition, protein, consumption of energy, niacin, riboflavin, and ascorbic acid are required. Jethi and Chandra (2013) found that energy consumption protein consumption and education, have positive and significant relationship with nutritional status whereas family size has negative association with nutritional status. So consumption of pulse is highly required for hilly states where malnutrition is prevalent especially for higher hills where malnutrition is more prevalent than mid hills and low hills Dutta *et al.* (2009).

The mountains are rich in resources, yet, limited access to current information and knowledge about good nutrition and health care practices and lack of crop diversity expose mountain people to high rates of malnutrition and disease. The under-nutrition in Hills is because of lack of dietary diversification. The cereal based cropping system incorporates more cereals in the diets of hill people, favoured especially for the energy it provides, resulting exclusion of more nutritious foods such as legumes and pulses, from the food busket. It was observed that the diets of susceptible populations are habitually devoid of protein leading to protein–energy deficiency; and essential micronutrients such as zinc calcium and iron, which further aggravating the problem.

Increasing production of pulses and improving its access will help reduce malnutrition among the poor, especially the women. Initiatives to enhance pulses production in Eastern Indian states has resulted in enhancing nutrition for the family, besides generating additional income, improving soil fertility and reducing migration. (Dogra *et al.*, 2014). To eradicate appearance of malnutrition, the policy needs not only to promote diverse diets but agricultural diversification. The pulses is such a class of food that has the power to eradicate malnutrition. For the importance of pulses as a primary source of protein (~ 21-26 per cent) and other essential nutrients, the United Nation has declared the year 2016 as the International Year of Pulses.

Nutritional Quality of Pulses

Pulses play a primary and essential role as a low-fat, high fibre source of protein, an essential component of traditional food baskets in most developing countries. Pulses are contributing about a volume of ~10 and 5 percent in the daily protein and energy intake repectively. It has profound importance for food and nutritional security in low income countries where the major sources of proteins are non-animal products. In India, from the ancient era many legumes and pulses have been consumed as part of a mostly cereal-based diet. They are consumed as whole grains or as split (called *dhals*). Economically, the pulses are relatively cheaper source of protein than that of milk, cheese, meat, fish cashew and almond *etc.*, hence important for the developing countries as accessability concern. If we consider the statistics, the share of food use in total consumption of pulses in the developing countries is over 75 per cent, compared to 25 per cent in the developed countries. Pulses are rich in protein (lysine) content the value is ~2 times that of cereals and several times that in root tuber (FAO 1968), so cereals and root tubers in combination with pulses can help to improve the protein intake (Kushwah *et al.*, 2002). Pulses are a great source of protein and fat. It contains about 18-32 per cent protein and about 1-5 per cent fat (Table 8.3). In children pulses consumption should be encouraged, particularly in those areas where animal protein is expensive and scant. Pulse consumption would surely help to provide necessary amino acids required for growth. As for example one cup serving of cooked lentils contains more than 15 g of fibre, meeting 60 per cent of our daily requirement. The fibre in the pulses may improve the health of heart by lowering down cholesterol levels (Burkitt and Trowell 1985). Besides protein and fibre, pulses are a significant source of vitamins and minerals, such as zinc, iron, folate, and magnesium, and taking half a cup of beans or peas a day can improve diet quality by increasing intakes of these nutrients. The vitamins present in considerably amount in pulses are thiamin, riboflavin, folic acid and pyridoxine, vitamin E and K. Pulses are considerably higher in calcium than most cereals and contain about 100 to 200 mg of calcium per 100 g of grain.

In addition, the phytochemicals like saponins, and tannins are available in pulses which have antioxidant and anti-carcinogenic properties, indicating that pulses may have significant anti-cancer effects (Mudryi *et al.*, 2014). Consumption of pulse also improves serum lipid profiles and positively reduces several other cardiovascular disease risk factors, such as inflammation, platelet activity and blood pressure (Mudryi *et al.*, 2014). Pulses are high in fibre and have a low glycemic index, which is beneficial to diabetic patients as it supporting in maintainance of healthy blood glucose and insulin levels.

Strategy for Enhancing Productivity of Pulse Crops in Hills

☆ The major problems related to productivity of pulses are technological hindrances as well as the lack of a managerial set-up to administer the landscape in hilly region. The lack of a supporting mechanism for the procurement and marketing of pulses has been a major impediment to the propagation of pulses in Hills. Incentivising pulse production through the price instrument will only work once the farmer is assured that the

Table 8.3: Proximate Composition of Pulse Grains (per 100 gm)

Pulses	Energy (KCal)	Protein (g)	Fat (g)	Carbohydrate (g)	Total Dietary Fibre (per cent)	Minerals (mg/100 g dried wt)												Vitamins/100 g				
						Fe (mg)	Zn (mg)	Ca (mg)	Mg (mg)	K (mg)	Na (mg)	Se (mg)	Thiamine (mg)	Riboflavin (mg)	Niacin (mg)	Pantothenic Acid (mg)	Vita B6 (mg)	Folate (µg)	Vita min C (mg)	Vita min E (mg)
Chick Pea	368	21.0	5.7	61	22.7	6.2	3.4	105	115	875	24	8.2	0.5	0.2	1.5	1.6	0.5	557	4.0	0.8
Pigeon Pea	342	21.7	1.49	62	15.5	5.2	2.7	130	183	1392	17	--	0.6	0.18	2.9	1.26	0.28	456	--	--
Urd	347	24.0	1.6	63.4	16.2	8.4	3.5	110	--	--	--	--	0.6	0.2	2.3	--	0.2	--	--	--
Moong Bean	345	25.0	1.1	62.6	16.3	6.7	2.7	132	189	1246	15	8.2	0.6	0.2	2.3	--	0.2	--	--	--
Lentil	346	27.2	1.0	60	11.5	7.5	4.7	56	122	955	6	8.2	0.8	0.2	2.6	2.12	0.54	479	4.4	0.3
Pea	345	25.1	0.8	61.8	13.4	4.4	3.0	55	115	981	15	1.6	0.7	0.2	2.9	1.8	0.2	274	1.8	0.3
Raj-mash	345	23.0	1.3	63.4	18.2	3.4	1.9	186	188	1316	18	12.9	0.53	0.22	2.08	0.79	0.4	399	4.6	--
Cow Pea	346	28.0	1.3	63.4	18.2	7.54	3.77	80.3	250	1450	23	--	0.94	0.22	2.36	1.39	0.44	545	--	--
Horse Gram	321	23.6	2.3	59.1	15.0	7.0	--	28.7	--	--	--	--	0.4	0.2	1.5	--	--	--	--	--
Moth Bean	330	24.0	1.5	61.9	--	9.6	--	--	--	--	--	--	0.4	0.09	1.5	--	--	--	--	--

Source: Pulses for Human health and Nutrition, ICAR-IIPR, Kanpur.

government will procure and distribute pulses along with rice and wheat through the Food Corporation of India and other State agencies.

☆ Low genetic yield of Indian pulses and their vulnerability to pests and diseases is a major hindrance to adoption of pulses by farmers. Being rain-fed, pulses often experience drought at critical growth stages. The lack of drought- and disease-resistant varieties of pulse seeds is alarming. Apart from these, public investment in agricultural research to develop high-yielding, short-duration strains of pulses, oilseeds and other horticultural crops has been exceptionally low. In such a scenario, public investment in research should be more and focused towards development of drought and disease resistant varieties. In such a condition public private partnership and convergenece approach can help to solve the problem (Mukherjee *et al.*, 2012a; Mukherjee and Maity 2015).

☆ At this juncture, government agencies have to act as an active role for rejuvenating the pulse technology extension in order to achieve the stipulated target. Public extension sytem has several advantages but have functional disadvantages too whereas the private extension are playing increasingly important role in rural knowledge systems, although complete privatisation is not feasible and advisable in country like India. So a converegent approach shall work better. The public extension should focus more on resolving policy issues and quality control issue which can establish a pragmatic mechanism to benefit resource poor pulse farmers of NWH. Private extension services are working in mode of viable business models. In pulse production upliftment a wide ranges of services are required and to ptovide that extension organisations need a large number of partner agencies capable of effectively supplementing and complementing their expertise. Right now there are pluralistic extensions organizations such as cooperatives, public, private, NGOs, Farmer Producer Organization *etc.* are working in this country. A strong value chain management is only be possible through effective convergance of resources and services.

☆ A strong extension strategy is required which shall be focused in production aspects. Pulse village concept needs to be implemented in hills. Identifying land for growing pulses must go hand in hand with promoting yield-augmenting and resource-saving technologies along with providing farmers better access to remunerative markets. Development of strong hub and spoke model like some private extension system have (Mukherjee *et al.*, 2011).Targeting large farmers can be another option. Large farmers and theprogressive have greater risk-absorbing capacity in case of inadvertent loss. Study depicts that marketing orientation, education, economic motivation, social participation, family size and extension agency contact are such factors can influence farmers progressiveness (Mukherjee *et al.*, 2012b and c).Large-scale and progressive farmers can be motivated towards pulse production. Although a comprehensive crop insurance scheme should be devised especially for pulse to cover losses due to unseasonal rains, risk of droughts, natural calamities and wild life damage which are the main problems now a days in planis (Mukherjee 2015) and hills too. In this respect *Pradhanmantri Fasal Bema Yojana* (prime minister crop insurance scheme 2016) can act better to cover these isuues under one single umbrella.

☆ Availability of critical inputs including seeds, bio-pesticides and micronutrients such as zinc should be at the door step of pulse village through PPP mode.

☆ The use of water harvesting and drip irrigation can additionally be promoted through subsidy for pulse-growing farmers.

☆ Dal mills and processing facilities should be encouraged within the vicinity of production areas, which will promote off-farm employment. A detailed study and pilot project may be warranted in initiating nitrogen credit for farmers. PPP in seed production, inputs, promotion and extension must be mapped out.

Conclusions

India has the potential to achieve much higher pulse productivity with available resource by utilizing improved technology. Special attention in terms of government policy initiatives are required particularly for the hilly states. Nutritional benefit outreach awareness programme should be created among the farmers to achieve higher pulses production. A new outlook on sustainability is required to be created to rejuvenate sustainable nutritional status for both soil and human being.

REFERENCES

Agricultural Statistics at a Glance (2014). Government of India Ministry of Agriculture Department of Agriculture and Cooperation Directorate of Economics and Statistics. Published in India by Oxford University Press YMCA Library Building, 1 Jai Singh Road, New Delhi 110 001, India.

Burkitt D P and Trowell H C (1985). *Refined carbohydrates in foods and disease. Some implications of dietary fiber*. Academic Press, New York.

Davies S and Stewart A (1987). *Nutritional Medicine*. Richard Clay Ltd., Bungay, Suffolk.

Dogra A, Sarker A, Hassan A, Sah P and Rizvi A H (2014). Consuming more pulses: Can it be a solution to fight malnutrition? LEISA India 16(4). Available online at http: //www.agriculturesnetwork.org/magazines/india/nutrition/consuming-more-pulses#sthash.ojstghKs.dpuf

Dutta A (1998). Environmental degradation and nutritional status of hill people: Some reflections. In: Singh V, Sharma ML, editors. *Mountain Ecosystems. A Scenario of Unsustainability*. New Delhi, India: Indus Publishing Company, pp. 199–202.

Dutta A, Pant K, Puthia R and Sah A (2009). Prevalence of undernutrition among children in the Garhwal Himalayas. *Food and Nutrition Bulletin*, 30: 77-81.

FAI (2006). Fertilizer Association, of India http: //www.faidelhi.org/statistical-database.htm

FAO (1968). FAO year book. FAO, Rome.

Infant RB and Cordeo R (1997). Status of children population living in Coche Islands. In: Fitzpatrick DW, Anderson JE, L'Abbé ML, editors. *Proceedings of the 16th International Congress of Nutrition*, 1997. Montreal, Canada, 27.

Jethi R and Chandra N (2013). Nutritional Status of Farm Women in Hills of Uttarakhand. *Indian Res. J. Ext. Edu.* 13 (3): 92-97.

Kushwah, A, Rajawat, P, Kushwah, H S (2002). Nutritional evaluation of extruded Faba bean (*Vicia faba* L.) as a protein supplement in cereals based diet in rats. *J. Exp Biol.* 40(1) 49–52.

Mudryi AN, Yu N and Aukema HM (2014). Nutritional and health benefits of pulses. *Appl Physiol Nutr Metab.* 2014 Nov; 39(11): 1197-204. doi: 10.1139/apnm-2013-0557. Epub 2014 Jun 13.

Mukherjee A, Maity A (2015). Public–private partnership for convergence of extension services in Indian agriculture. *Current Science.* 109 (9) : 1557-1563.

Mukherjee A, Bahal R, Roy Burman R, and Dubey SK (2012a). Conceptual Convergence of Pluralistic Extension at Aligarh District of Uttar Pradesh. *Journal Of Community Mobilization and Sustainable Development.* Jan-July 2012. **7** (1 and 2): 85-94.

Mukherjee A, Bahal R, Roy Burman R, Dubey SK, and Jha GK (2011). Effectiveness of Tata Kisan Sansar in Technology Advisory and Delivery Services in Uttar Pradesh. *Indian Research Journal of Extension Education.* 11 (3): 8-13.

Mukherjee A, Bahal R, Roy Burman R, and Dubey SK (2012b). Factors Contributing Farmers' Association in Tata Kisan Sansar: A Critical Analysis. *Indian Research Journal of Extension Education.* 12 (2): 81-86.

Mukherjee A, Bahal R, Roy Burman R, Dubey SK, and Jha GK (2012c). Constraints in Privatized Agricultural Technology Delivery System of Tata Kisan Sansar. *Journal of Global Communication.* 5 (2): 155-159.

Mukherjee A (2015). Prioritization of Problems in Integrated Agriculture: A Case of Rampur Village in Sub Humid Region of Eastern India. *Indian Res. J. Ext. Edu.* **15** (1): 53-59.

U.S. Department of Agriculture, Agricultural Research Service (2008). Published online at www.ars.usda.gov/ba/bhnrc/ndl. USDA National Nutrient Database for Standard Reference, release 21. Nutrient Data Laboratory home page, 2008.

Section D

Technological Intervention and Approaches for Sustainable Pulse Production

2018, *Climate Risks Management: Sustainable Pulse Production*
Editors: A K Srivastava and Yogranjan
Published by: **ASTRAL INTERNATIONAL PVT. LTD., NEW DELHI**   *Pages 181–203*

Chapter 9

Physiological Approaches for Faba Bean Breeding in Drought Prone Environments

M.A. Muktadir[1,3,4], U. Ijaz[2], A. Merchant[3],
A. Sadeque[1] and K.N. Adhikari[1,2]

[1]*Plant Breeding Institute, The University of Sydney,*
Narrabri, NSW 2390, Australia
[2]*Plant Breeding Institute, The University of Sydney,*
107 Cobitty Rd, Cobitty, NSW 2570
[3]*Centre for Carbon Water and Food, The University of Sydney,*
NSW 2570 Australia
[4]*Pulses Research Centre, Bangladesh Agricultural Research Institute,*
Gazipur 1701, Bangladesh

ABSTRACT

The ongoing rise in global temperatures owing to climate change is likely to aggravate the negative effects of hot and dry climatic conditions on faba bean farming. Programs aiming at genetic improvement of the drought resistance of this crop by both classical breeding and molecular breeding methods are hampered due to lack of high throughput screening methodology. Our review focused on adaptation mechanisms of drought namely avoidance, escape and tolerance, with highlighting on physiological traits such aswater potential, stomatal conductance, relative water carbon isotope discrimination, vapor pressure deficit and mesophyll conductance. Drought adaptation is necessary in regions experiencing transient droughts and can be screened by rapid test of stomatal characteristics followed by carbon isotope discrimination.Screening based on single trait is not feasible to improve yield in drought affected environments somewhat a recipe of different traits is required.

Keywords: Fababean, Drought resistance, Physiological attributes, Yield.

Introduction

Among the cool season grain legumes Faba bean (*Vicia faba* L.) ranked fourth. This crop originating in the Middle East has distributed its cultivation throughout Central Asia, Europe and North Africa. Faba bean was moved to China over 2,000 years ago *via* merchants along the Silk Road to South America in the Columbian period and more recently to Canada and Australia (Stoddard, 1991). The crop is now cultivated in more than 55 countries. The totalcultivated area globally was 2.25Mha with average annualproduction of 4.2 Mt dry seed over the last 5 years. Asia andAfrica account for 72 per cent of the area and 80 per cent of the productionof dry faba bean seed (FAOSTAT, 2015). Faba bean hassubstantial rotational advantage in cereal-based farming systems by assisting in control of root diseases, nematodes and weeds. Faba beanhave been shown tofix more than 80 per cent of their own nitrogen need under a wide range of condit ions (Somerville, 2002). Consequently, it is one of the best crops that can be used as green manure and one of the best bio factory of nitrogen byfixing 130 to 160 kg N/ha (Hoffmann *et al.*, 2007; Horst *et al.*, 2007). Due to its multiple uses, highnutritional value both for human and agricultural consumption, and ability to grow in diverse environments, it garners attention for sustainable agriculture in many areas.

Faba bean seeds are a good source of protein, fiber and are widely grown for food and feed (Duc, 1997). Faba beans are an important source of protein in those dry areas of developing countries most likely to be impacted by climate change. The protein content in faba bean is 24 ~ 35 per cent of the seed dry matter and, among other things is very rich in lysine – an essential amino acid in human diets as it cannot be synthesized by animals. Faba bean is one of the main sources of cheap protein for people in the Middle East, Latin America and Africa (consumed as dry or canned) and for livestock (mostly pigs and poultry) in many developed countries. Immature beans are also used as a vegetable.

Among legume species, Faba bean is sensitive to water deficit (Khan *et al.*, 2010). In irrigated farming, faba bean yield increased significantly compared to rainfed systems (Oweis, 2005) though most of production regions depend on rainfall and stored soil moisture for its growth and development. A meta analysis from 1980 to 2014 showed faba bean had the highest yield reduction (40 per cent) under the highest observed water reduction (*i.e.* >65 per cent). Although the amount of yield reduction varied among varieties,there were consistent positive linear relationships between observed in yield reduction (*i.e.*, ratio between yield during drought and during well-watered condition) and the corresponding observedwater reduction across different varieties (Daryanto *et al.*, 2015). Thus, breeding for drought tolerance is of utmost importanceto improve yield and yield stability, however progress has been relatively slow compared to other crops due to lack of efficient screening methods (Stoddard *et al.*, 2006). Wide genotypic variation in water stress response has been reported in this species (Abdelmula *et al.*, 1999; Amede *et al.*, 1999; Grzesiak *et al.*, 1997; Khan *et al.*, 2007; Link *et al.*, 1999; Ricciardi *et al.*, 2001) indicating the potential for breeding of drought-prone environments. Field phenotyping for drought responsesare costly and time-consuming and often produce unreliable output due to variable periods of drought length, timing and intensity (Khan *et al.*,

2007). In contrast, screening under controlled conditions can be reliable however a link to translate laboratory findings to the field, must be made in order for the breeding program to be successful (Passioura, 2012). Molecular and biotechnological tools are poorly developed for faba bean and the fruits of transgenic research developed in model plant species are not reaching this crop (Lavania *et al.*, 2014).

Plant Response Mechanism against Drought

Dry lands cover over 40 per cent of the world and are the home of around 2.1 billion poor people in nearly 100 countries (UN, 2011). Drought is the single most devastating environmental stress which influences the global crop productivity more than any other abiotic stress (Farooq *et al.*, 2009). In changing climatic conditions, models predict that under climate change effects, the frequency and intensity of droughts will increase in agricultural areas worldwide (Gornall *et al.*, 2010). To improve the crop performance and resilience against drought it is essential to understand the underlying mechanism of crop response and adaptation to drought. The key to drought adaptation for plant breeders and crop physiologists is tailoring the morphology, physiology and phenology of a crop to its environment in order to manage water economy.

A number of mechanisms operate singly or jointly to allow plants to adjust with drought stress. Therefore drought stress is demonstrated as a complex trait (Krishnamurthy *et al.*, 1996). Plant has its own way to alter against drought. So far three prominent mechanisms found on which plants adjust with drought conditions: Drought avoidance, Drought tolerance and Drought escape. However, crop plants use more than one mechanism at a time to cope with drought.

Drought Avoidance

Drought physiology ismainly controlled by the use of water through the stomata and uptake through the roots. Drought avoidance denotes mechanisms related to keeping high plant water status under water-limited conditions by reducing water loss through cuticle and/or exploiting water uptake through root characteristics. Stomatal closure is the key drought avoidance mechanism and is driven by abscisic acid (ABA) transported from the roots (Schachtman and Goodger, 2008). The guard cells are responsible to open and close stomata. Stomatal traits such as morphology (*e.g.*, density and size) and function (*e.g.*, stomatal conductance, g_s) are considered key elements of plant growth and water status in Faba bean (Khazaei *et al.*, 2013). Faba bean accessions having higher stomatal density produced lower yield and less resistance to water stress while lesser stomatal density was related with improved drought avoidance (Ricciardi, 1989). Suitable stomatal activity might play a key role for improving drought adaptation in Faba beans by reducing water loss and increasing transpiration efficiency (Darwish and Fahmy, 1997). (Khan *et al.* (2007) and Nerkar *et al.* (1981) found stomatal conductance (g_s) was the most important trait determining water use in relatively small sets of Faba bean accessions.

Cuticular wax accumulation hasalso been considered an important drought avoidance strategy, however, abundant wax was not always correlated with drought tolerance (Ristic and Jenks, 2002). Although accumulation of cuticular waxes significantly increased in soya bean plants exposed to water deficit (Kim *et al.*, 2007), variation in the wax content and in its role against drought stress has not yet been exposed in Faba bean.

Drought avoidance mechanisms are also driven by root characteristics and have been shown to be an important contributor to in many crop species (Blum, 2011; Monneveux and Ribaut, 2011) including legume species, such as groundnut, pigeon pea, cowpea, soya bean, common bean (Purushothaman *et al.*, 2013), chickpea (Kashiwagi *et al.*, 2006) and lentil (Sarker *et al.*, 2005; Singh *et al.*, 2013). The uptake of water can be maximized and dehydration postponed by appropriate root characteristics (Neil C. Turner, 2001). (Kashiwagi *et al.* (2005) proposed higher root biomass, root length density and root depth as the main drought avoidance traits for higher seed yields under drought conditions in chickpea while study of root characteristics in faba bean is still yet to investigate.

Drought Tolerance

Dehydration tolerance refers to the phenomenon of maintaining metabolic activity at low tissue water potential (surviving internal water deficits). Keeping high tissue turgor by osmotic adjustment (OA) antioxidant defense systems and changed dynamics of plant hormones are the major physiological stress-adaptive traits to minimize the detrimental effects of water deficit (Morgan, 1984). OA maintains cell water content by increasing the osmotic pressure that can be exerted by cells on their surroundings and thus increasing water uptake (Blum, 2011). OA has not been deeply investigated in Faba bean but has been demonstrated in many other legumes (Silvente *et al.*, 2012; Turner *et al.*, 2007). In chickpea, OA was not related to changes in carbohydrate composition or leaf gas exchange under drought conditions (Basu *et al.*, 2007). A wide variation was found in pea genotypes for OA when they were exposed to water stress with some producing considerable amounts of soluble sugars and the amino acid proline (Sánchez *et al.*, 1998). Infaba bean, a decline in osmotic potential (representing the accumulation of solutes) has been reported in limited lines under water stress conditions, but there was no evidence of osmotic adjustment (Amede *et al.*, 1999) and (Khan *et al.*, 2007) in the solutes quantified possibly representing a constitutive presence of solutes (Merchant, 2014). Given the potential for the regulation of osmotic potential to confer resilience against drought effects, a wider spectrum of faba bean germplasm needs to be surveyed.

Drought Escape

Drought escape is the ability of a crop to complete its life cycle before the onset of drought or unfavorable conditions. The matching of crop growth to the pattern of rainfall is one of the most important breeding goals for drought-prone environments particularly those prone to terminal drought. Early flowering is the most important phenological trait when terminal drought is likely. Consequently, early sowing along with early seedling vigor and better root establishment in dry environments allows rapid ground coverage and as a result reduces evaporation from the soil (Loss *et al.*, 1997). Although earliness allows crops to avoid terminal drought it can lead to a yield penalty under well watered conditions (Monneveux *et al.*, 2012). There is considerable variation in days to onset of flowering in faba bean accessions (Stoddard, 1993) and developmental responses to temperature and day length (Iannucci *et al.*, 2008; Patrick and Stoddard, 2010). Earliness is also a desirable attribute for often short season climates such as boreal zone (Stoddard and Hämäläinen, 2011). Combinations of thesecharacters are required to improve yield and its stability under drought situations. For example, combined selection for increased harvest index, earliness and phenotypic plasticity has been suggested as a method to stabilize legume yield under Mediterranean environment (Siddique *et al.*, 2003).

Detrimental Effects of Drought Stress on Faba Bean Cultivation

The world area under faba bean cultivation has declined from 5 million hectare in 1965 to less than half of this figure in 2007 due to unstable yields and biotic and abiotic stresses (Rubiales, 2010). Tangibleeconomic gains will be visible if the cultivation of this crop is extended to larger farming areas.Faba bean plants experience stressedby drought mainly towards the end of the growing season (Stoddard *et al.*, 2006). End-of-season water shortage, commonly called'terminal drought', often occurs in the Mediterranean-type environmentwhere faba beansgrow and yield better than other grain legumes (Loss *et al.*, 1997). During floweringand podding, the soil surface frequently dries whilesufficient water for root functioning is available atdepth but unreachable due shallow root system of faba beans compare to other legumes.De Costa *et al.* (1997) reported that shortage of soil water is a major limiting factor for increasing the production of faba bean and different cultivars are suited to different watering regimes, biomass production capacities of a determinate cultivar 'Tina' and an indeterminate cultivar 'Gobo' being highly sensitive to water stress. These authors further noted that water shortage affected the canopy size and reduced leaf initiation, expansion and longevity hampering biomass production and the indeterminate cultivar was more suitable for cultivation under water stress to obtain higher total biomass production than the determinate cultivar. Link *et al.* (1999) analyzed the ratio of yield under drought, to yield under irrigation among different genotypes of faba bean. Three cultivars and eight drought-tolerant breedinglines selected at Mediterranean region, which is knownfor its irregular water distribution and moderate moisture levels (~500mmrainfall). Therewere significant genotypic differences for seed yield under supplemented irrigation and rainfed conditions observed (Maalouf *et al.*, 2015). Significant variance in drought stress tolerance among the genotypes suggested the feasibility of genetic improvement fordrought tolerance by conventional breeding. Because of the high sensitivity of faba bean to drought stress, it is not possible to extend the cultivation of this crop to arid and semi-arid regions of middle eastwith low rainfall (100–200 mm), high soil water evaporation and deep water percolation. The unstable weather conditions in many regions of Saudi Arabia and the variability among locations, years and seasons severely affected the yield of faba bean crop (Alghamdi, 2007).Alghamdi (2009) analyzed 13 faba bean genotypes grown under three different watering regimes 13200, 7600 and 4800 m^3/ha representing heavy, moderate, and normal irrigation respectively during the crop growth season. This study revealed that faba bean genotypes exhibited large differences in their seed composition among different watering regimes. Al-Suhaibani (2009) assessed the influence of water regimes on faba bean cultivar Giza 957 at four different growth stages encompassing its entire life cycle. The results showed that water deficit significantly influenced seed yield and quality of faba bean under the arid environment of Saudi Arabia. The genetic variability, yield stability and correlations among yield, yield components and other vegetative traits were significantly different among genotypes for most of the agronomic characters. From the above account it emerges that drought stress conditions may either partially or completely damage the faba bean plants depending upon the severity of stress.

Tolerance to water deficit at three critical reproductive stages namely floral initiation, flowering and pod development was found to vary considerably in three contrasting faba bean genotypes ACC286, Fiord and Icarus affecting growth and yield (Mwanamwenge *et al.*, 1999). According to (Patrick and Stoddard, 2010; Sekara *et al.*, 2001) drought stress resulted in increased numbers of flowers, but decreased number of pods and seeds because of high rates of abscission in faba bean. Chbouki *et al.* (2005) studied the abscission process in three physiological types of faba bean and found precipitation to be the major factor limiting seed production. They noted that rainfall was highly variable from season to season and the decline in the amount of total annual rainfall resulted in unstable and decreased seed yields from 0.91 t ha^{-1} in 1994 to 0.4 t ha^{-1} in 1999. It was shown that the heavy flower loss in faba bean resulted from a combination of plant attributes and environmental stresses including drought.

Physiological Attributes Related of Drought Tolerance

Understanding the physiological basis of drought tolerance will indicate traits that can be used as indirect selection criteria for the development of cultivars adapted to drought conditions. The use of secondary plant characteristics related to enhanced production in water-limited environments has often been suggested to complement phenotypic selection (Boyer, 1982; Lafitte *et al.*, 2003; Richards, 2006). Carbon isotope discrimination ($\Delta^{13}C$) has been proposed as a physiological criterion to select drought tolerance in C_3 crops (Seibt *et al.*, 2008) due to its close association with water use efficiency (WUE). Good correlations with yield and water use efficiency have been found as a result of balance between photosynthetic activity and g_s (Condon *et al.*, 2004). Khan *et al.* (2007) found wide genetic variation for $\Delta^{13}C$ in faba bean and reported a negative correlation between transpiration efficiency and $\Delta^{13}C$ under well watered conditions. Leaf temperature has been shown to be a rapid and cost-efficient surrogate for measuring g_s and water status in several plant species (Blum, 2011) including faba bean (Khan *et al.*, 2010). Water use efficiency (WUE) has often been proposed as one the most important determinants of crop productivity under water-limited conditions and a target for plant drought adaptation. Blum (2009) argued however that selection for high WUE under drought conditions leads to dramatic losses in yield and drought adaptation. To overcome this problem, two layers of selection were suggested: one for higher $\Delta^{13}C$ and the other for shoot biomass (Krishnamurthy *et al.*, 2013). Efficient use of water *i.e.*, maximizing soil moisture capture through transpiration is the target for improvement of crop yield under drought conditions (Blum, 2009). Analysis of physiological responses to water deficit can identify sources of transpiration efficiency (TE) and water interception and furthermore provide efficient tools for use by breeders (Jackson *et al.*, 1996; Richards, 2006). Cultivars that show stability of performance under unfavorable environments should possess physiological components of tolerance to environmental stresses. Direct measurement of physiological traits for drought response is useful when a large number of genotypes can be measured (Wery *et al.*, 1994). Various physiological attributes of the faba bean, such as water use (Amede *et al.*, 1999), water potential (Karamanos and Papatheohari, 1999), stomatal characteristics (Bond *et al.*, 1994; Ricciardi, 1989) leaf temperature and carbon isotope discrimination ($\Delta^{13}C$) (Khan *et al.*, 2007) may be useful in developing selection techniques for drought avoidance or tolerance.

Screening Techniques

Canopy Temperature

The surface temperature of the canopy is related to the amount of transpiration resulting in evaporative cooling. A hand-held infrared thermometer (IRT) allows canopy temperature (CT) to be directly and easily measured remotely and without interfering with the crop. Canopy temperature measured by IRT as a screening technique for dehydration avoidance was first reported by (Blum *et al.*, 1982). Stomatal closure resulting from plant water deficit leads to decreased transpirational cooling and consequently increased leaf temperature, relative to well watered plants (Kramer and Boyer, 1995; O'Neill *et al.*, 2006). In the field, lower canopy temperature under drought stress conditions indicates a better capacity to take up water from deep soil and hence better water status (Blum, 2011). Canopy temperature can be measured by an infrared thermometer (IRT) as a rapid and cost-efficient alternative for preliminary screening for drought adaption under controlled and uniform conditions in faba bean (Khan *et al.*, 2010; Khan *et al.*, 2007). Canopy temperature has also been shown as an efficient field phenotyping tool for drought adaptation in other legume species e.gSoyabean (McKinney *et al.*, 1989), chickpea (Kashiwagi *et al.*, 2008) cowpea (Hall, 2012). Its assessment is non-destructive and large numbers of accessions can be evaluated at the vegetative stage.

Leaf Water Relations

Water stress affects several aspects of plant physiology such as gas exchanges, hormone relations and water relations. Attempts to correlate values of stomatal conductance and leaf water potential with particular environmental variables in the field are generally of only limited success because they are simultaneously affected by a number of environmental variables. For example, correlations between leaf water potential and either flux of radiant energy or vapor pressure deficit show a diurnal hysteresis. Physiological knowledge of stomatal functioning is adequate to provide a mechanistic model linking stomatal conductance to all these variables.

Leaf water potential (Ψ) and its twomain components; osmotic potential (Ψs) and turgor potential (Ψp) are useful as selection criteria for improving drought tolerance in cropplants. Leaf water potential valuates the water stress intensity sensed by leaves (Hsiao, 1973) and is recognized as an index for whole plant water status (Pantuwan *et al.*, 2002; Turner, 1982). It is considered as a reliable parameter for quantifying plant water stressresponse (Siddique *et al.*, 2000). In general, the maintenance of high Ψ determined by theinteraction of numerous plant mechanisms at both shoot and root levels is considered tobe associated with dehydration avoidance mechanisms (Levitt, 1980). Maintenance of leafturgor in the face of decreasing soil moisture has been emphasized as an important adaptation trait that contributes to drought tolerance (Hsiao *et al.*, 1976). (Jongdee *et al.*, 2002; Pantuwan *et al.*, 2002; Sibounheuang *et al.*, 2001) found that genotypes with high Ψ had less reproductive sterility and produced higher yield than genotypes with lower Ψ under drought stress conditions. Other reports suggest that plant metabolic processes are in fact more sensitive to turgor and cell volume than absolute water potential (Jones and Corlett, 1992).

Relative Water Content

A satisfactory basis for relating cellular water status to metabolism is relative water content (RWC) which is easily measured and a robust indicator of water status for comparison of tissues and species.It 'normalizes' water content by expressing it relative to the fullyturgid (hydrated) state (Lawlor and Cornic, 2002). Sinclair and Ludlow (1985) proposedthat leaf relative water content (RWC) is a better indicator of water status than was water potential (Ψ). RWC is a measure of relative change in cell volume, is the resultant ofcell turgor (Ψp) and osmotic potential (Ψs) and thus depends both on solute concentrationand cell wall rigidity and does not relate directly to cell volume (Kaiser, 1987; Lawlor, 1995). RWC as an integrative indicator of internal plant waterstatus under drought conditions has successfully been used to identify drought-resistantcultivars of barley (*Hordeumvulgare*) (Matin *et al.*, 1989), faba bean (Khan *et al.*, 2007) and common bean (França *et al.*, 2000).

Stomatal Conductance

Stomatal conductance has been proposed as a selectiontool and, when measured on multiple plants in a canopy, is equally effective as CT (Condon *et al.*, 2007).Stomatal conductance isdetermined by the combination of stomatal density, size andopening. While stomatal size and density are evaluated bymicroscopy of leaf surfaces (Grzesiak *et al.*, 1997) theaverage size of stomatal opening is hard to measure since itresponds rapidly to the environment.Stomatal conductance isefficiently measured with a porometer offering values that are integrated only over a small chamber space and short measurement period but nevertheless commonly reflect physiological patterns and behavior. (Ricciardi, 1989) found that accessions with higherstomatal density had lower seed yields and less resistance towater deficit, while low stomatal density was associated withbetter adaptation to stress conditions.Drought causes lower stomatal conductance and higher transpiration efficiency among tested genotypes. This confirms the physiological significance of lower stomatal conductance in regulating the water loss from leaves (Alghamdi *et al.*, 2014; Khan *et al.*, 2007).Variation for stomatal conductance measurements in a large population (402), suggesting a high potential for breeding for these traits (Khazaei, 2013).

Carbon Isotope Discrimination

Carbon isotope discrimination has been proposed as an indirect selection index for transpiration efficiency (TE) in C_3 species under water-limited environments (Condon *et al.*, 2004). The carbon in atmospheric CO_2 occurs as two stable isotopic forms: 99 per cent is ^{12}C and 1 per cent ^{13}C. The ^{13}C content of plant material is described in two ways; carbon isotope composition ($\Delta^{13}C$) and carbon isotope discrimination ($\Delta^{13}C$). $\Delta^{13}C$ is a measure of the $^{13}C/^{12}C$ ratio in the plant material relative to the value of the same ratio in the air that the plant consumed (Farquhar and Richards, 1984). Generally among C_3 plants plant material contains fractionally less ^{13}C and more ^{12}C than the atmospheric CO_2 (Condon *et al.*, 2006) attributable to both diffusional and biochemical fractionation processes. In C_3 plant species, variation in $\Delta^{13}C$ can result from variation in stomatal conductance and photosynthetic capacity which is large enough to generate substantial differences in TE (Condon *et al.*, 2006). The use of $\Delta^{13}C$ to evaluate TE has been well established in cereals and more recently extended to lucerne (*Medicago sativa* L.) (Johnson and Tieszen, 1994); common bean (*Phaseolus vulgaris* L.) (Zacharisen *et al.*, 1999); (Lockhart *et al.*, 2016; Smith *et al.*, 2016) soybean (*Glycine max* (L.) Merr.), (Clay *et al.*, 2003); chickpea (*Cicer arietinum* L.), (Khan *et al.*, 2004). Genetic variation for $\Delta^{13}C$ also exists in faba bean (Khan *et al.*, 2007), the youngest fully expanded leaf of six-week-old plants of drought-avoiding accession ILB 938/2 and cv. Me´lodie had significantly lower values of $\Delta^{13}C$ than those of the droughtsensitive inbred line Aurora/1 at adequate moisture supply. $\Delta^{13}C$ was negatively correlated with TE in controlled-environment conditions (Khan *et al.*, 2007). Measures of drought avoidance were correlated with yield maintenance as shown in an experiment using rain-exclusion shelters (Link *et al.*, 1999) has not confirmed that $\Delta^{13}C$ can be used to predict grain yield in field experiments in this species. $\Delta^{13}C$ is the "gold standard" against which other methods are compared but it is expensive so it is best kept as a final test for a few tens of lines after the thousands have been screened by more economical methods such as leaf temperature, canopy temperature. One of the effective trait not yet reported any faba bean breeding program is selection for high TE but widely used in cereals.

Mesophyll Conductance

There is a tight trade-off between uptake of CO_2 and water loss through stomata. According to the curvilinear relationship between net assimilation (A) and g_s, drought stresscauses reduced photosynthesis ultimately causing lower yield (Flexas *et al.*, 2004). During drought maintaining or increasing yield would require an increase A at a given g_s, *i.e.*, genotypic modifications in A-g_s relationship (Parry *et al.*, 2005). At leaf level, one way to achieve this will be to increase the mesophyll conductance to CO_2 (g_m) which regulates the diffusion of CO_2 from sub-stomatal cavities to the carboxylation site.

Using Fick's law, mesophyll conductance (g_m) has been defined as A/ (Ci–Cc) where Ci is the concentration of CO_2 in the intercellular space and Cc is the CO_2 concentration at the site of carboxylation. After entering through the stomata, CO_2 diffuses through air spaces, cell walls, cytosol and chloroplast envelopes and finally reaches the chloroplast stroma where it is fixed by Rubisco (Evans *et al.,* 2009; Evans and Von Caemmerer, 1996). CO_2 from photorespiration and mitochondrial respiration diffuses through the mitochondrial envelope, cytosol, and chloroplast membranes into the stroma and forms an additional source of CO_2 for photosynthesis. Together these components form the substomatal cavities to the stroma constitute g_m (Kaldenhoff, 2012).

The term mesophyll conductance; g_m (Harley *et al.,* 1992) is synonymous with internal conductance; g_i (Lloyd *et al.,* 1992), wall and liquid phase conductance and CO_2 transfer conductance (Caemmerer and Evans, 1991). Some papers *e.g.* (Eichelmann *et al.,* 2004; Laisk and Loreto, 1996; Samsuddin and Impens, 1979; Troughton and Slatyer, 1969) call 'mesophyll conductance' or 'internal conductance' to the initial slope of A versus Ci relationship. However, in this review as proposed by (Flexas *et al.,* 2008) the term mesophyll conductance is restricted to the diffusion of CO_2 through leaf mesophyll.g_m was originally assumed large enough to have a negligible impact on photosynthesis (Farquhar *et al.,* 1980) and was assumed to be constant throughout a leaf's lifespan (Evans and Von Caemmerer, 1996). More recent research suggests that g_m is finite and even may be of similar magnitude as g_s *i.e.* it significantly decrease the concentration of CO_2 at the site of carboxylation (Cc) relative to that in the intercellular space (Ci) thereby limiting photosynthesis (Flexas *et al.,* 2008). It was recently suggested by (Barbour *et al.,* 2010) that further improvements in A/gs could be accelerated by explicitly selecting for g_m.

Reduction of g_m under drought has been observed in several studies (Flexas *et al.,* 2002; Flexas *et al.,* 2006; Grassi and Magnani, 2005; Misson *et al.,* 2010). Ferrio *et al.* (2012) and Rancourt *et al.* (2013) observed that g_s responded rapidly to early signs of water stress and g_m followed afterwards. Ferrio *et al.* (2012) demonstrated a strong linear relationship between g_m and hydraulic conductance of the leaf lamina only below a certain threshold. Karatassiou *et al.* (2009) studied the drought adaptation of two annual legumes *i.e. Medicago minima* L. and *Onobrychisa equidentata,* and found that these two legumes exhibited a completely different photosynthetic behavior. The photosynthetic rate in *M. minima* was mainly depended on stomatal conductance while in *O. aequidentata* seems to be depended both on stomatal and mesophyll conductance.

Vapor Pressure Deficit

Leaf-to-air water vapor pressure deficit (VPD) is an important environmental factor that can affect stomatal functioning in higher plants. The response of plants to changes in VPD are multifaceted and encompass the range of physiological and chemical processes outlined above. The use of changes in VPD as a surrogate for drought effects offers some promise for rapid screening of individual phenotypes. Maintaining high photosynthetic activity when the vapor pressure deficit is high would obviously favor biomass accumulation, albeit at a high water cost since water use efficiency is inversely related to vapor pressure deficit. Therefore, genotypes capable of transpiring less at high vapor pressure deficit would save water at the cost of a lower carbon accumulation potential. However, over the long-term a parsimonious water-use strategy may increase mean yield if water is limited. This hypothesis is supported by modeling results in sorghum showing that imposing a maximum transpiration rate per day saves water, increases the transpiration efficiency and leads to a yield benefit in most years (Sinclair *et al.*, 2005). In drought-tolerant soybean there is evidence for physiological responses to water stress: at vapor pressure deficits above 2.0 kPa transpiration rates are flat or increase at relatively lower rates (Sinclair *et al.*, 2008). This trait limits soil moisture use at high vapor pressure deficit when carbon fixation has a high water cost, leaving more soil water available for subsequent grain filling. There are conflicting results about whether stomata respond to VPD or not. Soil water stress and leaf position are factors that may affect the stomatal response to VPD and can help to explain these conflicting results. When stomata do respond to VPD the mechanism causing such response is not well understood and two contrasting hypotheses have been proposed. The feed forward hypothesis states that stomatal conductance (g_s) decreases directly as VPD increases with abscisic acid (ABA) in the leaves probably triggering the response. The feedback hypothesis states that g_s decreases as VPD increases because of an increase in transpiration (E) that lowers the leaf water potential.

Stomata respond to VPD, when VPD increases stomata begin to close. Closure is the outcome of a decline in guard cell turgor but the link between VPD and turgor is poorly understood. As water stress developed, stomata regulated transpiration at ever decreasing values of VPD. Thus, stomatal sensitivity to VPD increased with increasing water stress.

The argument about the potential mechanism (s) involving stomatal responses to VPD (feedback or feed forward) is eventually the same argument about the mechanism that causes stomata to respond to soil water stress (Kramer, 1988; Passioura, 1988). The earlier view of leaf water potential as a measure of water stress in plants (Kramer, 1988; Raschke, 1975) has been null and void (Passioura, 1988; Sinclair and Ludlow, 1985) because of the proof that soil water instead of leaf water status governs stomatal function under drought conditions through chemical signals (among them ABA) that originate in the root system (Davies and Zhang, 1991; Liang *et al.*, 1997). The feedback theory as the mechanism that stomata respond to VPD was proposed based on the assumption that leaf water status was accountable for stomatal control (Monteith, 1995). The feed forward hypothesis assumed that if chemical signals are responsible for the stomatal response to soil water stress, then similar signals (*e.g.*, ABA) would also be produced in the leaves in response to VPD (Bunce, 1998; Farquhar, 1978). In the field, drought history of the crop may define the type of mechanism that causes stomatal response to VPD.

Conclusion

Faba bean is establishing asone of the major components of future cropping systems due its multipurpose use as food, feed and relative independence from soil nitrogen supply. Faba bean is more sensitive to drought than other legumes, such as chickpea, and will be more affected by future climate change effects. Improving drought tolerance has thus become an unavoidable subject of research for sustainable legume production. Lately there have been significant progresses made in drought, notably in cereals by selecting for plants with high transpiration efficiency (A/g_s). Genotypic variation in A/g_s has been observed in faba bean as well.Different physiological traits may be effective in selecting materialsfor tolerance to drought stress.Some of the traits like canopy temperature can be efficiently used for large volume accessions while carbon isotope discrimination is for low number of accession. The rich germplasm collection of fababean should be systematically maintained and utilized as aresource base of genes against tolerance to drought stresses. The genome size of faba bean is larger compare to other legumes and next-generation genotypingand sequencing technologies can aid the generation ofmolecular markers for genetic diversity assessment,demographic studies, construction of genetic linkage maps,QTL mappingenhance the pace of genetic improvement against drought stresses.Finally, no single trait is adequate to improve yield in drought-prone environments, rather a combination of characteristics is needed.

REFERENCES

Abdelmula, A. A., Link, W., Kittltz, E. v., and Stelling, D. (1999). Heterosis and inheritance of drought tolerance in faba bean (*Vicia faba* L). *Plant Breeding* **118**, 485-490.

Al-Suhaibani, N. (2009). Influence of early water deficit on seed yield and quality of faba bean under arid environment of Saudi Arabia. *Am-Eurasian J Agric Environ Sci.*, **5**, 649-654.

Alghamdi, S. S. (2007). Genetic behavior of some selected faba bean genotypes. *In* "African Crop Science Conference Proceedings", Vol. 8, pp. 709-714.

Alghamdi, S. S. (2009). Chemical composition of faba bean (*Vicia faba* L.) genotypes under various water regimes. *Pakistan Journal of Nutrition* **8**, 477-482.

Alghamdi, S. S., Al-Shameri, A. M., Migdadi, H. M., Ammar, M. H., El-Harty, E. H., Khan, M. A., and Farooq, M. (2014). Physiological and Molecular Characterization of Faba bean (*Vicia faba*L.) Genotypes for Adaptation to Drought Stress. *Journal of Agronomy and Crop Science*, n/a-n/a.

Amede, T., Kittlitz, E. V., and Schubert, S. (1999). Differential Drought Responses of Faba Bean "*Vicia faba* L" Inbred Lines. *Journal of Agronomy and Crop Science*, **183**, 35-45.

Barbour, M. M., Warren, C. R., Farquhar, G. D., Forrester, G., and Brown, H. (2010). Variability in mesophyll conductance between barley genotypes, and effects on transpiration efficiency and carbon isotope discrimination. *Plant, Cell and Environment*, **33**, 1176-1185.

Basu, P. S., Berger, J. D., Turner, N. C., Chaturvedi, S. K., Ali, M., and Siddique, K. H. M. (2007). Osmotic adjustment of chickpea (*Cicer arietinum*) is not associated with changes in carbohydrate composition or leaf gas exchange under drought. *Annals of Applied Biology*, **150**, 217-225.

Blum, A. (2009). Effective use of water (EUW) and not water-use efficiency (WUE) is the target of crop yield improvement under drought stress. *Field Crops Research*, **112**, 119-123.

Blum, A. (2011). Drought Resistance is it really a complex trait? *Functional Plant Biology*, **38**, 753-757.

Blum, A., Mayer, J., and Gozlan, G. (1982). Infrared thermal sensing of plant canopies as a screening technique for dehydration avoidance in wheat. *Field Crops Research*, **5**, 137-146.

Bond, D., Jellis, G., Rowland, G., Le Guen, J., Robertson, L., Khalil, S., and Li-Juan, L. (1994). Present status and future strategy in breeding faba beans (*Vicia faba* L.) for resistance to biotic and abiotic stresses. *In* "Expanding the Production and Use of Cool Season Food Legumes", pp. 592-616. Springer.

Boyer, J. S. (1982). Plant productivity and environment. *Science*, **218**, 443-448.

Bunce, J. (1998). Effects of humidity on short-term responses of stomatal conductance to an increase in carbon dioxide concentration. *Plant, Cell and Environment*, **21**, 115-120.

Caemmerer, S., and Evans, J. (1991). Determination of the average partial pressure of CO_2 in chloroplasts from leaves of several C3 plants. *Functional Plant Biology*, **18**, 287-305.

Chbouki, S., Shipley, B., and Bamouh, A. (2005). Path models for the abscission of reproductive structures in three contrasting cultivars of faba bean (*Vicia faba*). *Canadian Journal of Botany*, **83**, 264-271.

Clay, D., Clay, S., Jackson, J., Dalsted, K., Reese, C., Liu, Z., Malo, D., and Carlson, C. (2003). Carbon-13 discrimination can be used to evaluate soybean yield variability. *Agronomy Journal*, **95**, 430-435.

Condon, A. G., Farquhar, G. D., Rebetzke, G. J., Richards, R. A., and Ribaut, J. (2006). The application of carbon isotope discrimination in cereal improvement for water-limited environments. *Drought Adaptation in Cereals*, 171-219.

Condon, A. G., Reynolds, M. P., Rebetzke, G. J., Ginkel, M. V. A. N., Richards, R. A., and Farquhar, G. D. (2007). "Using stomatal aperture related traits to select for high yield potential in bread wheat," Springer Netherlands.

Condon, A. G., Richards, R. A., Rebetzke, G. J., and Farquhar, G. D. (2004). Breeding for high water-use efficiency. *Journal of Experimental Botany*, **55**, 2447-2460.

Darwish, D. S., and Fahmy, G. M. (1997). Transpiration decline curves and stomatal characteristics of faba bean genotypes. *Biologia Plantarum*, **39**, 243-249.

Daryanto, S., Wang, L., and Jacinthe, P. A. (2015). Global Synthesis of Drought Effects on Food Legume Production. *PLoS One***10**, e0127401.

Davies, W. J., and Zhang, J. (1991). Root signals and the regulation of growth and development of plants in drying soil. *Annual Review of Plant Biology*, **42**, 55-76.

De Costa, W., Dennett, M., Ratnaweera, U., and Nyalemegbe, K. (1997). Effects of different water regimes on field-grown determinate and indeterminate faba bean (*Vicia faba* L.). II. Yield, yield components and harvest index. *Field Crops Research*, **52**, 169-178.

Duc, G. (1997). Faba bean. *Field Crops Research*, **53**, 99-109.

Eichelmann, H., Oja, V., Rasulov, B., Padu, E., Bichele, I., Pettai, H., Möls, T., Kasparova, I., Vapaavuori, E., and Laisk, A. (2004). Photosynthetic parameters of birch (*Betula pendula* Roth) leaves growing in normal and in CO_2- and O_3-enriched atmospheres. *Plant, Cell and Environment*, **27**, 479-495.

Evans, J. R., Kaldenhoff, R., Genty, B., and Terashima, I. (2009). Resistances along the CO2 diffusion pathway inside leaves. *Journal of Experimental Botany*, **60**, 2235-2248.

Evans, J. R., and Von Caemmerer, S. (1996). Carbon dioxide diffusion inside leaves. *Plant Physiology*, **110**, 339.

Farooq, M., Wahid, A., Kobayashi, N., Fujita, D., and Basra, S. M. A. (2009). Plant drought stress: effects, mechanisms and management. *Agronomy for Sustainable Development*, **29**, 185-212.

Farquhar, G. (1978). Feedforward responses of stomata to humidity. *Functional Plant Biology*, **5**, 787-800.

Farquhar, G., and Richards, R. (1984). Isotopic composition of plant carbon correlates with water-use efficiency of wheat genotypes. *Functional Plant Biology*, **11**, 539-552.

Farquhar, G. D., Caemmerer, S. v., and Berry, J. A. (1980). A Biochemical Model of Photosynthetic CO2 Assimilation in Leaves of C3 Species.pdf>. *Planta*, **149**, 78-90.

Ferrio, J. P., Pou, A., FLOREZ-SARASA, I., Gessler, A., Kodama, N., Flexas, J., and RIBAS-CARBÓ, M. (2012). The Péclet effect on leaf water enrichment correlates with leaf hydraulic conductance and mesophyll conductance for CO_2. *Plant, Cell and Environment*, **35**, 611-625.

Flexas, J., Bota, J., Escalona, J. M., Sampol, B., and Medrano, H. (2002). Effects of drought on photosynthesis in grapevines under field conditions: an evaluation of stomatal and mesophyll limitations. *Functional Plant Biology*, **29**, 461-471.

Flexas, J., Bota, J., Loreto, F., Cornic, G., and Sharkey, T. (2004). Diffusive and metabolic limitations to photosynthesis under drought and salinity in C_3 plants. *Plant Biology*, **6**, 269-279.

Flexas, J., Ribas-Carbó, M., Diaz-Espejo, A., Galmés, J., and Medrano, H. (2008). Mesophyll conductance to CO_2: current knowledge and future prospects. *Plant, Cell and Environment*, **31**, 602-621.

Flexas, J., Ribas-Carbó, M., Hanson, D. T., Bota, J., Otto, B., Cifre, J., McDowell, N., Medrano, H., and Kaldenhoff, R. (2006). Tobacco aquaporin NtAQP1 is involved in mesophyll conductance to CO_2 *in vivo*. *The Plant Journal*, **48**, 427-439.

França, M. G. C., Thi, A. T. P., Pimentel, C., Rossiello, R. O. P., Zuily-Fodil, Y., and Laffray, D. (2000). Differences in growth and water relations among Phaseolus vulgaris cultivars in response to induced drought stress. *Environmental and Experimental Botany*, **43**, 227-237.

Gornall, J., Betts, R., Burke, E., Clark, R., Camp, J., Willett, K., and Wiltshire, A. (2010). Implications of climate change for agricultural productivity in the early twenty-first century. *Philos Trans R Soc Lond B Biol Sci*, **365**, 2973-89.

Grassi, G., and Magnani, F. (2005). Stomatal, mesophyll conductance and biochemical limitations to photosynthesis as affected by drought and leaf ontogeny in ash and oak trees. *Plant, Cell and Environment*, **28**, 834-849.

Grzesiak, S., Iijima, M., Kono, S., and Yamauchi, A. (1997). Differences in drought tolerance between cultivars of field bean and fieldpea. A comparison of drought-resistant and drought-sensitive cultivars. *Acta Physiologiae Plantarum*, **19**, 349-357.

Hall, A. E. (2012). Phenotyping cowpeas for adaptation to drought. *Frontiers in Physiology*, **3**.

Harley, P. C., Loreto, F., Di Marco, G., and Sharkey, T. D. (1992). Theoretical considerations when estimating the mesophyll conductance to CO_2 flux by analysis of the response of photosynthesis to CO_2. *Plant Physiology*, **98**, 1429-1436.

Hoffmann, D., Jiang, Q., Men, A., Kinkema, M., and Gresshoff, P. M. (2007). Nodulation deficiency caused by fast neutron mutagenesis of the model legume Lotus japonicus. *Journal of Plant Physiology*, **164**, 460-469.

Horst, I., Welham, T., Kelly, S., Kaneko, T., Sato, S., Tabata, S., Parniske, M., and Wang, T. L. (2007). TILLING mutants of *Lotus japonicus* reveal that nitrogen assimilation and fixation can occur in the absence of nodule-enhanced sucrose synthase. *Plant Physiology*, **144**, 806-820.

Hsiao, T. C. (1973). Plant responses to water stress. *Annual Review of Plant Physiology*, **24**, 519-570.

Hsiao, T. C., Acevedo, E., Fereres, E., and Henderson, D. (1976). Water stress, growth, and osmotic adjustment. *Philosophical Transactions of the Royal Society B: Biological Sciences*, **273**, 479-500.

Iannucci, A., Terribile, M., and Martiniello, P. (2008). Effects of temperature and photoperiod on flowering time of forage legumes in a Mediterranean environment. *Field Crops Research*, **106**, 156-162.

Jackson, P., Robertson, M., Cooper, M., and Hammer, G. (1996). The role of physiological understanding in plant breeding; from a breeding perspective. *Field Crops Research*, **49**, 11-37.

Johnson, R., and Tieszen, L. (1994). Variation for water-use efficiency in alfalfa germplasm. *Crop Science*, **34**, 452-458.

Jones, H., and Corlett, J. (1992). Current topics in drought physiology. *The Journal of Agricultural Science*, **119**, 291-296.

Jongdee, B., Fukai, S., and Cooper, M. (2002). Leaf water potential and osmotic adjustment as physiological traits to improve drought tolerance in rice. *Field Crops Research*, **76**, 153-163.

Kaiser, W. M. (1987). Effects of water deficit on photosynthetic capacity. *Physiologia Plantarum*, **71**, 142-149.

Kaldenhoff, R. (2012). Mechanisms underlying CO_2 diffusion in leaves. *Current Opinion in Plant Biology*, **15**, 276-281.

Karamanos, A., and Papatheohari, A. (1999). Assessment of drought resistance of crop genotypes by means of the water potential index. *Crop Science*, **39**, 1792-1797.

Karatassiou, M., Noitsakis, B., and Koukoura, Z. (2009). Drought adaptation ecophysiological mechanisms of two annual legumes on semi-arid Mediterranean grassland. *Sci Res Essays*, **4**, 493-500.

Kashiwagi, J., Krishnamurth, L., Singh, S., Gaur, P., Upadhyaya, H., Panwar, J., Basu, P., Ito, O., and Tobita, S. (2006). Relationships between Transpiration Efficiency and Carbon Isotope Discrimination in Chickpea (*C. arietinum* L). *SAT eJournal*, **2**.

Kashiwagi, J., Krishnamurthy, L., Upadhyaya, H., and Gaur, P. (2008). Rapid screening technique for canopy temperature status and its relevance to drought tolerance improvement in chickpea. *Journal of SAT Agricultural Research*, **6**, 105-4.

Kashiwagi, J., Krishnamurthy, L., Upadhyaya, H. D., Krishna, H., Chandra, S., Vadez, V., and Serraj, R. (2005). Genetic variability of drought-avoidance root traits in the mini-core germplasm collection of chickpea (*Cicer arietinum* L.). *Euphytica*, **146**, 213-222.

Khan, H., McDonald, G., and Rengel, Z. (2004). Zinc fertilization and water stress affects plant water relations, stomatal conductance and osmotic adjustment in chickpea (*Cicer arientinum* L.). *Plant and Soil*, **267**, 271-284.

Khan, H. R., Paull, J. G., Siddique, K. H. M., and Stoddard, F. L. (2010). Faba bean breeding for drought-affected environments: A physiological and agronomic perspective. *Field Crops Research*, **115**, 279-286.

Khan, H. u. R., Link, W., Hocking, T. J., and Stoddard, F. L. (2007). Evaluation of physiological traits for improving drought tolerance in faba bean (*Vicia faba* L.). *Plant and Soil*, **292**, 205-217.

Khazaei, H., Street, K., Santanen, A., Bari, A., and Stoddard, F. L. (2013). Do faba bean (*Vicia faba* L.) accessions from environments with contrasting seasonal moisture availabilities differ in stomatal characteristics and related traits? *Genetic Resources and Crop Evolution*, **60**, 2343-2357.

Khazaei, H. S., Kenneth Bari, Abdallah Mackay, Michael Stoddard, Frederick L. (2013). The FIGS (focused identification of germplasm strategy) approach identifies traits related to drought adaptation in Vicia faba genetic resources. *PloS one*, **8**, e63107-e63107.

Kim, K. S., Park, S. H., Kim, D. K., and Jenks, M. A. (2007). Influence of Water Deficit on Leaf Cuticular Waxes of Soybean (*Glycine max* [L.] Merr.). *International Journal of Plant Science*, **168**, 307–316.

Kramer, P. (1988). Changing concepts regarding plant water relations. *Plant, Cell and Environment*, **11**, 565-568.

Kramer, P. J., and Boyer, J. S. (1995). *"Water Relations of Plants and Soils,"* Academic press.

Krishnamurthy, L., Ito, O., and Johansen, C. (1996). Genotypic differences in root growth dynamics and its implications for drought resistance in chickpea. *Dynamics of Roots and Nitrogen in Cropping Systems of the Semi-Arid Tropics. Eds O Ito, C Johansen, JJ Adu-Gyamfi, K Katayama, JVDK Kumar Rao and TJ Rego*, pp. 235-250.

Krishnamurthy, L., Kashiwagi, J., Tobita, S., Ito, O., Upadhyaya, H. D., Gowda, C. L. L., Gaur, P. M., Sheshshayee, M. S., Singh, S., Vadez, V., and Varshney, R. K. (2013). Variation in carbon isotope discrimination and its relationship with harvest index in the reference collection of chickpea germplasm. *Functional Plant Biology*, **40**, 1350.

Lafitte, R., Blum, A., and Atlin, G. (2003). Using secondary traits to help identify drought-tolerant genotypes. *Breeding Rice for Drought-Prone Environments*, pp. 38-39.

Laisk, A., and Loreto, F. (1996). Determining photosynthetic parameters from leaf CO_2 exchange and chlorophyll fluorescence (ribulose-1, 5-bisphosphate carboxylase/oxygenase specificity factor, dark respiration in the light, excitation distribution between photosystems, alternative electron transport rate, and mesophyll diffusion resistance. *Plant Physiology*, **110**, 903-912.

Lavania, D., Siddiqui, M. H., Al-Whaibi, M. H., Singh, A. K., Kumar, R., and Grover, A. (2014). Genetic approaches for breeding heat stress tolerance in faba bean (*Vicia faba* L.). *Acta Physiologiae Plantarum*, **37**.

Lawlor, D. (1995). The effects of water deficit on photosynthesis. *Environment and Plant Metabolism. Flexibility and Acclimation*, pp. 129-160.

Lawlor, D., and Cornic, G. (2002). Photosynthetic carbon assimilation and associated metabolism in relation to water deficits in higher plants. *Plant, Cell and Environment*, **25**, 275-294.

Levitt, J. (1980). *"Responses of plants to environmental stresses. Volume II. Water, radiation, salt, and other stresses,"* Academic Press.

Liang, J., Zhang, J., and Wong, M. (1997). Can stomatal closure caused by xylem ABA explain the inhibition of leaf photosynthesis under soil drying? *Photosynthesis Research*, **51**, 149-159.

Link, W., Abdelmula, A. A., Kittli, E. V., Bruns, S., Riemer, H., and Stelling, D. (1999). Genotypic variation for drought tolerance in Vicia faba. *Plant Breeding*, **118**, 477-483.

Lloyd, J., Syvertsen, J., Kriedemann, P., and Farquhar, G. (1992). Low conductances for CO_2 diffusion from stomata to the sites of carboxylation in leaves of woody species. *Plant, Cell and Environment*, **15**, 873-899.

Lockhart, E., Wild, B., Richter, A., Simonin, K., and Merchant, A. (2016). Stress-induced changes in carbon allocation among metabolite pools influence isotope-based predictions of water use efficiency in Phaseolus vulgaris. *Functional Plant Biology*.

Loss, S. P., Siddique, K. H. M., and Tennant, D. (1997). Adaptation of faba bean (*Vicia faba* L.) to dryland Mediterranean-type environments III. Water use and water-use efficiency. *Field Crops Research*, **54**, 153-162.

Maalouf, F., Nachit, M., Ghanem, M. E., and Singh, M. (2015). Evaluation of faba bean breeding lines for spectral indices, yield traits and yield stability under diverse environments. *Crop and Pasture Science*, **66**, 1012.

Matin, M., Brown, J. H., and Ferguson, H. (1989). Leaf water potential, relative water content, and diffusive resistance as screening techniques for drought resistance in barley. *Agronomy Journal*, **81**, 100-105.

McKinney, N., Schapaugh, W., and Kanemasu, E. (1989). Selection for canopy temperature differential in six populations of soybean. *Crop Science* **29**, 255-259.

Merchant, A. (2014). The regulation of osmotic potential in trees. *In* "Trees in a Changing Environment", pp. 83-97. Springer.

Misson, L., LIMOUSIN, J., Rodriguez, R., and Letts, M. G. (2010). Leaf physiological responses to extreme droughts in Mediterranean Quercus ilex forest. *Plant, Cell and Environment*, **33**, 1898-1910.

Monneveux, P., Jing, R., and Misra, S. C. (2012). Phenotyping for drought adaptation in wheat using physiological traits. *Frontiers in Physiology*, **3**.

Monneveux, P., and Ribaut, J.-M. (2011). Drought phenotyping in crops: from theory to practice. *Available at Generation Challenge Program website www. generationcp. org*.

Monteith, J. (1995). A reinterpretation of stomatal responses to humidity. *Plant, Cell and Environment*, **18**, 357-364.

Morgan, J. M. (1984). Osmoregulation and water stress in higher plants. *Annual Review of Plant Physiology*, **35**, 299-319.

Mwanamwenge, J., Loss, S. P., Siddique, K. H. M., and Cocks, P. S. (1999). Effect of water stress during floral initiation, flowering and podding on the growth and yield of faba bean (Vicia faba L.). *European Journal of Agronomy*, **11**, 1-11.

Neil C. Turner, G. C. W. a. K. H. M. S. (2001). Adaptation of Grain Legumes (Pulses) to Water-Limited Environments.. *Advances in Agronomy*, **71**, 193–231.

Nerkar, Y. S., Wilson, D., and Lawes, D. A. (1981). Genetic variation in stomatal characteristics and behaviour, water use and growth of five *Vicia faba* L. genotypes under contrasting soil moisture regimes. *Euphytica*, **30**, 335–345.

O'Neill, P. M., Shanahan, J. F., and Schepers, J. S. (2006). Use of chlorophyll fluorescence assessments to differentiate corn hybrid response to variable water conditions. *Crop Science*, **46**, 681-687.

Oweis, T. H., Ahmed Pala, Mustafa (2005). Faba bean productivity under rainfed and supplemental irrigation in northern Syria. *Agricultural Water Management*, **73**, 57-72.

Pantuwan, G., Fukai, S., Cooper, M., Rajatasereekul, S., and O'Toole, J. (2002). Yield response of rice (*Oryza sativa* L.) genotypes to different types of drought under rainfed lowlands: Part 1. Grain yield and yield components. *Field Crops Research*, **73**, 153-168.

Parry, M., Flexas, J., and Medrano, H. (2005). Prospects for crop production under drought: research priorities and future directions. *Annals of Applied Biology*, **147**, 211-226.

Passioura, J. (1988). Root signals control leaf expansion in wheat seedlings growing in drying soil. *Functional Plant Biology*, **15**, 687-693.

Passioura, J. B. (2012). Phenotyping for drought tolerance in grain crops: when is it useful to breeders? *Functional Plant Biology*, **39**, 851.

Patrick, J. W., and Stoddard, F. L. (2010). Physiology of flowering and grain filling in faba bean. *Field Crops Research*, **115**, 234-242.

Purushothaman, R., Zaman-Allah, M., Mallikarjuna, N., Pannirselvam, R., Krishnamurthy, L., and Gowda, C. L. L. (2013). Root anatomical traits and their possible contribution to drought tolerance in grain legumes. *Plant Production Science*, **16**, 1-8.

Rancourt, G. T., Éthier, G., and Pepin, S. (2013). Threshold response of mesophyll CO_2 conductance to leaf hydraulics in highly transpiring hybrid poplar clones exposed to soil drying. *Journal of experimental botany*, ert436.

Raschke, K. (1975). Stomatal action. *Annual Review of Plant Physiology*, **26**, 309-340.

Ricciardi, L. (1989). Plant breeding for resistance to drought. I: Stomatal traits in genotypes of *Vicia faba* L. *Agricoltura Mediterranea*, **119**, 297-308.

Ricciardi, L., Polignano, G. B., and Giovanni, C. D. (2001). Genotypic response of faba bean to water stress. *Euphytica*, **118**, 39–46.

Richards, R. A. (2006). Physiological traits used in the breeding of new cultivars for water-scarce environments. *Agricultural Water Management*, **80**, 197-211.

Ristic, Z., and Jenks, M. A. (2002). Leaf cuticle and water loss in maize lines differing in dehydration avoidance. *Journal of Plant Physiology*, **159**, 645-651.

Rubiales, D. (2010). Faba beans in sustainable agriculture. *Field Crops Research*, **115**, 201-202.

Samsuddin, Z., and Impens, I. (1979). Photosynthesis and diffusion resistances to carbon dioxide in Hevea brasiliensis muel. agr. clones. *Oecologia*, **37**, 361-363.

Sánchez, F. J., Manzanares, M. a., Andres, E. F. d., Tenorio, J. L., and Ayerbe, L. (1998). Turgor maintenance, osmotic adjustment and soluble sugar and proline accumulation in 49 pea cultivars in response to water stress. *Field Crops Research*, **59**, 225-235.

Sarker, A., Erskine, W., and Singh, M. (2005). Variation in shoot and root characteristics and their association with drought tolerance in lentil landraces. *Genetic Resources and Crop Evolution*, **52**, 89-97.

Schachtman, D. P., and Goodger, J. Q. (2008). Chemical root to shoot signaling under drought. *Trends Plant Sci*, **13**, 281-7.

Seibt, U., Rajabi, A., Griffiths, H., and Berry, J. A. (2008). Carbon isotopes and water use efficiency: sense and sensitivity. *Oecologia*, **155**, 441-54.

Sekara, A., Poniedzialek, M., Ciura, J., and Jedrszczyk, E. (2001). The effect of meteorological factors upon flowering and pod setting of faba bean [*Vicia faba* L.] at different sowing times. *Vegetable Crops Research Bulletin*, **1**.

Sibounheuang, V., Basnayake, J., Fukai, S., and Cooper, M. (2001). Leaf-water potential as a drought resistance character in rice. *In* "Increased lowland rice production in the Mekong Region: Proceedings of an International Workshop held in Vientiane, Laos, 30 October-2 November 2000.", pp. 86-95. Australian Centre for International Agricultural Research (ACIAR).

Siddique, K., Loss, S., Thomson, B., and Saxena, N. (2003). Cool season grain legumes in dryland Mediterranean environments of Western Australia: significance of early flowering. *Management of agricultural drought: agronomic and genetic options*, 151-162.

Siddique, M., Hamid, A., and Islam, M. (2000). Drought stress effects on water relations of wheat. *Botanical Bulletin of Academia Sinica*, **41**.

Silvente, S., Sobolev, A. P., and Lara, M. (2012). Metabolite adjustments in drought tolerant and sensitive soybean genotypes in response to water stress. *PLoS ONE* **7**.

Sinclair, T., and Ludlow, M. (1985). Who taught plants thermodynamics? The unfulfilled potential of plant water potential. *Functional Plant Biology*, **12**, 213-217.

Sinclair, T. R., Hammer, G. L., and Van Oosterom, E. J. (2005). Potential yield and water-use efficiency benefits in sorghum from limited maximum transpiration rate. *Functional Plant Biology*, **32**, 945-952.

Sinclair, T. R., Zwieniecki, M. A., and Holbrook, N. M. (2008). Low leaf hydraulic conductance associated with drought tolerance in soybean. *Physiol Plant*, **132**, 446-51.

Singh, A. K., Bharati, R. C., Manibhushan, N. C., and Pedpati, A. (2013). An assessment of faba bean (*Vicia faba* L.) current status and future prospect. *African Journal of Agricultural Research*, **8**, 6634-6641.

Smith, M., Wild, B., Richter, A., Simonin, K., and Merchant, A. (2016). Carbon Isotope Composition Of Carbohydrates And Polyols In Leaf And Phloem Sap Of Phaseolus Vulgaris L. influences Predictions Of Plant Water Use Efficiency. *Plant and Cell Physiology*, pcw099.

Somerville, D. (2002). Honeybees in faba bean pollination. *Agnote DAI-128.*

Stoddard, E. L. (1991). Pollen Vectors and Pollination of Faba Beans in Southern Australia. *Australian Journal of Agricultural Research,* **42**, 1173-1178.

Stoddard, F. (1993). Limits to retention of fertilized flowers in faba beans (*Vicia faba* L.). *Journal of Agronomy and Crop Science,* **171**, 251-251.

Stoddard, F., and Hämäläinen, K. (2011). Towards the World's earliest maturing faba beans. *Grain Legume,* **56**, 9-10.

Stoddard, F. L., Balko, C., Erskine, W., Khan, H. R., Link, W., and Sarker, A. (2006). Screening techniques and sources of resistance to abiotic stresses in cool-season food legumes. *Euphytica,* **147**, 167-186.

Troughton, J., and Slatyer, R. (1969). Plant water status, leaf temperature, and the calculated mesophyll resistance to carbon dioxide of cotton leaves. *Australian Journal of Biological Sciences,* **22**, 815-828.

Turner, N. (1982). The role of shoot characteristics in drought resistance of crop plants. *Drought resistance in crops with emphasis on rice,* 115-134.

Turner, N. C., Abbo, S., Berger, J. D., Chaturvedi, S. K., French, R. J., Ludwig, C., Mannur, D. M., Singh, S. J., and Yadava, H. S. (2007). Osmotic adjustment in chickpea (*Cicer arietinum* L.) results in no yield benefit under terminal drought. *J Exp Bot,* **58**, 187-94.

UN (2011). Global drylands: A UN system-wide response. Prepared by the Environment management Group. 132 p.

Wery, J., Silim, S. N., Knights, E. J., Malhotra, R. S., and Cousin, R. (1994). Screening techniques and sources of tolerance to extremes of moisture and air temperature in cool season food legumes. *Euphytica,* **73**, 73-83.

Zacharisen, M., Brick, M., Fisher, A., Ogg, J., and Ehleringer, J. (1999). Relationships between productivity and carbon isotope discrimination among dry bean lines and F2 progeny. *Euphytica,* **105**, 239-250.

2018, *Climate Risks Management: Sustainable Pulse Production*
Editors: A K Srivastava and Yogranjan
Published by: **ASTRAL INTERNATIONAL PVT. LTD., NEW DELHI**   *Pages 205–220*

Chapter 10

Redefining Lentil (*Lens culinaris* M.) through Genomics for Nutritional Security: Status and Outlook

Yogranjan[1], A K Srivastava[1], Lalit M. Bal[1]
and Sudhakar P. Mishra[2]

[1]College of Agriculture, Jawaharlal Nehru Agricultural University,
Tikamgarh– 472 001, M.P.
[2]Division of Agricultural Biochemistry, Department of Crop Sciences,
MGCGVV, Chitrakoot, Satna – 485 780, M.P.
E-mail: yogranjan@gmail.com

ABSTRACT

Recent years witnessed an interest from the industrial sector in lentil for production of superior quality of pulse augmented with sound nutrients. However, despite its consumers' demand being widely felt, breeding programmes of lentil has not been complemented adequately with the innovative breeding and biotechnological tools. The crop has massive variability and several molecular marker systems have been used to increase selection efficiency for genotyping of several traits as well. Rich genetic diversity is available to ease the genomic advancements in the crop through detailed genetic linkage mapping. Wild species constitute a rich repertoire of genes for biotic and abiotic stresses besides protein quality traits and need to be utilized. In vitro manipulations and genetic transformation protocols through vector mediated gene transfer in lentil crop are in place providing scope for development of transgenic for desirable traits. This article reviews the benchmark interventions in the field of genomics of the crop and highlights further research thrust for promoting the same in an international arena.

Keywords: Pulse nutrients, Lentils, Genomics, Tissue culture, Marker assisted breeding.

Introduction

Lentil (*Lens culinaris* M.), a species of the family leguminosae and sub-family Papilionaceae is a self-pollinating diploid ($2n = 14$ chromosomes) legume crop that is cultivated throughout the world and is highly valued as a high protein food. The crop is supposed to be coevolved with wheat, barley and other pulses in the Near East arc about 8000 years ago (Cubero, 1981; Ladizinsky, 1979). The genus *Lens* comprises of seven taxa in four species (Ferguson and Erskine, 2001; Ferguson *et al.*, 2000). *Lens orientalis* is the presumed progenitor of cultivated *L. culinaris* and the two species are fully crossable and generate fully fertile progenies.

Lentils are primarily consumed as seeds after cooking for human consumption, while the remaining parts of the plant are used as animal feed in some developing countries. A number of lentil varieties have been commercially released throughout the world and are easily distinguishable on the basis of seed appearance and taste.

The total lentil cultivated area in the world is estimated around 4.34 million hectares with annual production and productivity of 4.95 million tons and 1260 Kgha^{-1} respectively (FAO, 2014). The crop is grown widely throughout the Indian Subcontinent, Middle East, Northern Africa and East Africa, Southern Europe, North and South America, Australia and West Asia (Erskine, 1997; Taylor *et al.*, 2006). The major lentil-growing countries of the world are Canada, India, Turkey, Australia, USA, Nepal, China, and Ethiopia. In India, Lentil production has always been important as it is the one of the most important rabi (winter) crops in the country.

Lentil crop is endowed with enormous agronomic values. The crop acts as an efficient green manure, and affects subsequent yield as well as protein content of winter wheat, and also the economic returns of the systems under no-till conditions. In this way, Lentils may have a proven potential to enhance the productivity of agroecosystems in dry areas, where water and nutrients are limited. On this accounts, generally lentil is produced on marginal lands that are relatively dry and without the benefit of fertilizer inputs or irrigation.

Food insecurity nowadays receives much attention, and several research efforts have contributed to remediate it also to a considerable extent, but the lack of nutritional security *i.e.* access to balanced ration is still unnoticed while the same is at alarming stage and equally devastating to the health and economic development of a country. Amidst the large proportion of malnourished population particularly in several developing countries in Asia, Lentils have proved to play vital roles in the nutritional security. Lentil grain is highly digestible and nutritious with sound protein and vitamins. Additionally lentil seed contains good to excellent amounts of macro- and micronutrients (Ca, P, K, Fe and Zn), vitamins (Niacin, Vitamin A, Ascorbic Acid and Inositol) and carbohydrates for balanced nutrition (Bhatty 1988; Savage 1988). It is also rich in lysine, an essential amino acid, found only at low levels in cereal protein and all with virtually no fat. Lentil carbohydrate has a low glycemic index and thus supposed to be a good food for diabetics. Lentils are also very good source of cholesterol-lowering fiber. They are of special benefit in managing blood-sugar disorders since their high fiber content prevents blood sugar levels from rising rapidly after a meal. Compared to other types of dried beans, lentils are relatively quick and easy to prepare. They readily absorb a variety of wonderful flavors from other foods and seasonings, are high in nutritional value and are available throughout the year.

Owing to their immense nutritive as well as agricultural values, exhaustive research has been in place at several levels for lentil improvement through conventional breeding. A number of breeding strategies were adopted to develop high-yielding lentil varieties suitable to fit in the cropping system because of its duration. Top priority was given to collection and evaluation of local and exotic lentil lines. At the same time, work on improved production packages, including pest and disease management, agronomic and cultural management also received due emphasis. Despite all these efforts of lentil improvement, an appreciable gain has not been materialized so far. The productivity of lentils remains quite low, around 1.7 tonnes per hactares, and large gap exists between their potential (2.4 tonnes per hactares) and actual yields (Ghanem *et al.*, 2015). The major reason for these low yields appears to be results of production on marginal lands in arid and semi-arid environments without irrigation, along with several abiotic and biotic constraints (Bejiga and Degago, 2000). The other constraints also restrict its productivity such as drought, *Fusarium* wilt, *Ascochyta* blight and rust. Rough estimates suggest that in India alone damage due to biotic and abiotic stresses causes losses that exceed 45 percent annually (Yadav *et al.*, 2007).

Classical improvement schemes comprise relatively long breeding cycles and are dictated by genetic complexity and the sensitivity of lentils to inbreeding depression (Tran and Credland, 1995). Applications of plant biotechnology over the past five decades have helped facilitate interspecies crosses and to augment and broaden the cultivated gene pool. Integrating genomic tools with conventional breeding methods holds the promising key to accelerate the progress of crop improvement and also the lentil production.

Plant biotechnology has traditionally encompassed the application of cell and tissue culture for crop improvement. From the mid-1980s, the development and application of transgenic plants was the dominant research activity associated with plant biotechnology. More recently research has expanded to include the application of genomics technologies. Members of the leguminaceae family, especially chickpea (*Cicer arietinum* L.), garden pea (*Pisum sativum* L.), soybean (*Glycin max* L.) and pigeonpea (*Cajanus cajan* L.) have been the subject of many of the developments in plant biotechnology, primarily due to their high propensity for growth and development in cell culture. Herein, we begin with a historical account of traditional tissue culture approaches and applications of cell culture, then review the past developments and new approaches for gene transfer via transformation. We also review the applications of genetic markers and genomics for lentil breeding.

Historical Perspectives of Lentils Biotechnology

Traditionally, lentil breeding involved classical plant breeding techniques of selection- hybridization -selection cycle among large seedling populations to have superior individuals with the desired combination of traits. These techniques have been successful in mainstreaming some of the easy-to-manage monogenic traits (Kumar *et al.*, 2016). However, in case of complex quantitative traits, the use of conventional approaches has not been proved effective. As most of the economic traits are complex, quantitative, and often influenced by environments and genotype–environment interaction, the genetic improvement of these traits becomes difficult. The main objectives of most lentils breeding programmes have involved the improvement in agronomic yield attributes and resistance to pests and diseases, while maintaining or improving traits such as protein contents and pulse quality. In the past few decades, plant biotechnology has contributed valuable solutions to some of the difficulties associated with conventional lentil breeding. These have related primarily to applications of cell and tissue culture. Modern tools of plant transformation and DNA markers offer new opportunities for both gaining improved efficiencies in selection of elite clones and enhancing genetic gain over time.

Gene transfer in lentil has been considered complicated and quite challenging because of its recalcitrant nature to *in vitro* regeneration (Gulati and Mc Hughen, 2003). Regeneration and transformation procedures of lentil could not be established as compared to the success reported in other grain legumes (Sarkar *et al.*, 2003). Very limited numbers of studies have been carried out and reported on *in vitro* plant regeneration through organogenesis and somatic embryogenesis in lentil. The first successful result of *in vitro* lentil regeneration from meristem tips had been reported by Bajaj and Dhanju (1979). Later Williams and McHughen (1986) reported a regeneration protocol for lentils, where hypocotyls and epicotyls were used as explants to induce callus. Saxena and King (1987) obtained whole plants from callus induced from embryonic axes, while Polanco *et al.*,(1988) reported multiple shoot formation from shoot tip, first node, and first pair of leaves. Similarly, plant regeneration was achieved through direct shoot organogenesis from intact seedling (Malik and Saxena, 1992), shoot tip (Bajaj and Dhanju, 1979; Polanco and Ruiz, 1988, 1997), cotyledonary node (Warkentin and Mc Hughen, 1993; Polanco and Ruiz, 1997; Gulati and Mc Hughen, 2001) cultures. Calli mediated indirect organogenesis were achieved from shoot meristems and epicotyls (William and Mc Hughen, 1986), axenic seedlings (Cambecedes *et al.*, 1991; Malik and Saxena,1992) and leaves (Polanco and Ruiz, 1988, 1997). A successful regeneration from protoplast culture from lentils leaf has also been reported ((Warkentin and Mc Hughen, 1993). The preferred choice of explants is often cultivar-dependent or based on preference of individual laboratories.

So far, there is no study reported on the direct somatic embryogenesis and subsequent plant regeneration in *Lens* cultivars. Most of these studies generally involve extensive manipulation of culture conditions to induce shoot regeneration and the frequency of shoot regeneration was not high. However, Saxena and King, (1987), reported indirect somatic embryogenesis from embryo-derived callus cultures of *Lens culinaris* Medik. cv. Laird. Halbach *et al.*,(1998) reported an efficient shoot regeneration from bisected embryos, when the cotyledons was kept attached to the embryo-axis. Ahmad *et al.*, 1997 and Ye *et al.*, 2000 reported successful protocols for the micropropagation of lentil and hybrids based on single node culture.

In order to achieve efficient plant regeneration, it is mendatory to define specific requirements of species and even within species, the genotypes (Dhir *et al.,* 1992), as it is not workable to have generalized culture medium for achieving totipotency in all plants or even in all cells of the same kind. Moreover, the type and balance of growth regulator needed for various morphogenetic responses vary from tissues to tissues and cell to cell depending on their metabolic status.

The *in vitro* studies in lentils mostly involve application of 1-Naphthaleneacetic acid (NAA) or Indole-3-acetic acid (IAA) at 2 mg/l as rooting media. However, Thidiazuron (TDZ) at low concentration in (MS) medium (Murashige and Skoog, 1962) induced multiple shoot formation (Malik and Saxena, 1992) while at higher concentration it caused a shift in regeneration from induction of shoots to majority of somatic embryos Polanco *et al.,*(1988). Warkentin and McHughen (1993) reported multiple shoot formation from cotyledonary nodes using benzyl adenine (BA). Malik and Rashid (1989) obtained multiple shoot formation from cotyledonary nodes on a medium with growth hormone Benzyl Adenine.

Singh and Raghuvansi (1989) reported that plants could be regenerated directly from nodal segment and shoot tip explants as well as from callus. On a medium containing kinetin, nodal segments and shoot tip explants produced multiple shoots without intervening by callus or root formation.A high frequency shoot organogenesis and somatic embryogenesis from the cotyledonary node explants of lentil in a single step of culturing them on a TDZ supplemented medium has been achieved (Chhabra *et al.,* 2008). These studies showed that cytokinins induced multiple shoot formation from different types of explants.

Embryo Rescue

The narrow genetic base of cultivated lentils restricts for the transfer of useful genes from diverse and distant sources. Embryo culture has been valuable for circumvention of interspecific incompatibility to a considerable extent. An extreme example of the use of embryo rescue to aid the recovery of interspecific hybrids in lentils has been reported (Cohen *et al.,* 1984; Ladizinsky *et al.,* 1979). Embryo culture in lentils has been used to rescue interspecific hybrids between *L. culinaris* and *L. orientalis*, and *in vitro* multiplication of F_1 hybrids has been used as a way of expanding hybrid populations (Ahmad *et al.,* 1997; Ye *et al.,* 2000).

Somaclonal Variation

During the course of regenerating plants from tissue cultures, the perceptions of the clonal integrity of the resulting plants have changed. The historical conception was of the view that all plants regenerated from somatic tissues are identical to the parent plant. However, Larkin and Scowcroft (1981) described the high frequency of phenotypically variant plants among tissue culture generated plant population and coined the term 'somaclonal variation' for their occurrence. Initially, it was seen as an inherent curse disturbing true-to-the-type traits of tissue culture generated plants while on one hand, somaclonal variation was considered as a new approach for generating novel variation in plants. As not all variants have a genetic basis, lines exhibiting phenotypic changes need to be grown over several field seasons to ensure stability of performance. Stable phenotypic changes of either heritable and / or epigenetic origin may arise through ploidy changes, chromosomal aberrations, gene amplification, activation of transposable elements, DNA methylation changes or point mutations. One of the potential benefits of somaclonal variation in plant genetic manipulation was seen to be the opportunity to create additional genetic variability in co-adapted, agronomically useful cultures without the need to resort to hybridization or the prod of transgenic plants.

Altaf *et al.*, 1999 found stable somaclonal variants till R_5 generation in lentil cultivar *viz*; Massor-85. The first selections of the variants were made from R_0 generation resulted from indirect organogenesis from shoot tip culture of Massor-85 cultivar. The improvement in yield through large seed size and more podded variants in the study were found significant contribution of lentil tissue culture. It has been recognized over time through a number of studies focusing somaclonal variation in other crops that the recovery of a somaclonal line exhibiting beneficial traits without other simultaneously arising negative attributes is very rare. Nowadays, the phenomenon of somaclonal variation is widely seen as an inherent negative feature of regeneration from cell culture and considered as something to be avoided. Strategies to minimize the impact of somaclonal variation on plant performance should be routinely implemented during other applications of cell culture for lentil improvement.

Genetic Transformation in Lentils

Genetic transformation offers an effective means of adding single genes to existing elite cultivar of any crop with no, or very minimal, disturbances to their genetic background. Several explants derived from lentil seedlings have been evaluated on their ability to express a foreign gene.

The first report of lentil transformation (Warkentin and McHughen, 1993) demonstrated the susceptibility of the crop to transformation by virulent strains of *Agrobacterium tumefaciens*. In the study, T-DNAs of four diverse strains of *Agrobacterium tumefaciens* (C58, Ach5, GV3111, and A281) were capable of inducing tumors at a high frequency on inoculated stems of lentil *in vivo*, and on excised shoot apices *in vitro*. Results of the study suggested that disarmed versions of any one of these strains could be suitable for the recovery of transgenic lentil plants from shoot apex explants. This progress was motivated by the advantages that transformation offered for the genetic improvement in lentils relative to the genetic limitations associated with traditional lentil breeding and led the quick adaptation of *Agrobacterium*-mediated gene transfer protocols for other explants as well as cultivars of the lentils throughout the world.

The most common reporter marker gene used for the selection of transformation events in lentil used to be β-glucuronidase based reporter system (GUS). Transient expression of the scorable marker gene GUS was achieved by *Agrobacterium* transfer and by particle bombardment methods (Maccarrone *et al.*, 1995). GUS expression has also been observed after inoculation of longitudinally sliced embryogenic axes of lentil with different *Agrobacterium* strains (Lurquin *et al.*, 1998). Transient expression of the scorable marker genes GUS and ALS and selectable marker gene *bla* was achieved by Khawar and Ozcan (2002), where they used *Agrobacterium* strain A281 for inoculation of several explants from 21 genotypes of lentil. Transformation frequency in *Agrobacterium*-mediated systems could be enhanced upon mechanical reinforcement of *Agrobacterium* cells into the tissues by means of sonication (Trick and Finer, 1997) or vacuum infiltration (Kapila *et al.*, 1997; Trieu *et al.*, 2000; Qing *et al.*, 2000).

Agrobacterium-mediated transformation results were reported by Sarker *et al.* (2003) through having successful regeneration from cotyledonary nodes, decapitated embryos, immature embryos and epicotyls. Histochemical assay confirmed that epicotyl explants exhibited highest transgene expression followed by decapitated embryos, and hence found to be more effective in formation of multiple shoots and were thus suggested as suitable explants for lentil transformation. Similar results were reported in a transformation study by Bayrac (2004), where regeneration was acieved through indirect organogenesis from different explants *viz;* peeled cotyledonary nodes, cotyledonary petioles, shoot tips and roots.

Dogan *et al.* (2005) compared tumour and root formation ability of different explants from several lentil cultivars after inoculation with *A. rhizogenes* and *A. tumefaciens* strains. The higher frequency of tumour formation from cotyledon node explants was observed as compared to shoot meristems.

Lentil transformation has also been achieved by direct DNA uptake, and success has been reported too for the generation of transgenic lentil however, *Agrobacterium*-mediated gene transfer is the preferred approach and is routinely performed in laboratories worldwide. Electroporation-mediated gene transfer into intact plant tissues was demonstrated in lentils (Chowrira *et al.*, 1996; Oktem *et al.*, 1999; Gulati *et al.*, 2002), where again transient expression of a chimeric *gus* reporter genes was used to monitor the uptake and expression of the introduced DNA in electroporated nodal axillary buds *in vivo*. Mahmoudian *et al.*,(2002), reported a rapid and convenient transient expression protocol first ever based on vacuum infiltration of *Agrobacterium* cells into lentil cotyledonary nodes. In the study nptII and GUS genes were used as selectable and scorable markers respectively. Similarly, although, in less extent, but reproducible levels of GUS expression could be obtained from several types of lentil explants, including shoot apices, epicotyls, and roots (Warkentin and McHughen, 1992). Liposomes mediated gene transformation was reported in lentil protoplasts and cotyledonary nodes (Maccarrone *et al.*, 1992). Chopra *et al.* (2011) developed a genotype independent *in vitro* regeneration system in lentils with adequate root induction. Transformation was achieved after regeneration using sonication-assisted A. tumefaciens (SAAT) transformation (first report of this method in lentil). Micro grafting has been extensively used to recover transformed plants during production of transgenic lentils. If an efficient and reproducible regeneration protocol for root induction from developing shoots could be established, the rigorous work of micrografting could be skipped.

The most common marker gene used for the selection of transformation events in lentil is the *npt*II gene conferring kanamycin resistance. This marker gene is highly effective and is used almost exclusively for lentil transformation. Another study has reported success with other marker genes such as acetolactate synthase gene (ALS) from tobacco conferring resistance to sulfonylurea herbicides (Gulati *et al.*, 2002).

Traits Conferred by Genetic Engineering in Lentils

The use of genetic engineering approaches has allowed the successful transfer of numerous transgenes into elite lentil cultivars by either *Agrobacterium*-mediated transformation or through direct DNA uptake. These transgenes have conferred several traits, covering pest and disease resistances to abiotic stress resistance. Traits successfully transferred to lentils by genetic transformation illustrate the immense potential of transformation for the genetic improvement in lentils.

Several fungal pathogens infects lentils such as ascochyta blight (*Ascochyta lentis*), *fusarium* wilt (*Fusarium oxysporum* f.sp. *lentis*), anthracnose (*Colletotrichum truncatum*), *stemphylium* blight (*Stemphylium botryosum*), rust (*Uromycesviciae-fabae*), collar rot (*Sclerotiun rolfsii*), root rot (*Rhizoctonia solani*), and white mold (*Sclerotinia sclerotiorum*) (Kumar *et al.*, 2013; Sharpe *et al.*, 2013). Classical plant breeding approach has been successful in developing some resistant varieties but the approach, being time consuming and having considerable influence of environment, is not of today's preferred choice. Application of biotechnology approaches to develop disease resistant varieties can contribute efficiently to solve or reduce these problems.

Hashem (2007) developed fungus-resistant lentils by transforming decapitated embryos with one cotyledon with *Ripgip* gene coding polygalacturonase inhibitory protein. The gene is accountable for conferring resistance against fungal pathogens. The study first ever demonstrated a marker free transformation system in legumes. In the marker-free transformation, low frequencies of transformed plants are recovered by PCR-screening of plants regenerated without selection after co-cultivation with *Agrobacterium*. In the study by Hashem (2007), *bar* gene was removed and PGIP gene was kept in T-DNA cassette before carrying out transformation. Fungus-resistant marker free plants were demonstrated via semi-quantitative polygalacturonase-inhibition assay. For achieving efficient transformation, wounding of explants, application of acetosyringone, vacuum infiltration and gradual selection were used. The transgene insertion and expression were confirmed through PCR, RT-PCR and Southern hybridisation, and the transgenes were segregated in Mendelian fashion.

For enhancing drought and salinity tolerance, Khatib *et al.* (2011) introduced *DREB1A* gene driven by the rd29A promoter into lentils' decapitated embryo explants followed by shoot regeneration from the apical meristems and cotyledonary buds via direct organogenesis. Putative transgenic explants were micro-grafted onto non-transformed rootstocks to establish transgenic plants. The transgene insertion and expression were confirmed through PCR and Southern hybridization. Expression of *DREB1A* gene in transgenic plants was induced by salt stress and was confirmed through RT-PCR.

Molecular Marker-Assisted Breeding

The use of molecular markers speeds up the selection process by alleviating time-consuming approaches of direct screening under field and/or greenhouse conditions. Molecular markers are particularly useful when target characters are polygenic or quantitative in nature. The prospective to map different Quantitative Trait Loci (QTLs) contributing to an agronomical trait and to identify linked molecular markers have paved the way to transfer several QTLs simultaneously and to pyramid QTLs for several agronomical traits in one improved cultivar. Various marker assisted selection (MAS) techniques such as isozymes, random amplified polymorphic DNA (RAPD), AFLP, inter simple sequence repeats (ISSR) and simple sequence repeats (SSR) *etc.* have been used in legume crop breeding to study genetic variability of legumes as well as in legumes in relation to biotic and abiotic stresses. Molecular markers technology enables the breeders to have the knowledge of the genetic control of specific resistance and/or tolerance in many legumes by providing information on the number, chromosomal location and individual or interactive effects of the QTLs involved.

Although the use of molecular markers has proved to be helpful for crop improvement, lentil lags behind other legumes in using these tools for resistance and/or tolerance to biotic or abiotic stresses. Large genome size (4063 Mbp/1C, Arumuganathan and Earle 1991), limited genomic resources (Tullu *et al.*, 2008; Mustafa *et al.*, 2009), lack of candidate genes, genetic complexity of most stress-related traits and the difficulty in identifying beneficial alleles are the main limiting factors in genomics enabled improvement in lentil.

Prior to the use of molecular markers, morphological and isozyme markers were accounted to be used to identify the first genetic linkages in lentil (Tadmor *et al.*, 1987; Vaillancourt and Slinkard, 1993; Zamir and Ladizinsky, 1984). After the advent of molecular marker technologies, restriction fragment length polymorphism (RFLP) markers, the non polymerase chain reaction (PCR) based marker, were the first reported molecular marker used in the construction of a lentil genetic linkage map (Havey and Muehlbauer, 1989). Two more extensive molecular linkage maps were reported, one using an intraspecific population (Kahraman *et al.*, 2004) and another by an inter-subspecific population (Dur´an *et al.*, 2004).

Thereafter, PCR based markers, such as random amplified polymorphic DNA (RAPD) were routinely and frequently used to study diversity, phylogeny and taxonomy of *Lens* (Ford *et al.*, 1997; Ferguson *et al.*, 2000; Sharma *et al.*, 1996), to develop linkage maps, and for tagging genes of interest (Chowdhury *et al.*, 2001; Eujayl *et al.*, 1998, 1999; Ford *et al.*, 1999; Tullu *et al.*, 2008) and to determine pathogen population structure (Ford *et al.*, 2000). Amplified fragment length polymorphism (AFLP) markers were also used to study linkage mapping in lentil (Dur´an *et al.*, 2004; Eujayl *et al.*, 1998; Hamwieh *et al.*, 2005; Kahraman *et al.*, 2004). Sharma *et al.* (1996) used AFLP to study genetic diversity in 54 lentil germplasm accessions. Tullu *et al.*, 2008 identified major gene as *LCt-2* as markers linked to specific traits *viz.*, resistance to anthracnose. Most relevant work has been carried out in lentil breeding for *Ascochyta*, *Colletotrichum* and *Fusarium* resistance under biotic stress while cold resistance under abiotic stress (Table 10.1). More recently, simple sequence repeat (SSR) or microsatellite Markers were used to construct lentil linkage maps (Hamwieh *et al.*, 2005). The ICARDA SSR library was developed from the genome of the Northfield cultivar (ILL5588) and was found to have $(CA)_n$ as the most abundant repeat type (Hamwieh *et al.*, 2005).Other molecular markers such as inter-simple sequence repeat (ISSR) and resistance gene analogue (RGA) markers were also used in lentil genome mapping (Dur´an *et al.*, 2004). Besides genomic based tools, proteomic approaches have also been applied to lentils to identify proteins involved in the response to different stresses (Fecht-Christoffers *et al.,* 2003). Molecular characterization in congruence with methodical screening of available germplasm would certainly be helpful in the identification and / or development of elite varieties with desired level of agronomic as well as nutritional traits.

Table 10.1: QTL Identified in *Lens culinaris* Associated to Biotic/Abiotic Stresses

Biotic/Abiotic Stresses	Gene(s)/ QTL	Associated Markers	Marker Type	References
Ascochyta lentis	ral1 (AbR1)	RV01and RB18	RAPD	Ford *et al.* (1999)
	ral2	UBC227$_{1290}$ and OPD-10$_{870}$	RAPD	Tar`an *et al.* (2003)
	AbR1	RB18 and RV01	RAPD	Chowdhury *et al.*, 2001
Colletotrichum trauncatum		OPD6$_{1290}$	RAPD	Tar`an *et al.* (2003)
	LCt-2	OPE061250 and UBC704700	RAPD	Tullu *et al.*, 2008
Fusarium oxysporum f.sp. lentis	FW	OPK15900	RAPD	Eujayl *et al.* (1998)
Cold stress	Frt	OPS16$_{750}$	RAPD	Eujayl *et al.* (1999)
		ubc808_3_ubc8073	SSR	
		ubc804_3	SSR	
		Cs48_1	RAPD	Kashraman *et al.* (2004)
		ubc808_12	SSR	
		E3M3	AFLP	

Conclusion

In the current world scenario, where population growth surpasses food supply, genomic tools aimed at overcoming severe biotic and abiotic stresses need to be adequately exploited. Lentils forms an important part of the Indian diet since it is a nutritionally-rich crop with several health promoting features, and its regular consumption would certainly have a significant impact on human health. Although lentil is a crop of choice of many farmers in the country because of its high agronomic value and low cost of cultivation, the average yield of lentils around the world is far below its physiological potential due to various biotic and abiotic stresses. In order to understand the molecular mechanisms involved in biotic and abiotic stress responses of lentils, wild or distant *lens* species could be used since they are adapted to grow in various environments. Wild crop relatives have been used by plant breeders for their undeniable benefits in providing a wide pool of potential genetic resources to improve the undesirable traits of modern crops (Hajjar and Hodgkin, 2007). Likewise, wild species in lentils too represent a great resource of genetic variability that can be used by breeders in identification of traits controlling the biotic and abiotic stress responses, and in improving the stress tolerance or resistance in breeding programs. Characterizing and understanding various plant phenotypes accurately in a short period of time has been rising to prominence in recent years. Although molecular breeding approaches require genotypic data for selection of breeding lines, development of molecular markers used in molecular breeding is still in need of a strong correlation with phenotypic data (McMullen *et al.*, 2009). Recent developments in high-throughput phenotyping will provide an opportunity for generating a fast, inexpensive, and massive collection of data that will together provide a reliable assessment of trait phenotypes for many of the underlying genotypes in a typical plant breeding population. High-throughput phenotyping for different stress tolerance traits would facilitate the discovery of new QTLs in lentils too.

REFERENCES

Ahmad, M., Fautier, A. G., McNeil, D. L., Hill, G. D. and Burritt. D. J. (1997). *In vitro* propagation of *Lens* species and their F1 interspecific hybrids. *Plant Cell Tiss. Organ Cult.* 47: 169–176.

Altaf N., Iqba J., and Ahmad M.S. (1999). Somaclonal variation in Microsperma lentil (*Lens culinaris* Medik). *Pakistan Journal of Biological Sciences*, 2(3): 697-699.

Arumuganathan K., Earle D.E. (1991). Nuclear DNA content of some important plant species. *Plant Mol Biol Rep*. 9: 208–218.

Bajaj, Y. P. S., Dhanju, M. S. (1979). Regeneration of plants from apical meristem tips of some legumes. *Current Science* 48: 906-907.

Bayrac, A. T. (2004). Optimization of a regeneration and transformation system for lentil (*Lens culinaris* M., cv. Sultan-I) cotyledonary petioles and epicotyls (Doctoral dissertation, Middle East Technical University.

Bejiga, G., Degago, Y. (2000). Region 4: Sub-Sahara Africa. In: Knight, R. (Ed.), Linking Research and marketing Opportunities for Pulses in the 21st Century.

Bhatty, R. S. (1988). Composition and quality of lentil (*Lens culinaris* Medik): a review. *Canadian Institute of Food Science and Technology Journal*, 21(2), 144-160.

Cambecedes, J., Duron, M., and Decourtye, L. (1991). Adventitious bud regeneration from leaf explants of the shrubby ornamental honeysuckle, *Lonicera nitida* Wils cv Maigrun: Effects of thidiazuron and 2,3,5- triiodobenzoic acid. *Plant Cell Rep* 10: 471-474.

Chhabra, G., Chaudhary, D., Varma, M., Sainger, M., and Jaiwal, P. K. (2008). TDZ-induced direct shoot organogenesis and somatic embryogenesis on cotyledonary node explants of lentil (*Lens culinaris* Medik.). *Physiology and Molecular Biology of plants*, 14 (4), 347-353.

Chopra, R., Prabhakar, A., Singh, N., and Saini, R. (2011). *In vitro* regeneration and sonication assisted *Agrobacterium tumefaciens* (SAAT) mediated transformation in Indian cultivars of lentil (*Lens culinaris* Medik.). In 5th Chandigarh Science Congress, Chandigarh, India (p. 10).

Chowdhury, M. A., Andrahennadi, C. P., Slinkard, A. E., and Vandenberg, A. (2001). RAPD and SCAR markers for resistance to acochyta blight in lentil. *Euphytica*, 118(3), 331-337.

Chowrira, G.M. Akella, V., Fuerst, P.E., and Lurquin, P.F. (1996) Transgenic grain legumes obtained by in planta electroporation-mediated gene transfer. *Mol. Biotechnol.* 5 85–95.

Cohen, D., Ladizinsky, G., Ziv, M., and Muehlbauer, F. J. (1984). Rescue of interspecific *Lens* hybrids by means of embryo culture. *Plant Cell, Tissue and Organ Culture*, 3(4), 343-347.

Cubero, J.I. (1981). Origin, taxonomy and domestication. *In*: C. Webb and G. Hawtin (Eds). Lentils, pp. 15–38. Commonwealth Agricultural Bureaux, Farnham, U.K.

Dhir, S. K., Dhir, S., and Widholm, J. M. (1992). Regeneration of fertile plants from protoplasts of soybean (*Glycine max* L. Merr.): genotypic differences in culture response. *Plant Cell Reports*, 11(5-6), 285-289.

Dogan D.Ý., Khawar K.M., Özcan S.E. (2005). *Agrobacterium* mediated tumor and hairy root formation from different explants of lentils derived from young seedlings. *International Journal of Agriculture and Biology*, 7: 1019-25.

Durán, Y., Fratini, R., Garcia, P., and De la Vega, M. P. (2004). An intersubspecific genetic map of Lens. *Theoretical and Applied Genetics*, 108(7), 1265-1273.

Erskine, W. (1997). Lessons for breeders from land races of lentil. *Euphytica*, 93(1), 107-112.

Eujayl, I., Baum, M., Powell, W., Erskine, W., and Pehu, E. (1998). A genetic linkage map of lentil (*Lens* sp.) based on RAPD and AFLP markers using recombinant inbred lines. *Theoretical and Applied Genetics*, 97(1), 83-89.

Eujayl, I., Erskine, W., Baum M., and Pehu, E. (1999). Inheritance and linkage analysis of frost injury in lentil. *Crop Sci* 39: 639–642.

Fecht-Christoffers, M. M., Führs, H., Braun, H. P., and Horst, W. J. (2006). The role of hydrogen peroxide-producing and hydrogen peroxide-consuming peroxidases in the leaf apoplast of cowpea in manganese tolerance. *Plant Physiology*, 140(4), 1451-1463.

Ferguson, M. E., Maxted, N., Slageren, M. V., and Robertson, L. D. (2000). A re-assessment of the taxonomy of *Lens* Mill.(Leguminosae, Papilionoideae, Vicieae). *Botanical Journal of the Linnean Society*, 133(1), 41-59.

Ferguson, M., and Erskine, W. (2001). Lentils (*Lens* L.). In Plant Genetic Resources of Legumes in the Mediterranean (pp. 125-133). Springer Netherlands.

Food and Agricultural Organization (FAO) (2014). FAO Statistical Database. Food and Agricultural Organization of the United Nations, USA.

Ford, R. E., Pang, E.C. K., and Taylor, P. W. J. (1997). Diversity analysis and species identification in *Lens* using PCR generated markers. *Euphytica*, 96(2), 247-255.

Ford, R., Pang, E.C.K., and Taylor, P.W.J.(1999). Genetics of resistance to ascochyta blight (*Ascochyta lentis*) of lentil and the identification of closely linked RAPD markers. *Theor Appl Genet* 98: 93–98.

Ford, R., Garnier-Géré, P. H., Nasir, M., and Taylor, P. W. J. (2000). Structure of Ascochyta lentis in Australia revealed with random amplified polymorphic DNA (RAPD) markers. *Australasian Plant Pathology*, 29(1), 36-45.

Ghanem, M. E., Marrou, H., Biradar, C., and Sinclair, T. R. (2015). Production potential of Lentil (*Lens culinaris* Medik.) in East Africa. *Agricultural Systems,* 137, 24-38.

Gulati A., and Mc Hughen, A. (2003). *In vitro* regeneration and genetic transformation of lentil. In: *Applied Genetics of Leguminosae Biotechnology* (Eds. Jaiwal PK and Singh RP), Kluwer Acad. Publ., The Netherlands, pp. 133-147.

Gulati, A., Schryer, P., and McHughen, A. (2002). Production of fertile transgenic lentil (*Lens culinaris* Medik.) plants using particle bombardment. *In Vitro Cell Dev. Biol. Plant.* 38 (2002) 316–324.

Gulati, A., Schryer, P., and McHughen, A. (2001). Regeneration and micrografting of lentil shoots. *In Vitro Cell. Dev. Biol. Plant* 37: 798–802.

Hajjar, R., and Hodgkin, T. (2007). The use of wild relatives in crop improvement: a survey of developments over the last 20 years. *Euphytica*, 156(1-2): 1-13.

Halbach, T., Kiesecher, H., Jacobsen, H. J., Kathen, A. (1998). Tissue culture and genetic engineering of lentil (*Lens culinaris* Medik.). In: *3rd European Conference on Grain Legumes*, Valladolid, Spain.

Hamwieh, A., Udupa, S.M., Choumane, W., Sarker, A., Dreyer, F., Jung C., and Baum, M. (2005). A genetic linkage map of *Lens* sp based on microsatellite and AFLP markers and the localization of *fusarium* vascular wilt resistance. *Theor Appl Genet* 110: 669–677.

Hashem, R. (2007). Improvement of lentil (*Lens culinaris Medik.*) through genetic transformation (Doctoral dissertation).

Havey, M.J., and Muehlbauer, F.J. (1989). Linkages between restriction fragment length. isozymes and morphological markers in lentil. *Theoretical and Applied Genetics* 77: 395-401.

Kahraman, A., Kusmenoglu, I., Aydin, N., Aydogan, A., Erskine W., and Muehlbauer, F.J. (2004). QTL mapping of winter hardiness genes in lentil. *Crop Sci* 44: 13–22.

Kapila, J., De Rycke, R., Van Montagu, M., and Angenon, G. (1997). An *Agrobacterium*-mediated transient gene expression system for intact leaves. *Plant Science*, 122(1), 101-108.

Khatib, F., Makris, A., Yamaguchi-Shinozaki, K., Kumar, S., Sarker, A., Erskine, W., and Baum, M. (2011). Expression of the DREB1A gene in lentil (*Lens culinaris* Medik. subsp. *culinaris*) transformed with the *Agrobacterium* system. *Crop and Pasture Science*, 62(6), 488-495.

Khawar, K. M., and Özcan, S. (2002). *In vitro* induction of crown galls by *Agrobacterium tumefaciens* super virulent strain A281 (pTiBo 542) in lentil (*Lens culinaris* Medik.). *Turkish Journal of Botany*, 26(3), 165-170.

Kumar, S., and Kaushik, N. (2013). Endophytic fungi isolated from oil-seed crop *Jatropha curcas* produces oil and exhibit antifungal activity. *PloS One*, 8(2), e56202.

Kumar, S., Rajendran, K., Kumar, J., Hamwieh, A., and Baum, M. (2016). Current knowledge in lentil genomics and its application for crop improvement. In *Crop Breeding: Bioinformatics and Preparing for Climate Change* (pp. 309-327). CRC Press.

Ladizinsky, G. (1979). Species relationships in the genus *Lens* as indicated by seed-protein electrophoresis. *Botanical Gazette*, 140(4), 449-451.

Larkin, P. J. and Scowcroft, W. R. (1981) Somaclonal Variation: A noval source of variability from cell cultures for plant improvement. *Theor. Appl. Genet.* 60: 197–214.

Lurquin, P. F., Cai, Z., Stiff, C. M., and Fuerst, E. P. (1998). Half-embryo cocultivation technique for estimating the susceptibility of pea (*Pisum sativum* L.) and lentil (*Lens culinaris* Medik.) cultivars to *Agrobacterium tumefaciens*. *Molecular Biotechnology*, 9(2), 175-179.

Maccarrone, M., Dini, L., Marizio, L.D., Giulio, A.D., Rossi, A., Mossa, G., and Frienzi-Agro, A. (1992). Interaction of DNA with cationic liposomes: ability of transfecting lentil protoplasts. *Biochem Biophys Res Comm* 186: 1417-1422.

Maccarrone, M., Veldink, G.A., Agrò, A.F., Vliegenthart, J.F.G. (1995). Lentil root protoplasts: a transient expression system suitable for coelectroporation of monoclonal antibodies and plasmid molecules. *Biochim. Biophys. Acta* 1243 136–142.

Mahmoudian, M. Yücel, M. Öktem, H.A. (2002). Transformation of lentil (*Lens culinaris* M.) cotyledonary nodes by vacuum infiltration of *Agrobacterium tumefaciens*. *Plant Mol. Biol. Rep.* 20: 251–257.

Malik, K. A., and Saxena, P. K. (1992). Thidiazuron induces high-frequency shoot regeneration in intact seedlings of pea (*Pisum sativum*), chickpea (*Cicer arietinum*) and lentil (*Lens culinaris*). *Functional Plant Biology*, 19(6), 731-740.

Malik, M. A., and Rashid A. (1989). Induction of multiple shoots from cotyledonary node of grain legumes: Pea and lentil. *Biol. Planta*,31: 230-232.

McMullen, M.D., Kresovich, S., Villeda, H.S., Bradbury, P., Li, H., Sun, Q., Flint-Garcia, S., Thornsberry, J., Acharya, C., Bottoms, C. and Brown, P. (2009). Genetic properties of the maize nested association mapping population. *Science*, 325(5941): 737-740.

Murashige, T. and Skoog, F. (1962). A revised medium for rapid growth and bioassays with tobacco tissue cultures. *Physiol. Plant.* 15: 473–497.

Mustafa, B. M., Coram, T. E., Pang, E. C. K., Taylor, P. W. J., and Ford, R. (2009). A cDNA microarray approach to decipher lentil (*Lens culinaris*) responses to *Ascochyta* lentis. *Australasian Plant Pathology*, 38(6): 617-631.

Oktem, H. A., Mahmoudian, M., and Yucel, M. (1999). GUS gene delivery and expression in lentil cotyledonary nodes using particle bombardment. *Lentil Experimental News Service.*

Polanco, M. C. and Ruiz, M. L. (1997). Effect of benzylaminopurine on *in vitro* and *in vivo* root development in lentil *Lens culinaris* Medik. *Plant Cell Rep.* 17: 22–26.

Polanco, M.C., Pelaez, M.I., and Ruiz, M.L. (1988). Factors affecting callus and shoot formation from *in vitro* cultures of *Lens culinaris* Medik. *Plant Cell, Tissue and Organ Culture* 15: 175-182.

Qing, C. M., Fan, L., Lei, Y., Bouchez, D., Tourneur, C., Yan, L., and Robaglia, C. (2000). Transformation of Pakchoi (*Brassica rapa* L. ssp. *chinensis*) by *Agrobacterium* infiltration. *Molecular Breeding*, 6(1), 67-72.

Sarker, R,H., Mustafa, B.M., Biswas, A., Mahbub, S., Nahra, M., Hashem, R., and Hoque M.I. (2003). *In vitro* regeneration in lentil (*Lens culinaris* Medik.). *Plant Tiss. Cult.* 13: 155-163.

Savage, G. P. (1988). The composition and nutritive value of lentils (*Lens culinaris*). In: *Nutrition Abstracts and Reviews* (Series A) (Vol. 58, No. 5, pp. 319-343).

Saxena, P. K., and King, J. (1987). Morphogenesis in lentil: plant regeneration from callus cultures of *Lens culinaris* Medik. via somatic embryogenesis. *Plant science*, 52(3), 223-227.

Sharma, S. K., Knox, M. R., and Ellis, T. H. (1996). AFLP analysis of the diversity and phylogeny of Lens and its comparison with RAPD analysis. *Theoretical and Applied Genetics*, 93(5), 751-758.

Sharpe, A.G., Ramsay, L., Sanderson, L.A., Fedoruk, M.J., Clarke, W.E., Li, R., Kagale, S., Vijayan, P., Vandenberg, A. and Bett, K.E. (2013). Ancient orphan crop joins modern era: gene-based SNP discovery and mapping in lentil. *BMC Genomics*, 14(1): 192.

Singh, R.K., and Raghuvanshi, S.S. (1989). Plantlet regeneration from nodal segments and shoot tip derived explants of lentil. *LENS Newsletter* 16: 33-35.

Tadmor, Y., Zamir, D., and Ladizinsky, G. (1987). Genetic mapping of an ancient translocation in the genus Lens. *Theoretical and Applied Genetics*, 73(6): 883-892.

Tar'an, B., L. Buchwaldt, A. Tullu, S. Banniza, T.D.Warkentin and Vandenberg, A. (2003). Using molecular markers to pyramid genes for resistance to ascochyta blight and anthracnose in lentil (*Lens culinaris* Medik). *Euphytica* 134: 223–230.

Taylor, P. W. J., Ades, P. K., and Ford, R. (2006). QTL mapping of resistance in lentil (*Lens culinaris* ssp. culinaris) to ascochyta blight (*Ascochyta lentis*). *Plant Breeding*, 125(5): 506-512.

Tran, B.M.D. and Credland, P.F. (1995) Consequences of inbreeding for the cowpea seed beetle, *Callosobruchus maculatus* (F.) (Coleoptera: Bruchidae). *Biological Journal of the Linnean Society*, 56: 483–503.

Trick, H. N., and Finer, J. J. (1997). SAAT: Sonication-assisted Agrobacterium-mediated transformation. *Transgenic Research*, 6(5), 329-336.

Trieu, A. T., Burleigh, S. H., Kardailsky, I. V., Maldonado-Mendoza, I. E., Versaw, W. K., Blaylock, L. A., and Weigel, D. (2000). Transformation of *Medicago truncatula* via infiltration of seedlings or flowering plants with *Agrobacterium*. *The Plant Journal*, 22(6): 531-541.

Tullu, A., Tar'an, B., Warkentin, T., and Vandenberg, A. (2008). Construction of an intraspecific linkage map and QTL analysis for earliness and plant height in lentil. *Crop Science*, 48(6), 2254-2264.

Vaillancourt, R. E., and Slinkard, A. E. (1993). Linkage of morphological and isozyme loci in lentil, *Lens culinaris* L. *Canadian Journal of Plant Science*, 73(4), 917-926.

Warkentin, T. D., and McHughen, A. (1992). *Agrobacterium tumefaciens*-mediated beta-glucuronidase (GUS) gene expression in lentil (*Lens culinaris* Medik.) tissues. *Plant Cell Reports*, 11(5), 274-278.

Warkentin, T. D. and McHughen, A. (1993): Regeneration from lentil cotyledonary nodes and potential of this explants for transformation by *Agrobacterium tumefaciens*. *Lens Newsletter* 20: 26-28.

Williams, D. J. and McHughen, A. (1986). Plant regenration of the legume *Lens culinaris* Medik (lentil) *in vitro*. *Plant Cell Tiss. Organ Cult*. 7: 149–153.

Yadav, S S., Rizvi, A. H., Manohar, M., Verma, A. K., Shrestha, R., Chen, C., Bejiga, G., Chen, W., Yadav, M. and Bahl. P. N.(2007). "Lentil growers and production systems around the world." In *Lentil*, pp. 415-442. Springer Netherlands.

Ye, G., McNeil, D. L., Conner, A. J., Hill, G. D. (2000). Improved protocol for the multiplication of lentil hybrids without genetic change by culturing single node explants. *SABRAO Journal of Breeding and Genetics* 32: 13-21.

Zamir, D., and Ladizinsky, G. (1984). Genetics of allozyme variants and linkage groups in lentil. *Euphytica*, 33(2), 329-336.

2018, *Climate Risks Management: Sustainable Pulse Production*
Editors: A K Srivastava and Yogranjan
Published by: **ASTRAL INTERNATIONAL PVT. LTD., NEW DELHI**   Pages **221–248**

Chapter 11

Breeding Faba Bean for Rust Resistance in Changing Environment

Usman Ijaz[1,2]*, Kedar N. Adhikari[2], M.A. Muktadir[2,3]
and Richard Trethowan[1,2]

[1]The University of Sydney, Plant Breeding Institute,
Cobbitty, NSW 2570, Australia
[2]The University of Sydney, I.A Watson Grain Research Centre,
Narrabri, NSW 2390, Australia
[3]The University of Sydney, Centre for Carbon Water and Food,
Brownlow hills NSW 2570 Australia
*E-mail: usman.ijaz@sydney.edu.au

ABSTRACT

Faba bean is one of the major cool season grain legumes and provides a good source of protein for human and animal consumption. It is considered as a staple food for Middle-East andNorth-African countries. It makes symbiotic association with the N-fixing bacteria and provides a natural source of fixed atmospheric nitrogen to the soil. Faba bean production is decreasing worldwide, due to various biotic (fungal and viral diseases) and abiotic (heat, frost and drought) factors. All of these factors are directly or indirectly related to seasonal shift due to fluctuating temperature and rainfall. Faba bean rust, Ascochyta blight and chocolate spot are the devastating diseases that can lead to the "no crop" losses. Good understanding on mechanism of disease resistance on host and pathogen virulence is needed to breed for disease resistance. Continuous crop breeding for developing better resistant cultivars is the only choice to ensure the sustainability in production. Environmental factors are the key elements that determine the epidemic development of disease. It is the reason why in some years a disease causes significant yield reduction, but negligible effect in other years. The aim of this review is to provide information on the current and past research on faba bean rust.Presence of different infection types, such as hypersensitive, slow-rusting and adult plant resistanceprovide good avenues for developing better resistance in the host. Proper utilisation of genetic factor of these responses combined with molecular markers will help in developing varieties with high level of rust resistance.Apart from host resistance, understanding on pathogen variation is equally important. Therefore a routine survey of pathogen variation is required to prevent possible epidemics caused by the development of virulent pathotypes.

Keywords: *Faba bean, Rust, Epidemiology, Screening, Resistance inheritance.*

Introduction

Grain legumes are edible group of leguminous crops providing protein, edible oil, food, feed andcredited for sustainable agriculture production because of their peculiar interface with nitrogen fixing bacteria (Torres *et al.*, 1993). Friedrich Kasimir Medikus (1736-1808) named faba bean as 'faba bona', but its authentic botanical name is*Vicia faba* as given by Linnaeus. It was known by a name "beans" that is still written in the 19th century dictionaries. Later on, it was renamed "field bean" that has been widely used in English. The name 'field bean' was often used for both *V. faba* and *Phaseolusvulgaris*causing confusion. Therefore, the Canadian breeders named it as 'faba bean' which is currently used widely (de la Vega *et al.*, 2011). Fababean (*Vicia faba* L.) also known as broad bean, fava bean or field bean is a major legumebelonging to family *fabaceae*. It has 2n = 12 chromosomeswith huge amount of repetitive DNA (Raina and Ogihara, 1995). It was domesticated in the Near East around 8,000 BC andthe oldest plant relics werefound in Jerico dated 6,000 BC. However, some authors (Cubero, 1973; 1974) speculate that faba bean could have been originated in the Near East based on accessions collected from this area.The origin of faba bean is still in debate because archaeological remainsand/or wild relatives have not yet been found (Cubero, 2011). It has been spread/acclimatised to wide geographical regionswith variable climatic and environmental conditions. Major faba bean growing countries include China, Ethiopia, UK and Australia comprising more than 80 per cent of the world's production (FAOSTAT, 2013).

Based upon seed morphology it is divided into four distinct groups: major, equina, minor and paucijuga (Cubero, 1974). Major seeds were evolved at South Mediterranean and China. Equina (medium seed) grown in the East and North Africa, while small seeded (paucijuga) varieties are cultivated in some parts of Ethiopia.Ingeneral faba beanhas been cultivated for human food and animal feed in crop rotation with winter cereals. It is an important legume in sustainable agriculture as its ability to fix atmospheric nitrogen is the highest among pulses.Faba bean has been extensively used for cytological studies due to its large chromosomes.

Faba bean is an annual winter legume requiring cool season for its growth and development (Duc, 1997). It is normally cultivated in northern latitude to warm temperate and subtropical zones. Some winter cultivars show photoperiod sensitivity and vernalisation requirement.Winter cultivars usually produce more tillers (4-6 stems/plant) as compared to spring cultivars (1-2 stem/plant)(Duc, 1997). Plants havea tap root system with secondary roots and bears nodule containing nitrogen fixing bacteria *Rhizobium leguminosarum*. Stem growth is indeterminate type, carrying a leaf up to more than tenth node. Flower is 2-3 cm long at anthesis with a typical papilionaceous structure. High variation for flower colour is present in faba bean such as white, brown, pink, purple and violet. Pods are short and erect in minor and paucijuga types (3-4 ovules/pod), but long and hanging in major (8-12 ovules/pod). Seed colour is also variable such as yellow, green, brown, beige, black or violet and sometimes seed carries brown spots or stripes around hilum.

Plant diseases are posing a major threat towards the sustainable crop production. Losses due to diseases are variable among geographic regions and year to year within the same location. Classic disease triangle explains the role of environment in plant disease as no virulent pathogen strain can induce disease on susceptible cultivars if climatic conditions are not favourable (Chakraborty *et al.*, 2000). Relationship between weather and disease are routinely used to forecast epidemics (Scherni and Yang, 1995). Environment change could be positive, negative or with no impact on individual plant disease cycle. Faba bean production is decreasing worldwide because of competitive pathogens of viruses, fungi, bacteria and nematodes prevail across all growing regions. Sustainable crop production requires better understanding of diseases in the changing climatic conditions triggered by global warming.

Understanding Disease Resistance

Genetic concepts of host-pathogen interaction keep changing since the rediscovery of Mendel's work (Hooker and Saxena, 1971). Parallel advancement in molecular genetics leads towards the better interpretation and understanding of data. Saga of scientific studies of disease resistance made progress in the field of genetics (Ellingboe, 1981). Cultivars of the crops showed differences in their degree of relative resistance and resistance can be transferred from one cultivar to other with the aid of breeding (Carleton, 1901). The first reference on the inheritance of resistance to pathogen was cited by Biffen (1905). He crossed a resistant wheat cultivar with a susceptible under natural epidemic field conditions. He collected data twice at early in development of rust epidemic and before maturity. He found three phenotypes: heavily rusted, intermediate and rust free (1:2:1) at rust developing stage and two types: heavily rusted and rust free at advanced plant stage (1:3). He reported the resistance was due to a recessive gene with partial dominance. Pathogenscontinue to evolve in competitive environment and show high variability in virulence (Barrus, 1911). Concept of physiological races has been established on the variability in *Puccinia graminis* f. sp. *tritici* by Stakman and Levine (1922). Physiological races are morphologically indistinguishable, but can be differentiated in their reaction to a set of host lines known as differentials. Concept of physiological races helps to understand why a resistant cultivar in one geographical zone becomes susceptible in other. Different physiological races of pathogens exist across environments. This discovery led plant breeders to think that resistant is another plant trait like awn vs awnless. This concept helped breeders to explain why a resistant cultivar becomes susceptible with a course of time. Plant breeders started finding new resistant sources from local or world host collections and transfer those genes to commercially adaptive cultivars (Ausemus, 1943). Some wild relatives and related host species are the source of rich gene bank, so breeders started interspecific cross to make future varieties' resistance more durable (Hayes *et al.,* 1920). The studies on the inheritance of plant pathogens were an obvious sequel. Discovery of genes controlling sexual incompatibilities in *Rhizopus* (Blakeslee, 1906) and *Schizophyllum* (Kniep, 1918) established Mendel's laws for fungi. Upon selfing a single isolate of *P. graminis* produced many new races (Craigie, 1930). This information explained origin of new pathogen races that were mysteryto breeders. A geographic race was a product of a collection of genetic variability, mutations, response to selection pressure and/or sexual incompatibilities. Presence of toxic substances proved to be plants arsenal against pathogens and invaders (Walker, 1924; Walker and Stahmann, 1955). Pathogen insensitivity to these compounds leads to the increase in virulence.

Flor laid the foundation of the modern genetics of host parasite relationship (Flor, 1946; 1947). He studied the inheritance of resistance/susceptibility (*Linumusitatissimum*) in host and virulence/avirulence in pathogen (*Melampsoralini*). Flor ascertained for each gene of resistance in a host there was a corresponding gene for avirulence in the pathogen. He formulated gene-for-gene hypothesis; the most suitable explanation to his findings (Flor, 1942; 1956;1971). He investigated progeny of different races of rust pathogen infected at lines of flex carrying different genes for resistance. He developed segregating population by crossing two different flex rust isolates and multiplied on 30 different host genotypes carrying single gene for rust reaction (Flor, 1955). He concluded that for successful resistance reaction genetic factors of both host and pathogens are involved. Specific plant pathogen interaction is determined by the resistant/avirulence gene products. This recognition mounts physiological defence response resulting hypersensitive cell death or accumulation of toxic molecules to pathogen which is called incompatible interaction. Absence of any of the gene products (resistant/avirulence) results compatible reaction and allows further growth of the pathogen. Thus loss of function either in resistant or avirulent gene by mutation or genetic recombination ended up with the loss of resistance. Mutation in avirulent gene elicits pathogen survival under selection pressure exerted by host resistant gene. Such events provoke disease epidemics and give rise to new pathogen races. Thus, genetic resistance against pathogen is highly race specific (Keller *et al.,* 2000). Late blight of potato (*Phytophthora infestans*) followed a pattern consistent with gene for gen hypothesis (Black, 1953; Toxopeus, 1956). Some genetic studies of apple scab disease was also an evident of the gene for gene hypothesis (Bagga and Boone, 1968).

Flor did not try to address the intermediate genotypes reactions. He focused on reaction types usually susceptible and resistant. Dealing with classification of intermediate reactions is more difficult because it is highly responsive to environmental variations (Ellingboe, 1981). Loegering (1966) presented an idea of "aegricorpus" defined as '*single living manifestation of specific genetic interaction in and between host and pathogen*'. This concept provided with a genetic framework of interactions between host and pathogen. He elucidated the necessity to investigate both host and pathogen genetic factors in determining the fate of interaction. The genetic basis of gene for gene interaction can be explained by a quadratic check first reported by (Martin and Ellingboe, 1976). Vanderplank (1978) published a simplest explanation of this idea illustrated in Figure 11.1. Resistance occurs only when a dominant R gene of a host interact with a dominant A gene of the pathogen. In all other cases, reaction is compatible, resulting susceptibility (Figure 11.1A). Mutation from R_1 to R_2 (Figure 11.1B) made the host resistant to parasite strain A_2. Therefore pathogen race A_1 has selective advantage over A_2 in the presence of resistance gene R_2 and *vice versa*. Gene product of A_1 did not recognise by the gene product of R_2. Therefore reaction ends up in compatibility with the fungal growth. To subjugate mutated virulence gene of pathogen the respective resistant host gene is needed (Figure 11.1B). These observations supported the idea that host-pathogen incompatibility is an active function (Shepherd and Mayo, 1972; Ellingboe, 1976). The disease resistance is frequently inherited as a dominant factor, but it can be recessive as well (Barrett, 1985; Adhikari *et al.*, 1999). Breeding favours the selection of dominant genes by early detection, whereas recessive genes are undetectable in F_1. Similarly dominant avirulence gene has been reported for pathogen, but it is not applicable under natural conditions like ascomycetes or bacteria which are haploid during infective stage. A molecular model of gene for gene interaction is reported by Staskawicz *et al.* (1995). Resistance occurs only when there is recognition between the products of avirulence gene (A) by resistant gene (R). Only the receptor-elicitor interaction results in specific recognition indicated by the hypersensitive response and disease resistance. Any alteration in gene products leads to the susceptible phenotype (Figure 11.2).

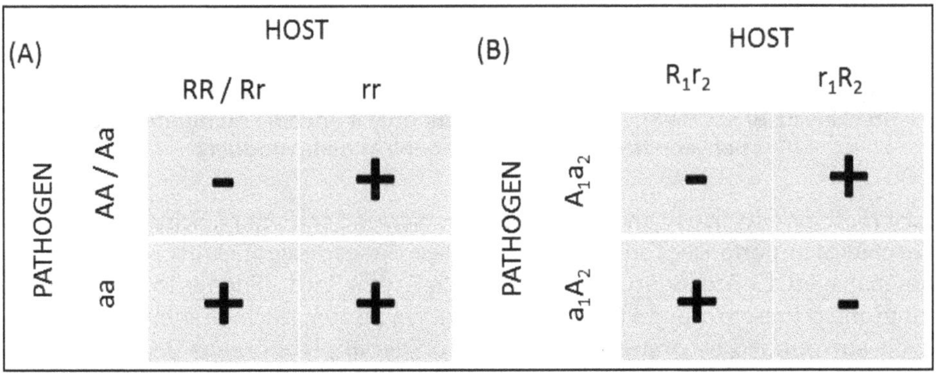

Figure 11.1: Quadratic Check Showing Interaction Types in Gene for Gene Reaction Modified from Keller *et al.* (2000).

A) incompatible reaction (-)between a dominant resistant gene R in a host and a dominant avirulence gene A in pathogen resulting in plant resistance. B) Two genetic loci of resistance (R_1 and R_2) in two plant cultivars corrsponding to two avirulence loci in two pathogen races (A_1 and A_2). Pathogen can grow when reciprocal dominant genes (R_1/A_2 and R_2/A_1) resulting compatible reaction (+).

Diseases of Faba Bean

Except increasing growing trend in Australia and Canada, the main reasons of declining faba bean cultivationis unreliable yield due to infestation of aerial fungi, soil borne pathogens, nematodes and viruses (Stoddard *et al.*, 2010).Major diseases caused by aerial fungi include Ascochyta blight (*Ascochyta fabae*Speg.), chocolate

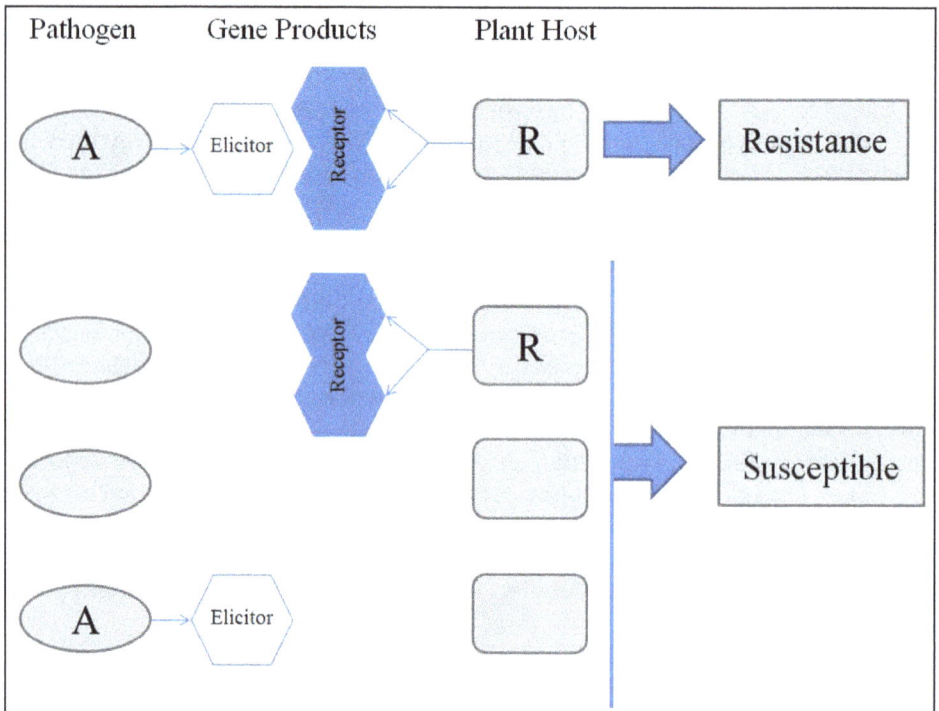

Figure 11.2: Molecular Model of the Gene for Gene Interaction Modified from Staskawicz *et al.* (1995). Resistance occurs only if specific recognition occurs between host (R) and pathogen (A) gene products.

spot (*Botrytis fabae*Sard.) and rust (*Uromyces viciae-fabae* (Pers.) J. Schrot.) and some are of local importance (Table 11.1). All of these diseases significantly reduce yield if not managed. Chocolate spot caused by *Botrytis fabae*sard., belongs to necrotrophic fungi and is present around the world. Primary symptoms include shiny dark brown spots surrounded by orange brown rings on aerial plant organs (leaves, flowers and stem). In the presence of mild temperature (15-22°C) and high humidity (>80 per cent) necrotrophic fungi start defoliating and killing the whole plant sometimes within two days (Harrison, 1988).Spores of *B. fabae*are visible through naked eye in senescent leave of diseased plants. Air born conidia will allow disease spread and new tissues produce more conidiophore to start another disease cycle. During off season, fungus survives on plant debris till onset of favourable condition. *B. fabae* also causes infection to some other alternate hosts including vetch (*Vicia sativa* L.), narbon bean (*Vicia narbonensis* L.) and lentils (*Lens culinaris* Medik.).

Ascochyta blightis a serious fungal disease of faba bean caused by *Ascochyta fabae*Speg. Characteristics symptoms include lesions of grey colour usually round in shape on leaves and elongated on stem. Elongated and slightly curved pycnidiospores can be seen easily inside lesion. In severe condition, the seed inside pods are stained resulting not only reduction in yield, but also in quality of the grain. The pathogen over winters on faba bean stubbles (Kaiser, 1997; Rubiales and Trapero-Casas, 2002).

Table 11.1: The Main Pathogenic Fungi on Faba Bean.
Table modified from Stoddard *et al.* (2010).

Name	Common Name	Parts Infected	Epidemiology	Distribution
Botrytis fabae Sard.	Chocolate Spot	Leaves, Flower, Stem and Seed	Mild temperature (15-22°C) High Humidity (>80 per cent)	Europe, North Africa, China and Australia
Ascochyta fabae Speg.	Ascochyta blight	Leaves, Stem, flowers and developing pods	Cooler temperature (15°C) during winter with high rainfall	Australia Mediterranean type Oceanic climates
Uromyces viciae-fabae (Pers.) J. Schro t.	Rust	Leaves, stems and pods	Warm (20-25°C) Humid (>80 per cent)	Worldwide
Peronospora viciae f. sp. fabae	Downey mildew	Leaves, stem and pods	Cool, Humid	Europe
Cercospora zonata	Cercospora leaf spot	Leaves stem	Cool, Humid	Australia China and Poland

Faba bean rust is a disease of worldwide distribution caused by a biotrophic fungus*Uromyces viciae-fabae* (Pers.) J. Schrot. It is considered as a major disease of faba bean in the Middle East, North Africa and sub-tropical region of Australia. All aerial plant parts are prone to rust infection. Rust epidemic occurs when climate becomes warm (20-25°C) with high humidity (>80 per cent) normally at the start of spring.If infected eraly in the season, rust epidemic can cause up to 70 per cent yield reduction (Liang, 1986; Rashid and Bernier, 1991).Infected plant parts (leaf, stem and pods) show oval uredial pustules of 1 mm in diameter. Telia developed beneath host epidermis appear to be dark brown to black in color. Resting teliospores are produced on infected plant debris in the areas with prolong dry summer. Under field conditions teliospore remain viable for two years and start germination when temperature becomes 17-22°C.

Peronosporaviciae f. sp. fabae is the causal agent of downy mildew in faba bean. Source of primary inoculation is soil born oospore that infects hypocotyl and upper root parts (Gaag and Frinking, 1997; Zinner *et al.,* 2008). Wet environmental conditions suits for the development of conidia that give rise to secondary infection cycle. Between winter and spring sown crops downy mildew survive on volunteer seedlings. Faba bean growers in Australia, China and Poland are concerned with the spread of*Cercosporazonata*. Lesions are of brown to grey in colour distinguished from Ascochyta blight by lack of typical pynidia and Chocolate spot by presence of zonation. It appears to infect other hosts including vetch, narbon bean and lentils (Kimber, 2011). *Cercospora* favours the same climatic conditions as Ascochyta and more prevalent invegetative phase (Hawthorne *et al.,* 2006).

Legumes are potentially vulnerable to the viral infection and about fifty viruses infecting faba bean have been reported (Makkouk *et al.,* 2003; Abraham *et al.,* 2006). During last twenty years eight persistent viruses causing yellowing, stunting and necrotic symptoms and ten non persistent viruses causing mosaics and molting in faba bean have been reported (Table 11.2). Faba bean necrotic yellow virus (FBNYV) is distributed Asian and African countries. Whereasbean leaf roll virus (BLRV), beet western yellow virus (BWYV), chickpea chlorosis dwarf virus (CpCDV) and milk vetch dwarf virus (MDV) has been found in China, Japan and subterranean clover stunt virus (SCSV) is in Australia only (Kumari and Van-Leur, 2011). Significant yield losses occur in plants showing yellowing and stunting symptoms. In 1991-92 FBNY epidemic resulted up to 90 per cent crop failure in Egypt. (Makkouk *et al.,* 1994). All these viruses are transmitted by aphid and are phloem limited except CpCDV. These viruses are neither transmitted by seed nor mechanically. BYMV is distributed worldwide but high incidence is observed in Egypt, Sudan and Coastal areas of Syria. These sites represented by warm winters that promote aphid populations. In addition some other viruses infecting faba bean found in specific countries under limited area such as tomato spotted wilt virus (TSWV) in China, Australia and broad bean necrosis virus (BBNV) in Japan. BBNV causes staining of seed coats that make undesirable for consumer. Disease control measures should be practical, affordable and efforts should be made in field of research, breeding and extension to cope with this challenge.

Table 11.2: The Reported Viruses Infecting Faba Bean.
Table modified from Kumari and Van-Leur (2011).

Viruses	Genome	Abbreviation	Distribution	Vector
Viruses causing yellowing/stunting/necrosis symptoms				
Beet western yellow virus	RNA	BWYV	Africa, Asia and Australia	Aphid
Bean leaf roll virus	RNA	BLRV	Worldwide	Aphid
Chickpea chlorosis dwarf virus	DNA	CpCDV	Africa and Asia	Leafhopper
Chickpea chlorosis stunt virus	RNA	CpCSV	Africa and Asia	Aphid
Faba bean necrotic yellow virus	DNA	FBNYV	Africa and Asia	Aphid
Milk vetch dwarf virus	DNA	MDV	Asia	Aphid
Soybean dwarf virus	RNA	SbDV	Africa, Asia and Australia	Aphid
Subterranean clover stunt virus	DNA	SCSV	Australia	Aphid
Viruses causing mosaics/mottling symptoms				
Alfalfa mosaic virus	RNA	AMV	Worldwide	Aphid
Bean Yellow Mosaic Virus	RNA	BYMV	Africa, Asia and Australia	Aphid
Broad bean mottle virus	RNA	BBMV	Africa and Asia	Beetles
Broad bean stain virus	RNA	BBSV	Africa, Asia and Europe	Beetles
Broad bean true mosaic virus	RNA	BBTMV	Africa, Asia and Europe	Beetles

Viruses	Genome	Abbreviation	Distribution	Vector
Broad bean wilt virus	RNA	BBWV	Asia and Australia	Aphid
Cucumber mosaic virus	RNA	CMV	Africa, Asia and Australia	Aphid
Pea early browning virus	RNA	PEBV	Africa	Nematode
Pea enation mosaic virus-1	RNA	PEMV-1	America, Asia and Europe	Aphid
Pea seed-borne mosaic virus	RNA	PSbMV	Africa, Asia and Australia	Aphid

Sometimes soil born nematodes alsostart damaging faba bean. *Ditylenchusdipsaci* (Kuhn) Filipjev is a stem nematode that infect more than 500 host species. Two separate oat race and giant race species infect the faba bean (Subbotin *et al.,* 2005; Kerkoud *et al.,* 2007). Infected plant showed symptomatic swelling or distortion of stem and yellowing of above surface plant parts. Root rot nematode (*Pratylenchusthornei* and*Melaoidogyn spp.*) is the problem of dry tropical and subtropical areas of India, Iraq and Egypt and Australia. Female breaks the cell of host root and develop a large multinuclear syncytia. Any damage to the root tissues ultimatelyaffects the nutrient uptake and exposes to secondary infection by fungi and bacteria. Plants appeared as stunted, yellow and wilted with damaged roots short black and rotten.

Uromycese Viciae-fabae

Rust fungus has served as an appropriate organism to gain subtle features of obligate biotrophic parasite. Rust fungus is a major group of living organisms comprising 100sof genera with about 7,000 species. *Puccinia* is the largest genus with 4,000 species followed by *Uromyces*with 600 species (Maier *et al.,* 2003).*U. viciae-fabae*senulato is a specie complexthat can infect faba bean, peas, lentils as well as more than 50 *Vicia* species and more than 20 *Lathyrus* species (Conner and Bernier, 1982a). It may be subdivided into three groups depending upon pathogenicity to faba bean, lentils and *Lathyrus* (Emeran *et al.,* 2005). This subdivision is further supported by the RAPD analysis (Emeran *et al.,* 2008).Physiological and cytological investigations started from 1970s (Abu-Zinada *et al.,* 1975) and continued to until 1990 because biochemical aspect became a new focus (Deising *et al.,* 1991).

Conner and Bernier (1982b) categorised faba bean rust reactions depending upon the colony size.On the basis of presence and absence of necrotic area, 16 races were identified (Emeran *et al.,* 2001). Thesame set of differentials were used to group 27 Spanish and Portuguese isolates into 15 physiological races (Del Mar Rojas-Molina *et al.,* 2007) with highest pathogenicity in Egyptian races.German botanist De Barry observed morphological and physiological alteration in faba bean rust pathogen after germination (Bary, 1866). The pathogen establishes infection structure to undertake decisive stages of pathogenesis: attachment, recognition, penetration, proliferation and nutrition. Our understanding towards biotrophy is still limited. Some characters pertinent to biotrophy include: developed infection structure, restricted secretory activity (lytic enzymes), extra haustorial matrix layer, host defence suppression and functional haustoria (Mendgen and Hahn, 2002).

Uromycese viciae-fabae is an autoecious and macrocyclic fungus. It completes all five spores stages in a single host (Mendgen and Deising, 1993).At the onset of spring dormant diploid (2n) teliospore startto germinate on plant residues andhaploid (n) basidiospores of different types (+ and -) are produced as a result of successive meiosis. After landing on a host, these haploid basidiosporesinvoke infection and give rise to pycnia.Dikaryotic pycniospores(n) from different mating types give rise to aecial primordia resulted by spermatization and dikaryotization (Voegele, 2006). Upon germination of aeciospore the major asexual urediosporesare formed in immense amount which cause frequent infection.Urediospores disperse by air and can travel to thousands of kilometres (Brown and Hovmoller, 2002). In fall uredia undergo through nuclei fusion and give rise to single cell diploid teliospore.Basidio-, aecio- and urediospores parasite the host with symptomatic infection while telio- and pycniospores do not.At the onset of summer, rust fungus survives as resting No information is available about this resting phase of fungi except few morphological features. During past two decade, most of the non-conventional studies were made upon urediospore to elucidate its cryptic behaviour. Analysis of infection structure made at both basidio- and urediospores allows comparison between both mono and dikaryotic spores on the same plant (Mendgen, 1997). Basidiospore is smooth and thin-walled as compare to the urediospore. Germ tube differentiated into well-developed appressorium, vesicle and haustorium in urediospores as compare to poorly differentiated structures in basidiospoere.

Epidemiology

If faba bean rust epidemics begin late in the season at the onset of pods filling, yield components are little affected by the infection, and losses usually range from 5 to 20 per cent (Sillero *et al.*, 2000; Sillero *et al.*, 2011). However, when the infection starts early in the season severe epidemics can occur and yield losses as high as 70 per cent has been reported by Rashid and Bernier (1991). In NSW, Australia rust appears from mid spring onward and favoured by warm temperature (Hawthorne *et al.*, 2012). Rust epidemics have been widely studied by many scientists under controlled environmental conditions (Sache and Vallavieille-Pope, 1993) because, it is difficult to access disease progress under interacting many biotic and abiotic factors.Joseph and Hering (1997) studied effects of interrelated environmental factors including leaf wetness, temperature and light that determines the fate of landed spore at leaf surface. Previous work on other fungi suggested that presence of moisture film is necessary for the successful spore germination and penetration._ ENREF_26Temperature range for the germination of *Uromycese viciae-fabae* sporeis 5-25°C with maximum at 20°C and none of the spore germinated at 30°C. A thin film of surface moisture is vital for spore germination and continuous moisture for 24 h give high infection response. During infection period these factors greatly influence the number of pustules. Only 60 per cent of the total number of pustules were produced at 10° C compared to 20°C even when moisture was maintained for 24h produced. This reduction response could not be explained by poor spore germination (Joshi and Tripathi, 2012). Negussie *et al.* (2005) found maximum infection of rust in lentils with dew period of 24h. Similar findings have been documented by Chauhan and Singh (1994) and Silva *et al.* (2001) with rust of pea (*Pisumsativum*) and common bean (*Phaseolus vulgaris*), respectively.

Effect of light on germination and infection establishment efficiently can be studied under laboratory conditions. Exposure to far red light wavelength (700-800 nm) induces delay in spore germination. This delay is proportionate to the intensity and exposure duration of light. Some studies upon other fungus such as *Puccinia graminis* reported a delay in spore germination when exposed with far red light (Lucas *et al.*, 1975). However, light inhibition in case of *Puccinia recondita* did not happen until spore become hydrated (Bromfield and Givan, 1963).Moreover, light and dark period also significantly affect the spore germination. Fungal spores upon exposure to alternative light and darkness starts germination upon 40 min dark period at 20°C. It is clear that *U. viciae-fabae* is different from other fungal genera because it did not show any response to all light wave lengths except far red (Chang *et al.*, 1973). Furthermore, the spores of *U. viciae-fabae*germinates during overcast day indicating that day time infection is possible if necessary surface moisture is present. This information, in combination with field metrological forecast can be utilised to sculpt epidemiological model (Vanderplank, 1963).

Screening Methods of Seedling and Adult Plant Resistance

Genetic resistance against any disease is the method of choice due to cost and environmental constraints. Durability of such resistance can be compromised due to sexual modifications in pathogen that makes resistant gene ineffective (Keen, 1990). The identification of new resistance sources and infection reaction with the present pathogen diversity should be kept under meticulous observations. This kind of information plays a key role in the success of breeding programme against disease control. Disease reaction of plants against fungal pathogens can be studied under both control and field conditions (Sillero *et al.*, 2006). Several components of rust infection have been studied under controlled environmental conditions (Dwivedi *et al.*, 2002; Sillero *et al.*, 2000;Sillero *et al.*, 2006). Time needed by a spore after germination to the successful sporulation is known as latent period (LP). Deep visual observation is needed to count number of pustules daily until it becomes constant. Infection frequency (IF) is an important resistance component and valued by counting number of pustules per unit area in same area where LP was estimated. Infection type (IT) is a key rust scoring method described above. Sillero and Rubiales (2002) used contrast phase microscope to measure colony size (CS). It can also be measured macroscopically on the basis of pustule diameter (Polignano *et al.*, 1990). Studies on these individual components are necessary to understand disease epidemiology. Prolonged LP, reduced IF, delayed spore production are important components of disease resistance (Habtu and Zadoks, 1995; Singh *et al.*, 2003).

Field Screening

Successful field fungal screening solely depends upon the uniformity of infection across the genotypes tested to avoid disease escape. Natural epidemics are so frequent in some locations that no artificial inoculation is needed (Tiwari *et al.*, 1997). Albeit, artificial inoculation is needed to ensure uniform disease spread under field conditions. For this purpose, plants are inoculated with spray containing aqueous suspension of rust of rust spores or dusting of rust spores with the inert carrying matter like pure talc powder (Tyagi and Srivastava, 1999; Timmerman-Vaughan *et al.*, 2000). In addition to inoculation, disease spreader rowsare planted around the field trial to increase disease pressure. These spreaders consist of highly susceptible genotypes which act as a source of inoculum or 'foci' of infection to adjacent entries (Tivoli *et al.*, 2006).Furthermore, plants should be inoculated after sunset to utilize high relative humidity needed for spore germination. Plots are misted withsprinklers to maintain thin water film over leaf surface. Different methods have been reported to measure quantitative rust resistance under field conditions (Adhikari *et al.*, 2016b). Disease severity (DS) can be calculated upon regular visual estimation of lead area under rust pustules. Genotypes differing in their rust reaction and their final disease severity can be compared by area under disease progress curve (AUDPC) or by epidemic growth rate (r) (Sillero *et al.*, 2000). Mean of AUDPC is modified in to relative value and expressed as percentage values. Infection type (IT) is studied through the amount of necrosis or chlorosis around the infection site and rate or size of individual sporulating colony. Different IT scales has been reported and used so far.

Glasshouse Screening

Growth chamber or glasshouse gives good control of environmental conditionsover field conditions. It allows testing plants both at seedling and adult stage against fungal pathogens. Controlled conditions are the most efficient methods for screening genotypes against fungal differentials and large scale segregating populations.Rust uredosporesare suspended in inert mineral oil and sprayed over the plants to ensure uniform distribution (Barsalobres-Cavallari *et al.*, 2009). Incubation chambers are used to ensure spore germination and plants incubated at $20°$ C at 100 per cent humidity for 24h in complete darkness (Ijaz and Adhikari, 2016). After inoculation, plants are shifted to growth chamber at $20°$C with 14-h photoperiod (Boyle and Walters, 2005).This technique resulted non-systemic very high infection rate (Pegg and Mence, 1970; Thomas and Kenyon, 2004). In order to multiply pathogen inoculum, plants are maintained at $20-22°$ C with high humidity for two weeks and humidity re-applied at the onset of sporulation. Collected spores can be stored at $-80°$C that remains viable for a year (Gill and Davidson, 2005).

Detached Leaf Test

Instead of using intact plant, detached plant organs provide an opportunity to study fungal differentials. Organs (leaves or stem) showed different reactions against pathogen (Kohpina *et al.*, 2000; Avila *et al.*, 2004). Faba bean screening by using detached leaf has been reported against Ascochyta blight (Kohpina *et al.*, 2000), chocolate spot (Tivoli *et al.*, 1986) and rust (Herath *et al.*, 2001). Young leaves collected from healthy plant 3-5 weeks after sowing can be used for discrimination of disease response. To study *B. fabae* benches of inoculated leaves are covered with polythene sheet at room temperature (20 ± 2°C) till the disease appear (Bouhassan *et al.*, 2004). Detached leaf test provides similar results for disease reaction in resistant and susceptible cultivars as field testing but no good correlation has been found for intermediate level of resistance (Tivoli *et al.,* 1986). Similarly, 5 to 8 young leaves from the top apex were excised to study reaction with faba bean rust (Herath *et al.,* 2001). Excised leaves were maintained in water medium with gibberellic acid under controlled environmental conditions. Rust spores dispersed in slurry with talc powder were applied at lower leaf surface. Disease development was monitored for fifteen days and infection type was accessed using 1-9 key scale of Stonehouse (1994). This scale measured surface covered with pustules, pustule size and pustule density. Pustule size was separately assessed using 1-5 scale of Davison and Vaughan (1963) as modified by Stavely (1984).

Methods of Disease Scoring

Stakman *et al.* (1962) reported first 0-4 IT scale where; 0 (no symptoms),; (necrotic flecks), 1 (minute pustule barely sporulate), 2 (necrotic patch surround small pustule), 3 (chlorotic halo surrounding pustules) and 4 (well-developed pustule). ITs from 0-2 are considered as resistant and 3-4 as susceptible. Another IT scoring scale ranging from 0 to 9 has been introduced by McNeal *et al.* (1971) where, IT 0-6 represents resistant and 7-9 susceptible. Some researchers usedcombined descriptive keys of disease severity and disease damage visible on rust infected leaves. One such work has been reported on faba bean by Bernier (1984) where individual with no pustule and minute non sporulating flecks, scored as 1 (highly resistant); few pustules covering < 1 per cent of leaf area and few or no stem pustule scored as 3 (resistant); pustules are common on the leaves covering 1-4 per cent leaf area, little defoliation and some pustules on stem scored as 5 (moderately resistant); pustules covering 4-8 per cent leaf are, defoliation and many stem pustules scored as 7 (susceptible) and highly susceptible individuals with severe defoliation and 8-10 per cent leaf area under sporulating rust pustules. These scales sketch a meticulous description between resistant and susceptible genotypes in comparison with check, but little information on the type of resistant. 1-9 scale of Subrahmanyam *et al.* (1995) has been used to investigate rust infection in groundnut where 1 = no disease; 3 = few pustules with poor sporulation; 5 = many pustules with moderate sporulation; 7 = pustules across the whole plant with leave withering; 9 = severely infected plant up to 100 per cent leaves withering.

Field screening is not an efficient mean of identifying resistance in many crops due to significant variation in disease severity and infection pressure. Therefore, field screening needs to be conducted over a number of seasons and locations to compensate for environmental variations and pathogen variation across and within the locations (Ryan, 1971; Stegmark, 1991; Thomas, 1992; Thomas and Kenyon, 2004). Site selection for the field screening plays vital role in the success of disease infestation. Continuous cropping of susceptible cultivars and their debris incorporation can create good infection site quickly and inexpensively. A reliable rust screening field needs to be developed as explained earlier. Percent value of infected area under disease cover is recognised as quantitative variable for data interpretation (Banyal and Tyagi, 1997; Tyagi, 1999). Tiwari *et al.* (1997) used 0-9 IT scoring scale where; 0-4 (20 per cent rust infected area) considered as resistant and 5-9 as susceptible. Pal and Brahmappa (1980) classified plants with no infection as resistant; less than 5 per cent infected foliage as moderately resistant, 6-25 per cent infected leaf area as moderately susceptible and over 26 per cent as highly susceptible. Some scale designed scoring according to the affected plant organs. Singh and Sokhi (1980) used 0-9 scale; 0 = no colony (highly resistant); 1 = traces of patches covering 1 per cent leaf area (resistant); 3 = few patches covering 1-10 per cent leaf area (moderately resistant); 5 = white powdery growth covering 11-25 per cent leaf area (resistant); 7 = powdery growth covering 26-50 per cent leaf area and affected pods (susceptible) and 9 = sever infection covering 51-100 per cent leaf area. A unique type of scaling ranges from 0 to 2 has been reported (Sharma, 1992). Healthy highly resistant plant scored with zero whereas, 0.5 = moderately resistant (25 per cent infection); 1 = moderately susceptible (50 per cent infection); 1.5 = susceptible (75 per cent infection) and 2 = highly susceptible (100 per cent infection). A scale 0-4 depending upon visually appeared micro- and macroscopic mycelium growth at leaf surface (Vaid and Tyagi, 1997).

Thomas and Camp (1997) studied faba bean response in reaction with *Peronosporaviciae*. For reaction assessment he established a sporulation index where, 0 = no visible infection; 1 = few flecks; 2 = large infected area with sparse sporulation; 3 = complete covered infected leaf with dense sporulation; 4 = total plant cover with denser sporulation. Systemic infections are more difficult to score under field conditions. Cycles of secondary infection generates severe disease infected leaves and it can easily be scored on the basis of visual infected (per cent) areas. Seven categories of scoring ng (0, 5, 10, 25, 50 and 100 per cent leaf affected area) wasintroduced by Dixon (1981). Taylor *et al.* (1989) also categorized the macroscopic reaction responses where, 0 = no visible symptoms; 1 = visible necrosis with no sporulation; 2 = limited sporangium growth with local necrosis; 3 = abundant sporulation confined to lesions; 4 = abundant sporangium production on whole plant. Semi quantitative rating scale based upon five classes reported by Buchwaldt *et al.* (2004). Highly resistant (HR) = no lesion on stem; resistant (R) = few superficial lesions; moderately resistant (MR) = 1-10 deep lesions; moderately susceptible (MS) = 15-20 deep lesions; susceptible = 25-30 deep lesions and highly susceptible (Joshi and Tripathi) = more than 30 deep lesions.

Mechanism and Inheritance of Rust Resistance in Faba Bean

Studies on the genetic bases of resistance in faba bean can be difficult either due to the absence of complete resistance (Hiratsuka, 1933) or lack of plants'genetic uniformity due to partial allogamy. The investigation conducted by Conner and Bernier (1982b) reported three race specific genes of rust resistance in seven faba bean inbred lines. On the basis of distinct virulence pattern they selected two single pustule rust isolates (SP3 and SP51). Using resistant and susceptible parents, hybridization was done to generate back crosses and F_2 generations. Three weeks after emergence plants of these generations were inoculated with SP3 and SP51 separately. Screening against two isolates allowed three homozygous genes (designated as *fr1, fr2 and fr3*) of resistance to be identified. The resistant phenotype was dominant in all the cases. Rashid and Bernier (1986a) reported the presence of five genes conditioning resistance against two rust isolates.

Uromycese viciae-fabae infected wide host range including *Vicia, Lathyrusand Pisum* genera (Conner and Bernier, 1982a). Conner and Bernier (1982c) identified seven genes conferring resistance to specific rust races in Manitoba, Canada. None of these genes survived enough to be called as durable resistance. Emeran *et al.* (2001) reported sixteen *U. viciae-fabae* pathotypes using infection type based upon presence and absence of necrosis. Slow rusting (quantitative resistance) has been considered to be durable (Bond *et al.*, 1994). Slow rusting or partial resistance has been reported in many other crops like field corn (Hooker, 1969), sweet corn (Groth *et al.*, 1983), wheat, barley (Wilcoxson, 1981) and faba bean (Conner and Bernier, 1982d). Incomplete resistance in faba bean has been reported earlier but hypersensitive resistance, where necrotic reaction occurs late resulting reduction of infection has been documented recently. Incomplete resistance associated with increase latent period, reduce colony size and decrease infection frequency.

Extensive variability of *U. viciae-fabae* required durability of faba bean rust resistance for successful crop cultivation (Singh and Sokhi, 1980). In slow rusting limited disease development occurs in relative to susceptible seedlings (Conner and Bernier, 1982d). Low yield losses of 1-2 per cent were found in slow rusting population as compared to 67 per cent in susceptible populations (Rashid *et al.*, 1991). Rashid and Bernier (1986b) examined more than two hundred faba bean lines from diverse origin against faba bean rust They found significant differences among genotypes for area under progress disease curve (AUDPC) and final rust severity. Low AUDPC in some accessions was not entirely due to slow rusting. Mixture of genotypes containing susceptible and resistant cultivars has been reported to be effective in slowing down the rust development (Browning and Frey, 1969). Information on genetic basis of these types of resistance is still scant. Low levels of hypersensitive response resulting intermediate infection type was reported (Sillero *et al.*, 2000). In another study, a race specific resistance governed by major genes has been reported (Avila *et al.*, 2003). Non-hypersensitive believed to be polygenic in nature and considered as durable. Recently, three major seedling rust resistance genes were found in Australian germplasm (Adhikari *et al.*, 2016b). The above studies indicated more than one source of resistance and avenues for deploying different strategies to prolong the resistance.

Breeding for Rust Resistance

Disease resistance is generally inherited as a dominant factor in natural populations (Barrett, 1985) as natural selection acts on them in the heterozygous phase. Resistance based on major hypersensitive genes is usually considered non-durable (Stuthman *et al.*, 2007) as described above in gene-for-gene hypothesis. Hypersensitivity has been observed among faba bean genotypes, where necrosis appears during the later stages of rust proliferation and restricts fungal growth by controlled cell death (Sillero *et al.*, 2000; Torres *et al.*, 2006). Avila *et al.* (2003) proposed that this type of hypersensitive response in faba bean is under the control of major genes. However, the longevity of major genes in faba bean is less well understood than in wheat, so the deployment of genes should be careful and strategic.

Gene pyramiding is the simultaneous deployment of more than one R gene in the same cultivar. SuchR gene pyramids can provide long-lasting resistance because multiple mutational events in *avr* genes are required to elicit susceptibility in the host. A particular gene combination (*Avr/R*) will remain effective as matching virulence is not widespread in the population. However, it is difficult to discriminate the effects of individual R genes in a complex pyramid unless virulent pathotypes are available (Pink, 2002) or specific DNA markers are used.

Several R gene pyramids have been achieved using marker-assisted selection (MAS) in rice (*OryzasativaL.*), wheat, cotton (*GossypiumhirsutumL*) and barley (*Hordeum vulgare* L.) (Table 1). The utility of MAS is limited by breeder access to R genes where the *avr* gene already exists in the population. Gene pyramiding provides an opportunity to achieve a 'clean crop' and ensures crop uniformity because all genes are deployed in a uniform genetic background. Nevertheless, R gene pyramids are vulnerable to virulence if the individual components of resistance are deployed singly in other cultivars grown in the same region (Parlevliet, 1997). The R gene in a pyramid exerts a strong unidirectional selection pressure against the matching virulence in a pathogen population. Therefore, pathogen monitoring to identify new virulent pathotypes is essential to minimizing risk through early warning. The number of R genes in a pyramid will determine the spread of emerging virulent pathotypes in a pathogen population (Kolmer, 1992; Mundt, 1991),but conflicting results have been reported. For example, Mundt (1991) reported that an R gene pyramid failed to provide resistance against stem rust in wheat. This breakdown of resistance started a debate over the utility of pyramiding R genes. The successful use of gene pyramiding against rust in faba bean has not been reported.

In comparison to domestic crops, wild species are more heterogeneous for resistance genes and thus avoid disease epidemics (Bevan *et al.*, 1993; Okamura and Ouchi, 2007). Genetic uniformity for quantitative and qualitative traits facilitates harvesting, processing and marketing, but it increases vulnerability to diseases and insect pests (Castro, 2001). Heterogeneous landraces are a common feature of subsistence farming and their lack of uniformity buffers insect and disease damage (Smithson and Lenne, 1996) by minimising selection pressure on the avirulent allele in the pathogen population. Multiline development (Browning and Frey, 1969) and cultivar mixtures (Wolfe and Barrett, 1980) have been proposed to keep genetic diversity as mean of prolonging Multiline is a composite of agronomically similar genotypes that differ forcertaintraits, such as resistance to different pathogen races (Jensen, 1952). Wolfe (1985) defined a cultivar mixture as a "mixture of cultivars that vary for many characters including disease resistance, but have sufficient similarities to be grown together". A modern multiline is a composite of genetically uniform near isogenic lines (NILs) differing only for a disease-resistance gene (Smithson and Lenne, 1996). Developing these isolines is time consuming, so the breeder may prefer to deploy them singly as individual cultivars. In contrast, mixtures are relatively easy to make and use. Considerable research has shown that cultivar mixtures reduce disease incidence in cropping systems.

On the basis of field observations, it was reported that crop heterogeneity exerts little unidirectional selection pressure on complex pathotypes (Wolfe and

Barrett, 1980; Chin and Wolfe, 1984;Chin *et al.*, 2001). Nevertheless, the frequency of complex pathotypes was found to be higher in mixtures than in pure crop stands (Huang *et al.*, 1994; Lannou, 2001) because many *R* genes exert high pressure on the corresponding avirulent pathotypes. This breakdown of resistance in multilines and mixtures changes the frequency of virulent pathotypes, thus eventually eroding resistance further. This issue can be resolved by replacing susceptible genotypes with resistant lines in the mixture. Furthermore, mixtures of cultivars can be synthesized to control the spread of more than one disease (Wolfe, 2000; Gurr *et al.*, 2003).

Early genetic studies in crops were based on morphological, cytological and biochemical evaluation (Xu, 2010). However, DNA markers such as restriction fragment length polymorphism (RFLP), amplified fragment length polymorphism (AFLP), random amplified polymorphic DNA (RAPD), simple sequence repeats (SSR) and single nucleotide polymorphisms (SNP) linked to traits can assist gene tracking throughout selection and breeding (Collard *et al.*, 2005). Markers that are closely linked (less than 5 cM of genetic distance) with the gene of interest reduce the probability of recombination so that the markers remain linked with the gene of interest can provide reliable selection in crop improvement (Jiang, 2013). Use of flanking markers increases the probability of success. Marker assisted selection (MAS) is a practical and effective way to pyramid rust resistance genes if the genes and closely linked markers are available (Servin *et al.*, 2004). MAS has been successfully employed to pyramid multiple leaf rust resistant (*Lr*) genes in common wheat (Tiwari *et al.*, 2008; Moullet *et al.*, 2009; Moullet *et al.*, 2010; Chhuneja *et al.*, 2011) and stripe rust in barley (Castro *et al.*, 2003), but not yet in faba bean.

Association mapping and linkage mapping are powerful tools used in identify QTLs and genes responsible for a particular phenotype (Collard and Mackill, 2008). Limited information is available on the development and use of molecular markers in faba bean. Three RAPD markers, namely $OPD13_{736}$, $OPL18_{1032}$ and $OPI20_{900}$ were mapped in coupling phase to the rust resistance gene *Uvf-1* in faba bean line 2N52 (Avila *et al.*, 2003). Two additional markers linked to the gene in repulsion at distances of 9.9 cM ($OPP02_{1171}$) and 11.5 cM ($OPR07_{930}$) were also reported. The line 2N52 used in this study was resistant to seven other rust isolates (2, 4, 5, 8, 9, 10 and 13) (Emeran *et al.*, 2001). Many other genotypes showing this type of resistance response have been reported (Sillero *et al.*, 2000; Adhikari *et al.*, 2016a). The next step is to pyramid these R genes using MAS to produce a line with potentially higher levels of longer lasting resistance.

Conclusion

Faba bean is one of the oldest crops grown for human consumption, animal feed and sustainable nitrogen fixation.Despite a wide range of benefits, the area under crop cultivation is reducing annually. The useful genetic variation identified for various diseases should be used to develop cultivars with multiple disease resistance to ensure stable yield. Faba bean rust is an emerging challengeto researchers and farmers worldwide. Recent understanding of the disease, mechanism of resistance and better gene deployment aided by molecular tools is in progress. More efforts are needed to find new resistant sources and broad adoption of new improvements in marker technology to ensure sustainability of the crop. Furthermore, comprehensive programme is needed to monitor seasonal shift, change in temperature and rainfall to predict rust epidemics. This prediction will be helpful in adoption of preventive measures to minimize such damages. Breeding rust resistance in faba bean should involve finding new sources of resistance,characterisation of resistant genes,deployment of effective resistant genes to construct pyramid or mixture aided with molecular tools can ensure rust free crop culture.

REFERENCES

Abraham, A. D., Menzel, W., Lesemann, D. E., Varrelmann, M. and Vetten, H. J. (2006). Chickpea chlorotic stunt virus: a new polerovirus infecting cool-season food legumes in Ethiopia. *Phytopathology* 96(5): 437-446.

Abu-Zinada, A. A. H., Cobb, A. and Boulter, D. (1975). An electron-microscopic study of the effects of parasite interaction between Vicia faba L. and Uromyces fabae. *Physiological plant pathology* 5(2): 113-118.

Adhikari, K. N., McIntosh, R. A. and Oates, J. D. (1999). Inheritance of the stem rust resistance phenotype Pg-a in oats. *Euphytica* 105(2): 143-154.

Adhikari, K. N., Zhang, P., Sadeque, A., Hoxha, S. and Trethowan, R. (2016a). Single independent genes confer resistance to faba bean rust (Uromyces viciae-fabae) in the current Australian cultivar Doza and a central European line Ac1655. *Crop and Pasture Science* 67(6): 649-654.

Adhikari, K. N., Zhang, P., Sadeque, A., Hoxha, S. and Trethowan, R. (2016b). Single independent genes confer resistance to faba bean rust (Uromyces viciae-fabae) in the current Australian cultivar Doza and a central European line Ac1655. *Crop and Pasture Science.*

Ausemus, E. (1943). Breeding for disease resistance in wheat, oats, barley and flax. *The Botanical Review* 9(4): 207-260.

Avila, C., Sillero, J., Rubiales, D., Moreno, M. and Torres, A. (2003). Identification of RAPD markers linked to the Uvf-1 gene conferring hypersensitive resistance against rust (Uromyces viciae-fabae) in *Vicia faba* L. *Theoretical and Applied Genetics* 107(2): 353-358.

Avila, C. M., Satovic, Z., Sillero, J. C., Rubiales, D., Moreno, M. T. and Torres, A. M. (2004). Isolate and organ-specific QTLs for ascochyta blight resistance in faba bean (Vicia faba L). *Theoretical and Applied Genetics* 108(6): 1071-1078.

Bagga, H. and Boone, D. (1968).Genes in Venturia inaequalis controlling pathogenicity to crabapples. In *Phytopathology*, Vol. 58, 1176- and : AMER PHYTOPATHOLOGICAL SOC 3340 PILOT KNOB ROAD, ST PAUL, MN 55121.

Banyal, D. and Tyagi, P. (1997). Role of climatic factors in the development of powdery mildew of pea. *Indian Journal of Mycology and Plant Pathology* 27(1): 64-66.

Barrett, J. (1985).The gene-for-gene hypothesis: parable or paradigm. In *LINN. SOC. SYMP. SER. 1985*.

Barrus, M. F. (1911). Variation of varieties of beans in their susceptibility to anthracnose. *Phytopathology* 1(6): 190-195.

Barsalobres-Cavallari, C. F., Severino, F. E., Maluf, M. P. and Maia, I. G. (2009). Identification of suitable internal control genes for expression studies in Coffea arabica under different experimental conditions. *BMC molecular biology* 10(1): 1.

Bary, A. (1866). *Morphologie und physiologie der pilze, flechten und myxomyceten*. W. Engelmann.

Bernier, C. C. (1984). *Field manual of common faba bean diseases in the Nile Valley*. International Center for Agricultural Research in the Dry Areas.

Bevan, J., Clarke, D. and Crute, I. (1993). Resistance to Erysiphe fischeri in two populations of Senecio vulgaris. *Plant Pathology* 42(4): 636-646.

Biffen, R. H. (1905). Mendel's laws of inheritance and wheat breeding. *The Journal of Agricultural Science* 1(01): 4-48.

Black, W. (1953). III.—A Genetical Basis for the Classification of Strains of Phytophthora infestans. *Proceedings of the Royal Society of Edinburgh. Section B. Biology* 65(01): 36-51.

Blakeslee, A. F. (1906). *Zygospore germinations in the Mucorineae*. Druck von A. Hopfer.

Bond, D., Jellis, G., Rowland, G., Le Guen, J., Robertson, L., Khalil, S. and Li-Juan, L. (1994).Present status and future strategy in breeding faba beans (*Vicia faba* L.) for resistance to biotic and abiotic stresses. In *Expanding the Production and Use of Cool Season Food Legumes*, 592-616: Springer.

Bouhassan, A., Sadiki, M. and Tivoli, B. (2004). Evaluation of a collection of faba bean (Vicia faba L.) genotypes originating from the Maghreb for resistance to chocolate spot (Botrytis fabae) by assessment in the field and laboratory. *Euphytica* 135(1): 55-62.

Boyle, C. and Walters, D. (2005). Induction of systemic protection against rust infection in broad bean by saccharin: effects on plant growth and development. *New phytologist* 167(2): 607-612.

Bromfield, K. and Givan, C. V. (1963).Light inhibition of uredospore germination in puccinia recondita. Army biological labs frederick md.

Brown, J. and Hovmoller, M. (2002). Aerial dispersal of pathogens on the global and continental scales and its impact on plant disease. *Science* 297: 537–541.

Browning, J. A. and Frey, K. J. (1969). Multiline cultivars as a means of disease control. *Annual Review of Phytopathology* 7(1): 355-382.

Buchwaldt, L., Anderson, K., Morrall, R., Gossen, B. and Bernier, C. (2004). Identification of lentil germ plasm resistant to Colletotrichum truncatum and characterization of two pathogen races. *Phytopathology* 94(3): 236-243.

Carleton, M. A. (1901). *Emmer: A grain for the semiarid regions.* US Department of Agriculture.

Castro, A. (2001). Cultivar mixtures. *Plant Health Instr. doi* 10.

Castro, A. J., Capettini, F., Corey, A., Filichkina, T., Hayes, P. M., Kleinhofs, A., Kudrna, D., Richardson, K., Sandoval-Islas, S. and Rossi, C. (2003). Mapping and pyramiding of qualitative and quantitative resistance to stripe rust in barley. *Theoretical and Applied Genetics* 107(5): 922-930.

Chakraborty, S., Tiedemann, A. and Teng, P. (2000). Climate change: potential impact on plant diseases. *Environmental Pollution* 108(3): 317-326.

Chang, H.-S., Calpouzos, L. and Wilcoxson, R. D. (1973). Germination of hydrated uredospores of Puccinia recondita inhibited by light. *Canadian Journal of Botany* 51(12): 2459-2462.

Chauhan, R. and Singh, B. (1994). Effect of different durations of leaf wetness on pea rust development. *Plant Disease Research* 9(2): 200-201.

Chhuneja, P., Vikal, Y., Kaur, S., Singh, R., Juneja NS Bains, S., Berry, O., Sharma, A., Gupta, S., Charpe, A. and Prabhu, K. (2011). Marker-assisted pyramiding of leaf rust resistance genes Lr24 and Lr28 in wheat (*Triticum aestivum*). *Indian Journal of Agricultural Sciences* 81(3): 214.

Chin, K., Chavaillaz, D., Kaesbohrer, M., Staub, T. and Felsenstein, F. (2001). Characterizing resistance risk of *Erysiphe graminis* f. sp. *tritici* to *strobilurins*. *Crop Protection* 20(2): 87-96.

Chin, K. and Wolfe, M. (1984). Selection on Erysiphe graminis in pure and mixed stands of barley. *Plant Pathology* 33(4): 535-546.

Collard, B., Jahufer, M., Brouwer, J. and Pang, E. (2005). An introduction to markers, quantitative trait loci (QTL) mapping and marker-assisted selection for crop improvement: the basic concepts. *Euphytica* 142(1-2): 169-196.

Collard, B. C. and Mackill, D. J. (2008). Marker-assisted selection: an approach for precision plant breeding in the twenty-first century. *Philosophical Transactions of the Royal Society B: Biological Sciences* 363(1491): 557-572.

Conner, R. and Bernier, C. (1982a). Host range of Uromyces viciae-fabae. *Phytopathology* 72(6): 687-689.

Conner, R. and Bernier, C. (1982b). Inheritance of rust resistance in inbred lines of Vicia faba. *Phytopathology* 72(12): 1555-1557.

Conner, R. and Bernier, C. (1982c). Race identification in Uromyces viciae-fabae. *Canadian Journal of Plant Pathology* 4(2): 157-160.

Conner, R. and Bernier, C. (1982d). Slow rusting resistance in Vicia faba. *Canadian Journal of Plant Pathology* 4(3): 263-265.

Craigie, J. H. (1930). An experimental investigation of sex in the rust fungi.

Cubero, J. (1973). Evolutionary trends in *Vicia faba* L. *Theoretical and Applied Genetics* 43(2): 59-65.

Cubero, J. I. (1974). On the evolution of Vicia faba L. *Theoretical and Applied Genetics* 45(2): 47-51.

Cubero, J. I. (2011). The faba beans: a historic perspective. *Grain legumes* 56: 5-7.

Davison, A. D. and Vaughan, E. K. (1963). A simplified method for identification of races of Uromyces phaseoli var. phaseoli. *Phytopathology* 53(4): 456- and.

de la Vega, M. P., Torres, A. M., Cubero, J. I. and Kole, C. (2011). *Genetics, genomics and breeding of cool season grain legumes*. CRC Press.

Deising, H., Jungblut, P. R. and Mendgen, K. (1991). Differentiation-related proteins of the broad bean rust fungus Uromyces viciae-fabae, as revealed by high resolution two-dimensional polyacrylamide gel electrophoresis. *Archives of microbiology* 155(2): 191-198.

Del Mar Rojas-Molina, M., Rubiales, D., Prats, E. and Sillero, J. C. (2007). Effects of phenylpropanoid and energetic metabolism inhibition on faba bean resistance mechanisms to rust. *Phytopathology* 97(1): 60-65.

Dixon, G. R. (1981).Downy mildew of peas and beans. In *The Downy Mildews*, 497–514 London.: Academic Press.

Duc, G. (1997). Faba bean (*Vicia faba* L.). *Field Crops Research* 53(1–3): 99-109.

Dwivedi, S. L., Pande, S., Rao, J. N. and Nigam, S. N. (2002). Components of resistance to late leaf spot and rust among interspecific derivatives and their significance in a foliar disease resistance breeding in groundnut (*Arachis hypogaea* L.). *Euphytica* 125(1): 81-88.

Ellingboe, A. (1976).Genetics of host-parasite interactions. In *Physiological plant pathology*, 761-778: Springer.

Ellingboe, A. H. (1981). Changing concepts in host-pathogen genetics. *Annual Review of Phytopathology* 19(1): 125-143.

Emeran, A., Roman, B., Sillero, J. C., Satovic, Z. and Rubiales, D. (2008). Genetic variation among and within Uromyces species infecting legumes. *Journal of phytopathology* 156(7-8): 419-424.

Emeran, A., Sillero, J., Niks, R. and Rubiales, D. (2005). Infection structures of host-specialized isolates of Uromyces viciae-fabae and of other species of Uromyces infecting leguminous crops. *Plant Disease* 89(1): 17-22.

Emeran, A., Sillero, J. and Rubiales, D. (2001).Physiological specialisation of Uromyces viciae-fabae. In *Proceedings, 4th European Conference on Grain Legumes, Cracow, Poland*, Vol. 263.

FAOSTAT (2013).World statistics on faba bean. Rome, Italy.: Food and Agriculture Organization of the United Nations. Available at http: //faostat.fao.org/.

Flor, A. H. (1955). host-parasite interaction in flax rust - its genetics and other implications. *Phytopathology* 45: 680-685.

Flor, H. (1942). Inheritance of pathogenicity in Melampsora lini. *Phytopathology* 32: 653-669.

Flor, H. (1946). Genetics of pathogenicity in *Melampsora lini. J. agric. Res* 73(11-12): 335.

Flor, H. (1947). Inheritance of reaction to rust in flax. *J. agric. Res* 74(9-10): 241.

Flor, H. (1956). The complementary genic systems in flax and flax rust. *Advances in genetics* 8: 29-54.

Flor, H. H. (1971). Current status of the gene-for-gene concept. *Annual Review of Phytopathology* 9(1): 275-296.

Gaag, D. and Frinking, H. (1997). Survival, germinability and infectivity of oospores of *Peronospora viciae* f. sp. fabae. *Journal of Phytopathology* 145(4): 153-157.

Gill, T. S. and Davidson, J. A. (2005). A preservation method for Peronospora viciae conidia. *Australasian Plant Pathology* 34(2): 259-260.

Groth, J., Davis, D., Zeyen, R. and Mogen, B. (1983). Ranking of partial resistance to common rust (Puccinia sorghi Schr.) in 30 sweet corn (*Zea mays*) hybrids. *Crop Protection* 2(2): 219-223.

Gurr, G. M., Wratten, S. D. and Luna, J. M. (2003). Multi-function agricultural biodiversity: pest management and other benefits. *Basic and Applied Ecology* 4(2): 107-116.

Habtu, A. and Zadoks, J. (1995). Components of partial resistance in Phaseolus beans against an Ethiopian isolate of bean rust. *Euphytica* 83(2): 95-102.

Harrison, J. (1988). The biology of Botrytis spp. on Vicia beans and chocolate spot disease-a review. *Plant Pathology* 37(2): 168-201.

Hawthorne, W., Bretag, T., Raynes, M., Davidson, J., Kimber, R., Nikandrow, A., Matthews, P. and Paull, J. (2006).Faba Bean Disease Management Strategy for Southern region GRDC 2004.

Hawthorne, W., Kimber, R., Davidson, J., Paull, J., Brand, J. and H., R. (2012).Faba bean disease management strategy -Southern Region. 1-8 (Ed P. Australia). Southern Pulse Bulletin.

Hayes, H. K., Parker, J. H. and Kurtzweil, C. (1920). Genetics of rust resistance in crosses of varieties of Triticum vulgare with varieties of *T. durum* and *T. diococcum. J. Agric. Res.* 19(5): 23-42.

Herath, I., Stoddard, F. and Marshall, D. (2001). Evaluating faba beans for rust resistance using detached leaves. *Euphytica* 117(1): 47-57.

Hiratsuka, N. (1933). Studies on Uromyces fabae and its related species. *Japanese Journal of Botany* 6(3): 329-379.

Hooker, A. (1969). Widely based resistance to rust in corn. *Iowa State Univ Ext Spec Rep.*

Hooker, A. and Saxena, K. (1971). Genetics of disease resistance in plants. *Annual review of genetics* 5(1): 407-424.

Huang, R., Kranz, J. and Welz, H. (1994). Selection of pathotypes of *Erysiphe graminis* f. sp. *hordei* in pure and mixed stands of spring barley. *Plant Pathology* 43(3): 458-470.

Ijaz, U. and Adhikari, K. (2016).Understanding the genetics of seedling rust resistance in faba bean. 21 PBA - Summer Newsletter: Pulse Breeding Australia.

Jensen, N. F. (1952). Intra-varietal diversification in oat breeding. *Agron. J* 44: 30-34.

Jiang, G.-L. (2013). Molecular markers and marker-assisted breeding in plants.

Joseph, M. and Hering, T. (1997). Effects of environment on spore germination and infection by broad bean rust (Uromyces viciae-fabae). *The Journal of Agricultural Science* 128(01): 73-78.

Joshi, A. and Tripathi, H. (2012). Studies on epidemiology of lentil rust (Uromyces viciae fabae). *Indian Phytopathology* 65.

Kaiser, W. J. (1997). Inter-and intranational spread of ascochyta pathogens of chickpea, faba bean, and lentil. *Canadian Journal of Plant Pathology* 19(2): 215-224.

Keen, N. (1990). Gene-for-gene complementarity in plant-pathogen interactions. *Annual review of genetics* 24(1): 447-463.

Keller, B., Feuillet, C. and Messmer, M. (2000).Genetics of disease resistance. In *Mechanisms of resistance to plant diseases*, 101-160: Springer.

Kerkoud, M., Esquibet, M., Plantard, O., Avrillon, M., Guimier, C., Franck, M., Léchappé, J. and Mathis, R. (2007). Identification of Ditylenchus species associated with Fabaceae seeds based on a specific polymerase chain reaction of ribosomal DNA-ITS regions. *European journal of plant pathology* 118(4): 323-332.

Kimber, R. B. E. (2011).Epidemiology and management of cercospora leaf spot (Cercospora zonata) of faba beans (*Vicia faba*). Faculty of Sciences, University of Adelaide.

Kniep, H. (1918). Uber die Bedingungen der Schnallenbildung bei den Basidiomyceten. *Flora* 111: 380-395.

Kohpina, S., Knight, R. and Stoddard, F. (2000). Evaluating faba beans for resistance to ascochyta blight using detached organs. *Animal Production Science* 40(5): 707-713.

Kolmer, J. (1992). Enhanced leaf rust resistance in wheat conditioned by resistance gene pairs with Lr13. *Euphytica* 61(2): 123-130.

Kumari, S. G. and Van-Leur, J. A. G. (2011).Viral diseases infecting faba bean (Vicia fabaL.). In *GRAIN LEGUMES*, Vol. 56, 24-26.

Lannou, C. (2001). Intrapathotype diversity for aggressiveness and pathogen evolution in cultivar mixtures. *Phytopathology* 91(5): 500-510.

Liang, X. (1986). Faba bean diseases in China. *FABIS Newsletter, Faba Bean Information Service, ICARDA* (15): 49-51.

Loegering, W. Q. (1966). The relationship between host and pathogen in stem rust of wheat. *Hereditas* 2: 167-177.

Lucas, J. A., Kendrick, R. E. and Givan, C. V. (1975). Photocontrol of fungal spore germination. *Plant physiology* 56(6): 847-849.

Maier, W., Begerow, D., Weiß, M. and Oberwinkler, F. (2003). Phylogeny of the rust fungi: an approach using nuclear large subunit ribosomal DNA sequences. *Canadian Journal of Botany* 81(1): 12-23.

Makkouk, K., Kumari, S., Hughes, J. d. A., Muniyappa, V. and Kulkarni, N. (2003). Other legumes. In *Virus and Virus-like Diseases of Major Crops in Developing Countries*, 447-476: Springer.

Makkouk, K., Rizkallah, L., Madkour, M., El-Sherbeeny, M., Kumari, S., Amriti, A. and Sohl, M. (1994). Survey of faba bean (*Vicia faba* L.) for viruses in Egypt. *Phytopathologia Mediterranea* 33(3): 207-211.

Martin, T. and Ellingboe, A. (1976). Differences between compatible parasite/host genotypes involving the Pm4 locus of wheat and the corresponding genes in Erysiphe graminis f. sp. tritici [Fungal diseases]. *Phytopathology*.

McNeal, F., Konzak, C. F., Smith, E., Tate, W. and Russell, T. (1971). *A uniform system for recording and processing cereal research data.* USDA-ARS.

Mendgen, K. (1997). *The uredinales.* Springer.

Mendgen, K. and Deising, H. (1993). Infection structures of fungal plant pathogens–a cytological and physiological evaluation. *New Phytologist* 124(2): 193-213.

Mendgen, K. and Hahn, M. (2002). Plant infection and the establishment of fungal biotrophy. *Trends in plant science* 7(8): 352-356.

Moullet, O., Fossati, D., Mascher, F., Guadagnolo, R. and Schori, A. (2009).Use of marker-assisted selection (MAS) for pyramiding two leaf rust resistance genes,(Lr9 and Lr24) in wheat. In *Int. Conf."Conventional and Molecular Breeding of Field and Vegetable Crops*, 24-27.

Moullet, O., Fossati, D., Mascher, F., Guadagnolo, R. and Schori, A. (2010). Use of marker-assisted selection (MAS) for pyramiding leaf rust resistance genes (Lr9, Lr24, Lr22a) in wheat. *Tagungsband der 60. Jahrestagung der Vereinigung der Pflanzenzüchter und Saatgutkaufleute Österreichs 24.-26. November 2009, Raumberg-Gumpenstein*: 143.

Mundt, C. (1991). Probability of mutation of multiple virulence and durability of resistance gene pyramids: Further comments. *Phytopathology* 81(3): 240-242.

Negussie, T., Pretorius, Z. and Bender, C. (2005). Effect of some environmental factors on *in vitro* germination of urediniospores and infection of lentils by rust. *Journal of phytopathology* 153(1): 43-47.

Okamura, H. and Ouchi, M. (2007). resistance to erysiphe fischeri in two populations of senecio vulgaris. *Plant Pathology*.

Pal, A. and Brahmappa, R. (1980). RD and Ullasa, BA 1980. Field resistance of pea germplasm to powdery mildew (*Erysiphe polygoni*) and rust (*Uromyces fabae*). *Plant Disease* 64: 1085-1086.

Parlevliet, J. E. (1997).Durable resistance. In *Resistance of crop plants against fungi.*, 238-253 (Eds H. H. Hartleb, R. Heitefuss and H. H. Hoppe). Jena, Germany: Gustav Fisher.

Pegg, G. and Mence, M. (1970). The biology of Peronospora viciae on pea: laboratory experiments on the effects of temperature, relative humidity and light on the production, germination and infectivity of sporangia. *Annals of Applied Biology* 66(3): 417-428.

Pink, D. A. (2002). Strategies using genes for non-durable disease resistance. *Euphytica* 124(2): 227-236.

Polignano, G., Casulli, F. and Uggenti, P. (1990). Resistance to Uromyces viciae-fabae in Ethiopian and Afghan faba bean entries. *Phytopathologia Mediterranea* 29(3): 135-142.

Raina, S. and Ogihara, Y. (1995). Ribosomal DNA repeat unit polymorphism in 49 Vicia species. *Theoretical and Applied Genetics* 90(3-4): 477-486.

Rashid, K. and Bernier, C. (1986a). The genetics of resistance in Vicia faba to two races of Uromyces viciae-fabae from Manitoba. *Canadian Journal of Plant Pathology* 8(3): 317-322.

Rashid, K. and Bernier, C. (1986b). Selection for slow rusting in faba bean (Vicia faba L.) to Uromyces viciae-fabae. *Crop Protection* 5(3): 218-224.

Rashid, K., Bernier, C. and Conner, R. (1991). Genetics of resistance in faba bean inbred lines to five isolates of Ascochyta fabae. *Canadian Journal of Plant Pathology* 13(3): 218-225.

Rashid, K. Y. and Bernier, C. C. (1991). The effect of rust on yield of faba bean cultivars and slow-rusting populations. *Canadian Journal of Plant Science* 71(4): 967-972.

Rubiales, D. and Trapero-Casas, A. (2002). Occurrence of Didymella fabae, the teleomorph of Ascochyta fabae, on faba bean straw in Spain. *Journal of Phytopathology* 150(3): 146-148.

Ryan, E. (1971). Two methods of infecting Peas systemically with Peronospora pisi, and their application in screening cultivars for resistance. *Irish Journal of Agricultural Research* 10(3): 315-322.

Sache, I. and Vallavieille-Pope, C. d. (1993). Comparison of the wheat brown and yellow rusts for monocyclic sporulation and infection processes, and their polycyclic consequences. *Journal of Phytopathology* 138(1): 55-65.

Scherni, H. and Yang, X. (1995). Interannual variations in wheat rust development in China and the United States in relation to the *El Niño*/Southern Oscillation.

Servin, B., Martin, O. C. and Mézard, M. (2004). Toward a theory of marker-assisted gene pyramiding. *Genetics* 168(1): 513-523.

Sharma, N. (1992). Evaluation of varietal susceptibility in pea to *Erysiphe polygoni*. *Tests of agrochemicals and cultivars*.

Shepherd, K. and Mayo, G. (1972). Genes conferring specific plant disease resistance. *Science* 175(4020): 375-380.

Sillero, J., Fondevilla, S., Davidson, J., Patto, M. V., Warkentin, T., Thomas, J. and Rubiales, D. (2006). Screening techniques and sources of resistance to rusts and mildews in grain legumes. *Euphytica* 147(1-2): 255-272.

Sillero, J., Moreno, M. and Rubiales, D. (2000). Characterization of new sources of resistance to Uromyces viciae-fabae in a germplasm collection of *Vicia faba*. *Plant Pathology* 49(3): 389-395.

Sillero, J. and Rubiales, D. (2002). Histological characterization of resistance to Uromyces viciae-fabae in faba bean. *Phytopathology* 92(3): 294-299.

Sillero, J. C., Rojas-Molina, M. M., Emeran, A. A. and Rubiales, D. (2011). Rust resistance in faba beans. *Grain legumes* 56: 27-28.

Silva, S. R., Rios, G. P. and Silva, S. C. (2001). Influence of genetic resistance and wetness period on infection and lesion development of common bean rust. *Fitopatologia Brasileira* 26(4): 726-731.

Singh, A., Dwivedi, S., Pande, S., Moss, J., Nigam, S. and Sastri, D. (2003). Registration of rust and late leaf spot resistant peanut germplasm lines. *Crop Science* 43(1): 440-441.

Singh, S. and Sokhi, S. (1980). Pathogenic variability in Uromyces viciae-fabae. *Plant Dis* 64: 671-672.

Smithson, J. B. and Lenne, J. M. (1996). Varietal mixtures: a viable strategy for sustainable productivity in subsistence agriculture. *Annals of Applied Biology* 128(1): 127-158.

Stakman, E. C. and Levine, M. N. (1922). The determination of biological forms of Puccinia graminis on Triticum spp. *Min. Agric. Exp. Stn. Bull.*: 8-10.

Stakman, E. C., Stewart, D. M. and Loegering, W. Q. (1962). Identification of physiologic races of Puccinia graminis var. tritici. *Identification of physiologic races of Puccinia graminis var. tritici.*

Staskawicz, B. J., Ausubel, F. M., Baker, B. J., Ellis, J. G. and Jones, J. D. (1995). Molecular genetics of plant disease resistance. *SCIENCE-NEW YORK THEN WASHINGTON-*: 661-661.

Stavely, J. (1984). Pathogenic specialization in Uromyces phaseoli in the United States and rust resistance in beans. *Plant Disease* 68(2): 95-99.

Stegmark, R. (1991). Comparison of different inoculation techniques to screen resistance of pea lines to downy mildew. *Journal of Phytopathology* 133(3): 209-215.

Stoddard, F., Nicholas, A., Rubiales, D., Thomas, J. and Villegas-Fernández, A. (2010). Integrated pest management in faba bean. *Field crops research* 115(3): 308-318.

Stonehouse, J. (1994). Assessment of Andean bean diseases using visual keys. *Plant Pathology* 43(3): 519-527.

Stuthman, D., Leonard, K. and Miller-Garvin, J. (2007). Breeding crops for durable resistance to disease. *Advances in agronomy* 95: 319-367.

Subbotin, S. A., Madani, M., Krall, E., Sturhan, D. and Moens, M. (2005). Molecular diagnostics, taxonomy, and phylogeny of the stem nematode Ditylenchus dipsaci species complex based on the sequences of the internal transcribed spacer-rDNA. *Phytopathology* 95(11): 1308-1315.

Subrahmanyam, P., McDonald, D., Waliyar, F., Reddy, L., Nigam, S., Gibbons, R., Rao, V. R., Singh, A., Pande, S. and Reddy, P. (1995). *Screening methods and sources of resistance to rust and late leaf spot of groundnut. Information Bulletin no. 47.* International Crops Research Institute for the Semi-Arid Tropics.

Taylor, P. N., Lewis, B. G. and Matthews, P. (1989). Pathotypes of Peronospora viciae in Britain. *Journal of Phytopathology* 127(2): 100-106.

Thomas, J. and Kenyon, D. (2004).Evaluating resistance to downy mildew (Peronospora viciae) in field peas (*Pisum sativum* L.) and field beans (Vicia fabae L.). In *5th European Conference on Grain Legumes*, 81-82 Dijon, France.

Thomas, J. E. (1992).Evaluation of cultivar resistance in field pea (*Pisum sativum* L.) and field bean (Vicia faba L.) to some soil soil borne and foliar fungal pathogens. In *Proc. 1st European Conference on Grain Legumes*, 77-78 Angers, France.

Thomas, J. E. and Camp, J. A. (1997).Assessment of Resistance to Downy Mildew (Peronospora viciae) in seedlings of field bean cultivars. In *Proc. International Food Legume Research Conference III, Abstracts*, pp. 158 Adelaide, Australia.

Timmerman-Vaughan, G., Frew, T. and Weeden, N. (2000). Characterization and linkage mapping of R-gene analogous DNA sequences in pea (*Pisum sativum* L.). *Theoretical and Applied Genetics* 101(1-2): 241-247.

Tivoli, B., Baranger, A., Avila, C. M., Banniza, S., Barbetti, M., Chen, W., Davidson, J., Lindeck, K., Kharrat, M. and Rubiales, D. (2006). Screening techniques and sources of resistance to foliar diseases caused by major necrotrophic fungi in grain legumes. *Euphytica* 147(1-2): 223-253.

Tivoli, B., Berthelem, D., Le Guen, J. and Onfroy, C. (1986). Comparison of some methods for evaluation of reaction of different winter faba bean genotypes to Botrytis fabae. *Fabis Newsletter* 16: 46-51.

Tiwari, K., Penner, G., Warkentin, T. and Rashid, K. (1997). Pathogenic variation in Erysiphe pisi, the causal organism of powdery mildew of pea. *Canadian Journal of Plant Pathology* 19(3): 267-271.

Tiwari, R., Kumar, Y., Saharan, M. and Mishra, B. (2008). Marker assisted approach for incorporating durable rust resistance in popular Indian wheat cultivars.

Torres, A., Weeden, N. and Martin, A. (1993). Linkage among isozyme, RFLP and RAPD markers in Vicia faba. *Theoretical and Applied Genetics* 85(8): 937-945.

Torres, A. M., Roman, B., Avila, C. M., Satovic, Z., Rubiales, D., Sillero, J. C., Cubero, J. I. and Moreno, M. T. (2006). Faba bean breeding for resistance against biotic stresses: towards application of marker technology. *Euphytica* 147(1-2): 67-80.

Toxopeus, H. (1956). Reflections on the origin of new physiologic races in Phytophthora infestans and the breeding for resistance in potatoes. *Euphytica*

5(3): 221-237.

Tyagi, M. (1999). Generation mean analysis of adult plant resistance to powdery mildew and rust disease in field pea. *PLANT DISEASE RESEARCH* 14(2): 171-174.

Tyagi, M. and Srivastava, C. (1999). Inheritance of powdery mildew and rust resistance in pea. *Annals of Biology* 15: 13-16.

Vaid, A. and Tyagi, P. (1997). Genetics of powdery mildew resistance in pea. *Euphytica* 96(2): 203-206.

Vanderplank, J. (1978). *Genetic and molecular basis of plant pathogenesis.* Springer-Verlag.

Vanderplank, J. E. (1963). Plant diseases: epidemics and control. *Plant diseases: epidemics and control.*

Voegele, R. T. (2006). Uromyces fabae: development, metabolism, and interactions with its host Vicia faba. *FEMS microbiology letters* 259(2): 165-173.

Walker, J. and Stahmann, M. (1955). Chemical nature of disease resistance in plants. *Annual Review of Plant Physiology* 6(1): 351-366.

Walker, J. C. (1924). On the nature of disease resistance in plants. *Trans. Wis. Acad. Sci. Arts, letters* 2: 225-247.

Wilcoxson, R. (1981). Genetics of slow rusting in cereals. *Phytopathology* 71(9): 989-993.

Wolfe, M. (1985). The current status and prospects of multiline cultivars and variety mixtures for disease resistance. *Annual Review of Phytopathology* 23(1): 251-273.

Wolfe, M. S. (2000). Crop strength through diversity. *Nature* 406(6797): 681-682.

Wolfe, M. S. and Barrett, J. (1980). Can we lead the pathogen astray? *Plant Disease* 64(2).

Xu, Y. (2010). *Molecular plant breeding.* CABI.

Zinner, H., Knippel, E., Tech, H.-J., Osten, R., Erfurt, G. and Schnell, M. (2008). survival, germinability and infectivity of oospores of *Peronospora viciae* f. sp. *fabae. Journal of Phytopathology.*

2018, *Climate Risks Management: Sustainable Pulse Production*
Editors: A K Srivastava and Yogranjan
Published by: **ASTRAL INTERNATIONAL PVT. LTD., NEW DELHI**   Pages 249–255

Chapter 12

Gene Action and Biochemical Genetics of Powdery Mildew (*Erysiphe polygoni* D.C.) Resistance in Mungbean [*Vigna radiata* (L.) Wilczek]

Nidhi Pathak[1], Manish K. Mishra[2] and M.N. Singh[3]

[1]*Department of Plant Breeding and Genetics,*
JNKVV College of Agriculture, Jabalpur, M.P.
[2]*AICRP-Sesame, JNKVV, College of Agriculture,*
Tikamgarh, M.P.
[3]*Department of Genetics and Plant Breeding,*
Institute of Agricultural Sciences, BHU, Varanasi, U.P.

ABSTRACT

Powdery mildew of mungbean is caused by Erysiphe polygoni (D.C.). It is the most devastating fungal disease of rabi mungbean in South-East Asia as well as late kharif sown crop in Northern part of the country and may cause yield loss up to 20-40 per cent. The mode of inheritance of powdery mildew resistance in mungbean was observed to be inconsistence and varied from cross to cross. Under most of the case, powdery mildew is under the control of monogenic dominant gene or two major dominant genes (Pm1, Pm2). On the contrary, the resistance to powdery mildew in mungbean is also reported to be governed by a single recessive gene. Inheritance of powdery mildew resistance was mostly governed by additive and additive × additive gene interaction. Powdery mildew resistant genotypes contained lower levels of reducing, non-reducing and total sugars than susceptible ones. However, resistant genotypes contained high level of total phenols, peroxidase and polyphenol oxidase activities as compared to susceptible genotype.

Keywords: *Mungbean, Powdery mildew, Gene action, QTLs and Total phenols.*

Introduction

Food legumes, being the major sources of protein, minerals and vitamins, are of prime importance in vegetarian diet as well as animal feed in our country (Salunkhe *et al.*, 1985). On an average, pulses contain 20-25 per cent protein in their dry seeds, which is almost 2.5-3.0 times more than the value normally found in cereals. Thus, the food legumes ensure nutritional security to the poor masses of the people in the country (Chaturvedi and Ali, 2002). The commonly grown pulses in India are chickpea, pigeonpea, cowpea, fieldpea, greengram, urdbean, lentil, mothbean, and frenchbean. Among these pulses, greengram or mungbean [*Vigna radiata* (L.) Wilczek] is an ancient and well known leguminous crop of Asia. India is the largest producer of mungbean in the world, followed by China and Myanmar. The crop is fetching wider acceptance in India, because of its short life cycle fitting well in intercropping with most prominent rice and wheat-based cropping system, besides being a potential nitrogen fixer. With the advances and affordability of genome sequencing technologies, generation of genomic resources of this valuable pulse crop would certainly be of use for diverse breeding activities including studies on crop domestication and species divergence. The genus *Vigna* is pantropical and comprises of about 150 species, most of which are found in Africa and Asia. Out of these, only seven (five Asiatic and two African) species are cultivated as pulse crops of which Asiatic group consists of Mungbean, Urdbean (*V. mungo*), Mothbean (*V. aconitifolia*), Ricebean (*V. umbellata*) and Adzukibean (*V. angularis*). The *Vigna* species are adapted to a wide range of agro-climatic conditions and consequently widely distributed in the tropics and sub-tropics. They may also be grown on poor soils without supplemental nitrogen and thus being advantageous to sustainable agriculture (Anishetty and Moss, 1987). Mungbean is thought to be native of India and Central Asia. According to Vavilov (1926), mungbean originated in India and *V. sublobata* (Roxb.) which occurs in a wild state in the sub-Himalayan region, is considered as the progenitor of mungbean. Mungbean is a strictly self-pollinated crop and belongs to family *Leguminosae* and sub-family *papilionaceae* having somatic chromosome number of 2n=22. The mungbean proteins are quite simple, highly digestible, free from flatulent effects and consequently, recommended as a medicinal diet for the patient. Mungbean is a short duration, annual, erect or sub-erect plant, sometimes slightly twining at the tips. The leaves are alternate, trifoliate, ovate and dark or light green. The inflorescence is an axillary raceme bearing yellow flowers and the keel is spirally coiled with a horn-like appendage (Rachie and Roberts, 1974). Pods are long, slender, containing 6-10 globose, round or elongated seeds which are mostly green but sometimes yellow, tawny brown, black or mottled in colour. The hilum is flat having epigeal germination (Bailey, 1970). With the increase in irrigation facilities, the area under mungbean cultivation has registered an increasing trend in recent years. For example, during 1971-72, the area, production and productivity of mungbean were 1.84 million hectares, 0.56 million tonnes and 306 kg/ha, while the year 2011-2012 recorded the area, production and productivity of mungbean as 3.43 million hectares, 1.71 million tonnes and 498 kg/ha, respectively (Anonymous, 2013). The lower productivity in green gram is mainly attributed to low genetic yield potentiality, indeterminate growth habit, canopy architecture, low partitioning efficiency, cultivation in marginal land including biotic and abiotic stresses. Among biotic stresses, powdery mildew, mungbean yellow mosaic virus (MYMV) and *Cercospora* leaf spot (CLS) are the major diseases of greengram. Powdery mildew caused by *Erysiphe polygoni* D.C. is the most devastating fungal disease of rabi mungbean in South-East Asia (Park and Yang, 1978) as well as late *kharif* sown crop in Northern part of the country. The powdery mildew disease usually appears on 35-40 days old crop (Khare *et al.*, 1998), during flowering and pod formation stage (Arjunan *et al.*, 1976). The fungus attacks on all parts of the plant except roots. The initial symptoms are faint, slightly dark areas developing over the leaf, later turning into small, white powdery spots. These spots enlarge, coalesce and develop into a complete coating by a white to dirty-white powdery growth consisting of mycelium and conidia. In epidemic form, the fungus covers all parts of the plant with white powdery growth, thereby adversely affecting the photosynthetic activity of the plant. Consequently, defoliation takes place and pods are not formed and if formed they bear subnormal seeds (Singh 1980), which in turn, reducing the yield as well as market price, causing enormous economic loss to the farmers. The extent of yield losses due to this disease in mungbean are reported to the tune of 20-40 per cent (Fernandez and Shanmugasundaram, 1988) and even 100 per cent, when the disease occurs at the seedling stage (Reddy *et al.*, 1994) and also reduces the yield up to 21 per cent when all the leaves are covered with mildew at flowering time (Soria and Quebral, 1973).

Gene Action in Powdery Mildew

The mode of inheritance of powdery mildew resistance in mungbean was observed to be inconsistent and varied from cross to cross. Anonymous (1980) reported that resistance to powdery mildew in mungbean is under the control of monogenic dominant gene, whereas Reddy *et al.* (1994) and Reddy (2009) observed the involvement of two major dominant genes (Pm1, Pm2) in donor genotypes for effecting résistance. On the contrary, Kalia and Sharma (1988), reported that the resistance to powdery mildew in mungbean is governed by a single recessive gene. In addition, Yohe and Poehlman, (1975), Sorajjapinun *et al.* (2005) and Kasettranan *et al.* (2009) observed the prevalence of additive gene effects for governing resistance, while Gawande and Patil (2003) noticed that resistance to powdery mildew is governed by more than one gene with both additive and dominance gene actions. Hegde *et al.* (1994) observed the major role of additive and additive × additive gene interaction for the inheritance of powdery mildew resistance. Involvement of three QTLs (Young *et al.*, 1993) and one QTL (Chaitieng *et al.*, 2003 and Humphry *et al.*, 2003) for powdery mildew resistance in mungbean was also on record. These reports are usually based on studies involving few parents with limited F_2 population. The study based on large number of parents/F_2 population is expected to provide more reliable information on these aspects in mungbean.

Biochemical Genetics of Disease Resistance

Biochemical constituents play an important role in the process of development of high yielding and disease resistance genotypes in any crop plants. Rathi *et al.* (1986) studied changes in different biochemical parameters like total phenolic content, protein pattern, polyphenol oxidase, peroxidase and isozymes of peroxidase in sterility mosaic resistant and susceptible pigeonpea varieties at different growth stages both under inoculated and uninoculated conditions. Resistant variety was characterized by the presence of specific isoperoxidase and proteins but only little difference was recorded between resistant and susceptible varieties with respect to preformed or induced total phenolics and peroxidase activity. The activity of polyphenol oxidase increased substantially in susceptible variety following infection. Kalia and Sharma (1988) observed that the resistant cultivars contained higher levels of phenolics and phenol- oxidizing enzymes than the susceptible ones. A further study of their F_1s, F_2s and backcross progenies suggested a higher heritability for all biochemical traits. The correlation coefficients between the biochemical parameters and the disease index were also high. Both additive (d) and dominant (h) components were found to contribute to the inheritance of these constituents. Bhattacharya and Shukla (2000) observed activities of monophenolase, o-diphenolase, catalase, phenol concentrations and powdery mildew severity in leaves at different growth stages. Levels of phenols were similar in resistance and healthy plants of susceptible cultivar along with higher activities of o-diphenolase and catalase. The higher accumulations of phenolic compounds in diseased plants were due to non-conversion to their respective quinones. Garain *et al.* (2003) observed only CLS and MYMV infection during pre-*kharif*, whereas powdery mildew severity was more during post-*kharif* season. Fifteen mungbean cultivars were selected for resistance to the disease. Higher phenol content was observed in resistant than susceptible cultivars. A positive correlation between stomatal frequency and susceptibility to CLS was observed. Gawande and Patil (2004) observed that powdery mildew resistant genotypes contained lower levels of reducing, non-reducing and total sugars than susceptible ones. However, higher levels of potash were found in resistant as compared to susceptible genotypes. The levels of sugars and potash were found to be reduced after infection. Gawande and patil (2006) observed that the powdery mildew resistant genotypes show higher total phenols, peroxidase and polyphenol oxidase activities as compared to susceptible genotypes. Further, the degree of increase in peroxidase and polyphenol oxidase activities in resistant genotypes after infection was found higher than in the susceptible one. The level of total phenol and total sugar were found to be more in resistant cultivar than susceptible one against anthracnose disease in grapes (Bhavani and Jindal, 2001).

Promising Genotypes against Powdery Mildew

A number of powdery mildew resistant genotypes have been reported by several authors. Anonymous (1974) screened 12 Indian greengram varieties against powdery mildew. However, 11 genotypes *viz.*,M-163,P1363152, PLM 187, MT 32/2B,

PLM 224, PLM 731, PLM 857,PLM 863,PLM 873, PLM 944 and PLM 4060 showed resistant reaction against powdery mildew. Patil and Moghe (1993) screened 10 varieties of greengram against powdery mildew and observed that four varieties *viz.*, AKM-5801, BM-M4, DM-85188 and TARM-2 were free from the disease. Singh *et al.* (2007) screened 215 mungbean germplasm accessions against powdery mildew under natural epiphytotic conditions. Of these, only DMG-1030 exhibited resistant reaction while, DMG 1045, DMG 1130-2, LM 164, and LM 421 and CO 4 showed moderately resistant reaction to powdery mildew. Rana *et al.* (2013) analyzed 57 accessions of peas that showed resistant reaction for three consecutive years in field screening but only 14 accessions originating from 10 countries showed resistant reaction in laboratory screening against the four most prevalent isolates of *E. pisi* collected from different places in the area of experiment. Germplasm lines showed both complete and incomplete levels of resistance and variable reactions to different isolates. There was sufficient genetic diversity and agronomic superiority in the resistant accessions *e.g.* EC598655, EC598878, EC598704, IC278261, and IC218988, which may serve as useful genetic material to plant breeders for breeding high yielding and powdery mildew resistance varieties in pea.

REFERENCES

Anishetty M and Moss H, (1987). *Vigna* genetic resources current status and future plans. *Proc. of the Second Intl. Symp.*, 13-18.

Anonymous, (1974). Mungbean, AVRDC progress report, pp. 28-44.

Anonymous, (1980). AVRDC Progress Report, Shanhua, Taiwan.

Anonymous, (2013). Coordinator report presented by PC (MULLaRP) during mungbean/urdbean at TNAU. Coimbatore.

Arjunan G, Vidyasekharan P and Kolandaisawmy S, (1976). How to combat disease of greengram. *Farmer and Parliament*, 12: 17-18.

Bailey LH, (1970). Manual of Cultivated Plants, (Rev.) *MacMillan.* New York,1116 p.

Bhattacharya A, and Shukla P, (2000). Changes in activity of some phenol related enzvme in fieldpea leaves infected with powdery mildew under rainfed and irrigated condition. *Ind. J. Agric. Res.,* 34(3): 147-151.

Bhavani SAV and Jindal PC, (2001). Biochemical resistance of grape genotypes against anthracnose. *Indian J Agric. Res.* 35(1): 44-47.

Chaitieng B, Laosuwan P and Wongkaew S, (2003). Inheritance of powdery mildew resistance in mungbean (*Vigna radiata* (L.) Wilczek). *J. Agric. Sci.,* 36(1): 73-79.

Chaturvedi SK and Ali M, (2002). Poor mans meat need fresh fillip. *The Hindu Survey I. Agric.,* p. 63.

Fernandez GCJ and Shanmugasundaram S, (1988). The AVRDC mungbean improvement program: the past, present and future. In: McLean BT (ed) Mungbean: proceeding of the second international symposium. Asian Vegetable Research and Development Center, Taiwan, pp. 58–70.

Garain PK, Dutta S, Roy S, Bhattacharya PM, and Gayen P, (2003). Remove from marked records response of mungbean germplasm against some important foliar diseases in pre and post *kharif* season under *terai* agroecological region of West Bengal. *J. of Mycopathol. Res.* 2(41): 201-203.

Gawande VL and Patil JV, (2003). Genetics of powdery mildew (*Erysiphe polygoni* D. C.) resistance in mungbean [*Vigna radiata* (L.) Wilczek]. *Crop Prot.*, 22: 567-571.

Gawande VL and Patil JV, (2006). Genetics of biochemical traits associated with powdery mildew (*Erysiphe polygoni* D.C.) resistance in mungbean [*Vigna radiata* (L.) Wilczek]. *J. Genet.* and Breed., 60: 59-66.

Gawande VL and Patil JV. (2004). Biochemical genetics of powdery mildew (*Erysiphe polygoni* D.C.) resistance in mungbean. *SABRAO J. Breed. Genet.*, 36(2): 63-72.

Hegde VS, Parameshwarappa R and Goud JV, (1994). Genetics of some quantitative characters in mungbean [*Vigna radiata* (L.) wilczek]. *Mysore J. Agric. Sci.*, 28(3): 204-208.

Humphry ME, Magner T, McIntyre CL, Aitken EA and Liu CJ (2003). Identification of a major locus conferring resistance to powdery mildew (*Erysiphe polygoni* D.C.) in mungbean [*Vigna radiata* (L.) Wilczek] by QTL analysis. *Genome*, 46(5): 738-744.

Kalia P and Sharma SK, (1988). Biochemical genetics of powdery mildew resistance in pea. *Theor Appl Genet*, 76: 795-799.

Kasettranan W, Somta P and Srinives P, (2009). Genetics of the resistance to powdery mildew disease in mungbean [*Vigna radiata* (L.) Wilczek]. *J. Crop Sci. Biotech.* (March) 12(1): 37 - 42.

Khare N, Lankpale N and Agarwal KC, (1998). Epidemology of powdery mildew of mungbean in Chattisgarh region of Madhya Pradesh. *J. Mycol. Pl. Path.*, 28: 5-10.

Park HG and Yang CN, (1978). The mungbean breeding programme at the Asian Vegetable Research and Development Centre. First Intl. Symp. *Proc. Univ. Philippines*, 1977, pp. 214–21.

Patil BG and Mohge PG, (1993). Reaction of mung, urd, rajma and kulti varieties to powdery mildew. *J. Maharastra Agric. Univ.*, 18: 467-468.

Rachie KO and Roberts LM, (1974). Grain legumes of the lowland tropics. *Adv. Agron*, 26: 129–132.

Rana JC, Banyal DK, Sharma KD, Sharma MK, Gupta SK and Yadav SK, (2013). Screening of pea germplasm for resistance to powdery mildew. *Euphyt.*, 189: 271–282.

Rathi YPS, Bhatt A and Singh US, (1986). Biochemical changes in pigeonpea (*Cajanus cajan* (L.) Millsp.) leaves in relation to resistance against sterility mosaic disease. *J. Biosci.*, 10(4): 467-474.

Reddy KS, (2009). Identification and inheritance of a new gene for powdery mildew resistance in mungbean [*Vigna radiata* (L.) Wilczek]. *Plant Breed.*, 128: 521-523.

Reddy KS, Pawar SE and Bhatia CR, (1994). Inheritance of powdery mildew (*Erysiphe polygon* D. C.) resistance in mung [*Vigna radiata* (L.) Wilczek]. *Theor. Appl. Genet.*, 88: 945-948.

Salunkhe DK, Kadam SS and Chavan Jk, (1985). Chemical composition, In: (eds). Post-harvest Biotechnology of Food legumes. CRC press Inc: Boca Rabon, FL. pp29-52.

Singh DP, (1980). Inheritance of resistance to yellow mosaic virus in blackgram [*Vigna mungo* (L.) Hepper]. *Theor. Appl. Genet.* 57: 233–235.

Singh RA, Kumar S, Gupta S and Singh BB, (2007). Powdery mildew resistant sources in mungbean and their reaction to yellow mosaic disease. *J. of Food Leg.*, 20(1): 98-99.

Sorajjapinun W, Rewthongchum S, Koizumi M and Srinives P, (2005). Quantitative inheritance of resistance to powdery mildew disease in mungbean [*Vigna radiata* (L.) Wilczek]. *SABRAO J. Breed. Genet.*, 37(2): 91-96.

Soria JA and Quebral FC, (1973). Occurrence and development of powdery mildew on mungbean. *Philippine Agric.* 57: 158-177.

Vavilov NI, (1926). Studies on the origin of cultivated plants. *Trudy Nyuro Bike Bot.,* 16: 139-248.

Yohe JM and Poehlman JM, (1975). Regressions, correlations and combining ability in mungbean. *Tropic. Agric.*, 52: 343-352.

Young ND, Danesh D, Menancio-Hautea D and Kumar L, (1993). Mapping oligogenic resistance to powdery mildew in mungbean with RFLPs. *Theor. and Appl. Genet.*, 87(1): 243-249.

Section E

Extension Approaches, Government Planning and Policies for Enhancing Pulse Production

2018, *Climate Risks Management: Sustainable Pulse Production*
Editors: A K Srivastava and Yogranjan
Published by: **ASTRAL INTERNATIONAL PVT. LTD., NEW DELHI**   *Pages 257–274*

Chapter 13

Agriculture Knowledge Information System for Effective Dissemination of Pulse Production Technology

Kamini Bisht[1], Prasanta Mishra[2], and A.A. Raut[3]

[1]*College of Agriculture, Tikamgarh, JNKVV, Madhya Pradesh*
[2]*Associate Professor, AAU, Jorhat, Assam*
[3]*ICAR-Agricultural Technology Application Research,*
Jabalpur, Madhya Pradesh

ABSTRACT

Farmers perceive pulses as having a lower cost benefit ratio vis-à-vis cereals and other crops. This mindset needs to be changed by sensitizing farmers that pulses can be a profitable venture, with use of modern technology. Besides improved varieties, fertilizer and inputs, the key to agricultural development lies in the mind, heart and hands of farmers. To assist the farmers in these changing contexts, new strategies and innovative solutions are urgently required. Technological interventions help to increase agricultural productivity, cut production costs, and lower consumer prices but the benefits depend on how the technology is transferred. Potentially all actors in the public and private sectors are involved in the creation, diffusion, adaptation and use of all types of knowledge relevant to agricultural production. Policies decision has been impacted the pulse production. The agricultural Innovation System should be adopted keeping farmer as a central focus and it may be suitable for accelerating pulses production in India.

Keywords: Adoption, Agricultural innovation, Technological intervention, ICT tools.

Agriculture is the backbone of Indian economy from ancient times. With the vision of food sufficiency, Government of India kept agriculture as the primary focus area for first five year plan. However, Green revolution in late sixties proved to be a major boost in agriculture production. It gave rise to productivity of different crops. Most of our food production strategies have been largely confined to the cereals, as these crops constitute the staple food of the majority of Indian population. This has relegated the grain legumes to further lower priority. Having realized this problem in recent past, the Government of India laid great emphasis on increasing the pulse production and considers it as one of the national priority. In the existing cropping systems, pulses fit well due to its short duration, low input, minimum care required and drought tolerant nature. The existing level of technology employed in terms of agronomic practices, time of operation, and input use are essential for achieving higher yield and return. Diffusion and adoption of modern technologies, high yielding varieties, dedicated efforts of farmers, extension personnel and scientists and also programmatic support of Central and State Governments have all contributed significantly from 50.82 million tons in 1950-51 to land mark achievement of 259.29 million tons of food production in 2010-11.

Adoption of crop technologies is essential feature of agriculture-led development strategies. "Key to agricultural development lies in the mind, heart and hands of farmers". Farmers are ultimate decision makers about an innovation introduced in their systems. They are heterogeneous and differ in various characteristics like education, experience in cultivation, farm size, annual income, media participation, extension contact, economic motivation, scientific orientation *etc*. Their receptivity to different agricultural innovativeness will vary depending on their personal socio-economic and psychological attributes. Given the range of its agro-ecological setting and produces, Indian agriculture is faced with a great diversity of needs, opportunities and prospects.

The National seminar of agricultural extension 2009 background note states that sustaining growth rate and achieving the required food grain production of 320 million tons by 2025 would be a herculean task considering some of the challenges like non-expanding land, depleting soil and water resources, adverse impact of climate change, rising cost of production, diminishing agriculture labour availability and farmers' reduced interest in agriculture (NSAE, 2009). If India is to respond successfully to these challenges and also to achieve accelerated growth there is needs to have greater use of modern information and communication technology among researchers, extension personnel, farmers and other stakeholders. Further, the agricultural extension requires paradigm shift from top-down, blanket recommendation of technological packages towards providing producers with the knowledge and understanding with which they solve their own location specific problems. Continuous two-way interaction among the farmers, agricultural scientists and extension personnel is the most critical missing component of agricultural extension (Chatterjee and Prabhakar, 2009). To assist the farmers in these changing contexts, new strategies and innovative solutions are urgently required which in turn will require technological support.

Pulse Importance and Production Scenario

Pulses have been cultivated since time immemorial and were among the first crops cultivated date back during the bronze age. Peas, chickpeas, lentils and faba beans were cultivated as far back as 11,000 years ago. The ancient Egyptians considered lentils to be an emblem of life and had temples dedicated to them. Hippocrates prescribed lentils to patients suffering from liver ailments. Four of the most distinguished Roman families were named after pulses: Fabius (faba bean), Lentulus (lentil), Piso (pea) and Cicero (chickpea). In India, pulses continue to be majour source of dietary protein and consumed in traditional forms *viz., dals, sambar, idli, vada, laddu* and *pakoras.*

In India pulses are cultivated on marginal lands under rain fed conditions. Only 15 per cent of the area under pulses has assured irrigation. The major pulse producing states are Madhya Pradesh, Uttar Pradesh, Maharashtra, Rajasthan, Karnataka and Andhra Pradesh which together share 80 per cent of the total area and 81.1 per cent of the total production (Expert Committee Report on Pulses, 2000).

Pulse Availability vis-à-vis Nutritional Sustainability

Pulses are an important and relatively inexpensive source of protein for human and animal nutrition in India. Their importance as builder and restorer of soil fertility in arid areas is well recognized. Pulses meet most of the protein needs of the Indian population. Pulses are also important for agriculture and livestock farming too. The hay and straws of the pulses are rich in amino acids and makes valuable cattle feed. Moreover, almost all pulses, being leguminous plants, their cultivation improve soil fertility by fixing nitrogen into the soil. Thus, the nitrogen depleted from the soil by cereals cultivation can be replenished by proper crop rotations with pulses cultivation. Thus pulses play an important role in food and nutritional security and environmental sustainability.

The per capita availability of pulses, the major source of protein for Indians, has fallen by less than half since independence. The average per capita availability of wheat and rice together has increased appreciably during this period. In recent years, it has begun to sound irrelevant. Pulses have the dubious distinction of being the single major food item that has declined sharply in terms of net availability. This is the consequence of rising population and stagnant pulse production over the past three decades.

Initiatives to Enhance Pulse Production

In an estimate by the World Health Organisation, every Indian needs 80 gram of pulses every day to meet his or her protein needs. This implies that by the year 2018, India will require 38 million tonnes of pulses to meet its population's nutritional requirements. In the past four years, there has been significant increase in pulse consumption averaging around 50 g due to somewhat higher production and larger imports. The increasing mismatch between production and consumption of pulses has resulted in larger imports of pulses in recent year with imports in 2012-13 (Apr-Mar). Despite being world's largest producer of pulses, only small exports of pulses are taking place from India, both because of restrictions on exports and the high domestic demand. (NCAP, 2014).

Research and Extension Noticing the continuous decline in pulses productivity, an All India Coordinated Pulses Research Project (AICRP) was initiated in 1965 to undertake a nation-wide research effort on pulses. Subsequently, launching of Intensive Pulse Development Programme (IPDP) during 4th Five Year Plan (1969-74), Centrally Sponsored National Pulses Development Programme (NPDP) during 7th FYP (1985-90), Integrated Scheme of Oil Seeds, Pulses and Maize (ISOPOM) during 10th FYP (2002-07) and National Food Security Mission (NFSM) during 11th FYP (2007-12) have been milestones in improving the pulses production in the country.

The research on pulse at the national level is undertaken by the Indian Institute of Pulses Research (IIPR, Kanpur) and also through the centres of All India Coordinated Research Projects on Pulses located at various State Agricultural Universities. The IIPR, Kanpur has evolved a number of varieties of pulses. Similarly under the AICRPs, location-specific improved production technologies and new varieties are developed by different centres attached to the State Agricultural Universities. Many State Departments, public sector organisations and international organisations have also been supporting research on pulses through their institutes.

The National Pulse Development Project (NPDP) launched by the Government of India during the 7th five year plan and continued during the 8th five year plan. Later in 1990, the NPDP had been brought under the purview of Technology Mission on Oilseeds and Pulses (TMOP). The major thrust area of the project were to provide support on seed production (breeder, foundation and certified), distribution of seed mini-kits, organization of frontline/block demonstrations on improved varieties and technology, integrated pest management (IPM), farmers' training, distribution of critical input like biofertilisers, sprinkler sets, weedicides, gypsum/pyrite and farm implements.

In order to give the much needed fillip to pulse production, the government has included pulses in the NFSM (along with wheat and rice) since the launch of NFSM in October 2007. The mission provides assistance for laying cluster demonstrations, distribution of Certified Seed, Micro-Nutrients, Plant and soil protection material, liming of acidic soils, improved implements and machinery, introduction of IPM, INM and capacity building of the farmers.

Pulse Panchayat is an integrated approach in establishing a sustainable production, value addition and marketing system by M.S. Swaminathan Research Foundation. The initiative implemented by a Farmer Producer Company in Tamil Nadu, is moving towards achieving self sufficiency in pulse production. The Pulse Biopark, based on the value chain analysis, is a pilot project, implemented by the Illuppur Agriculture Producer Company Limited (IAPCL) with the technical support of MSSRF and other stakeholders.

The Pulse Biopark has significantly enhanced the pulse cultivating farmers share in the consumer rupee. This has also reduced the post harvest losses significantly. These approaches will bridge the supply-demand gap and have significant importance in Grain Legume research and development.

The Pulse Panchayat movement has demonstrated that innovative approaches with knowledge management enhancement, through multi-stakeholder platforms and policy making networks, are key to achieving self sufficiency in pulse production. For example, grow more food campaign is an initiative supported by Tata Chemicals and Rallis-India to promote the cultivation and availability of pulses in India. The Grow More Food Campaign aims to bring together those communities associated with the production of pulses, including the academic, agricultural and government sector. Documenting such existing and innovative value chains for different pulses, will enable Scaling up, Sales and Sustaining (3 S) the Pulse production.

Determinants of Adoption of Technology toward Higher Production

The adoption of agricultural innovation in developing countries attracts considerable attention because it can provide the basis for increasing production and income. Small scale farmers' decisions to adopt or reject agricultural technologies depend on their objectives and constraints as well as cost and benefit accruing to it (Million and Belay, 2004). Hence, farmers will adopt only technologies that suit their needs. Several factors influence adoption of agricultural technologies.

In a study on Technological gap in adoption of the improved cultivation practices by the soybean growers, Kumar (2009) reported that technological gap among farmers in respect of the recommended cultivation practices were application of zinc sulphate, seed inoculation with rhizobium culture, seed treatment with captan or thiram, potassic fertilizer and plant protection measures. Further, Varma (2011) reported that the important package of practices on which they were having high technological gap were; precaution in using seed and sowing management followed by plant protection management, fertilizer management, weed management, storage management harvesting management, field preparation, and irrigation management.

As far as adoption level regarding improved pulse production technology was considered, the respondents sequentially reported selection of suitable high yielding varieties, followed by time of harvesting, use of culture, control of insect/pest, disease management, precaution before storage, application of fertilizers, seed treatment, selection of land, crop rotation, time of sowing/seed rate, land preparation, depth of sowing/spacing, whereas weed management, irrigation management and sowing method were having less adoption level. Regarding change of productivity after conducting the Front Line Demonstration (FLD), most of the FLD beneficiaries were having low productivity change after adopting the new production technologies. The status of the beneficiaries regarding level of education, size of land holding, area under pulses crops, occupation, risk preference, extension participation, mass media contact, extent of knowledge and contact with development agencies had significant association in respect of adoption of pulses production technologies (Patidar, 2011).

Haque *et al.* (2014) mentioned that diseases and pest infestation, lack of good quality seed, lack of knowledge about improved technologies were the major constraints to mungbean cultivation. The adoption of any technology/innovation in cultivation practices depends on various factors such as awareness about practices, extent of change agencies efforts, complexity of practices, timely availability of inputs, characteristics of farmers *etc.* Under situational, technological, economical, market-related, extension and institutional constraints were concerned. Varma (2011) reported the constraints mentioned by the respondents. Among them distance of market from village, irregular supply of electricity, poor transport facilities for crop produce, lack of knowledge about seed treatment, no knowledge about recommended doses of fertilizers, no knowledge about improved varieties, lack of fund to purchase agriculture inputs, complex procedure of bank loan, high cost of inputs like seed, fertilizer and bio-fertilizers, irregular supply of seed, fertilizers and pesticides, lack of knowledge about market value of product, less contact with RAEOs, lack of knowledge about communication media, lack of technological knowledge, unavailability of seed, fertilizer through government agencies and lack of co-operative societies were the major constraints reported by the respondents (Varma, 2011). The lack of supporting mechanism for the procurement and marketing of pulses has been a major impediment to the production of pulses. Low genetic yield of Indian pulses and their vulnerability to pests and diseases is a major hindrance of adoption of pulses by farmers. The risk of low productivity and income is too high for farmers to bear (Aivalli, 2015).

A wide technological gap with respect to seed inoculation with rhizobium culture, seed treatment with captan or thiram, application of plant protection measures and weedicide application was observed. Since, these practices are important from the point of increasing production and net return, it warrants the attention of extension workers and scientists to intensify their efforts in these areas where wide gap observed and appropriate educational activities like organising trainings, demonstrations, exhibitions, field days *etc.* should be undertaken to reduce the technological gap.

Extension Intervention for Dissemination of Technology

In India, there is a continued need for extension to focus on grains, pulses and oilseeds in lagging areas, while at the same time covering high-value products in the supply chains that already exist or are being formed. Infrastructure such as electricity and roads is less well reducing the relative effectiveness of extension, however well-conceived. Farm extension services play an important role in agrarian development (Birner *et al.*, 2006; Anderson, 2007). This renewed interest in farm extension is linked to the rediscovery of the role that the farm sector can play in reducing persistent rural poverty (World Bank, 2007). Extension services are expected to facilitate the farming community to access backward and forward inputs and service, educate the farmers on better farm management practices.

The introduction of T and V extension management system in the mid-1970s was one of the significant developments that influenced extension practices. Approaches to agricultural extension in India and worldwide continue to evolve. Since the Green revolution in 1960s and the acknowledged unsustainability of the T and V programme (Anderson *et al.*, 2006), agricultural extension with its focus on increasing production via technology transfer, has adopted decentralised participatory, and demand driven approaches in which accountability is geared towards the users (Swanson, 2009). Extension went through distinctive stages overtime evolving with national priorities (Singh and Swanson, 2006).

When the public sector is involved in technology transfer the overlap or concordance in the interests of producers, extension agents and researchers may be limited and this may substantially lower the effectiveness of the research and development process (Mandell 1990). Röling (1988) argue that research and extension should not be seen as separate processes involving distinct institutions which must somehow be linked. FAO and the World Bank refer to this larger system as Agriculture Knowledge and Information System (AKIS). The AKIS framework conceives extension as a tool linking multiple players in the development process.

Framework for Sustainable Agricultural Knowledge and Information System

Farmers and rural people are partners within the agricultural knowledge system, and as the main subsystems for achieving a more flexible, productive and sustainable agriculture. For upgrading the effectiveness of current agricultural knowledge and information systems toward sustainability, new opportunities are emerging. These opportunities are as follow: 1) Advances in the agricultural sciences are crucial, but other advances are also needed (such as precision farming); 2) Relationships are changing between different subsystems (accurately and synergistically relationships of subsystems); 3) Communication and information technologies are advancing rapidly (Integrating Indigenous Knowledge Systems (IKSs), in the SAKIS); 4) New concepts are emerging for farmers' participation in the learning, problem solving and problem posing processes (such as Participatory Action and Learning Method (PALM). Therefore, "networking" and "synergy" processes among different subsystems increase the effectiveness of total SAKIS (Figure 13.1).

Factors Affecting Technology Transfer in Pulses

Technological interventions help to increase agricultural productivity, cut production costs, and lower consumer prices. The benefits depend on how the technology is transferred, the speed of transfer, and the degree of government

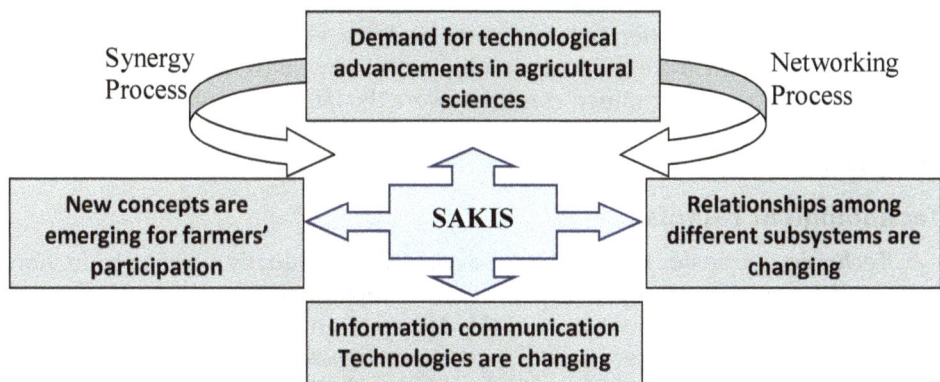

Figure 13.1: Framework for Sustainable Agricultural Knowledge and Information System.
(*Source*: Adapted from Hashemi, 2011).

policy influence on technology transfers. There are number of factors that influence the extent of adoption of technology such as farmers' use of technologies can be influenced by socio-economic, cultural, political factors and characteristics or attributes of technology (Figure 13.2).

Socio-economic
Education level
Agro-ecological condition
Community interaction
Availability of resources

Cultural
Language
Beliefs
Values
Gender roles

Successful Technology Transfer

Technological
Attributes of technology
Perception of technology
Learning of technology
Availability of technology

Political
Polices
Leadership
Investment
Programme funding
Availability of resources

Figure 13.2: Factors Influencing Technology Transfer to Farmers.

Socio-economic Factors

Extension planners face some difficult choices because of the need to respond to the diverse technology and information needs of farmers from many different zones and, at the same time, to satisfy a requirement for extensive countrywide coverage of the rural population. Also socio-psychological traits of farmers are important, their age, education, income, family size, tenure status, credit uses are positively related to adoption. The personal characteristics of extension worker such as credibility, relationship with farmers, emphatic ability, sincerity, resourcefulness, persuasiveness, and development orientation and the bio-physical environment influences the adoption. The conditions of the farm include its location, availability of resources and other facilities such as roads, markets, transportation, rainfall distribution, soil type, water, services, and electricity. For instance, farmers whose farms were irrigated were the earliest adopters of new high yielding varieties of pulses, while those under rained condition were the late adopters. The innovation diffuses slowly wherever the input cost is high, yield is uncertain and product price is low.

Technological Factors

Technology transfer helps increase agricultural productivity, cut production costs, and lower consumer prices. The benefits depend on how the technology is transferred, the speed of transfer, and the degree of government policy influence on technology transfers. There is a big gap between research station technology and farmers' technology, which has resulted in low yields The existing technologies in production of pulses have the potential of doubling production at national level without increasing area if farmers adopt the recommended package of practices. Farmers' perception of technology is important in adoption and diffusion. Farmers lack awareness and knowledge regarding most of improved production technologies related to pulse crops including, high yielding varieties, seed treatments, integrated crop management practices *etc.* Technological intervention should be concurrent with training of farmers in domain of intervened technology.

Cultural Factors

Cultural differences among farmers, as well as differences in community, also need to be taken into account. In particular, these are reflected in land-use strategies and gender roles. Socio-cultural factors are leading constraints to the effectiveness of extension. Beliefs, values, tradition and language differences can impede the communication of improved technology unless they are taken into account. Tribal farmers, for example, will require different types of subject-matter expertise, and extension interventions to promote pulse production and thus need to use different strategies to transfer technology to them than to permanent field agriculturalists. The resource endowments of different categories of farmers also affect technology adoption levels. Therefore, extension interventions with focus on cultural factors and compatible technologies may be more appropriate.

Political Factors

The political component of an agricultural technology system can enable or limit extension in ways beyond the reach of extension functionaries. Policy-making bodies of government set development goals and objectives such as self sufficiency in pulses and nutritional security. Governments set policies on consumer and producer commodity prices, subsidies for inputs, credit availability and import substitution for self-sufficiency. Political shifts at the national level often result in changes in schemes and programmes. All these impact on farmers influence their production decisions.

Constraints of Growth in Area Production and Productivity of Pulses

India produces nearly a dozen varieties of pulse crops. At the world level, it is the biggest cultivator of pulses. But, the gap between actual and potential yield remains very high even between farmer's yield of pulses and front line demonstration plot yield. Thus, challenges faced in improving the level of production of pulses three decades ago, still exist and the situation has not improved despite the government adopting a mission mode approach under the Technology Mission on Oilseeds and Pulses since 1990-91. A few constraints are described below:

☆ **Socio-economic constraints:** Farmers engaged in cultivation of pulses are mostly small and marginal. A majority are in areas with poor banking infrastructure. They have poor resource base and lack risk-bearing capacity. Delivery of credit to such farmers is also not hassle-free. There is lack of marketing network in remote areas. Procurement of produce by a dedicated agency is virtually non-existent or in-effective.

☆ **Technological constraints:** Pulses are grown under varied agro-climatic conditions in the country. Production technology for a pulse crop has to be soil type/region specific. The region specific pulse production technology including crop varieties with traits relevant to prevailing biotic and abiotic stresses may be emphasised. Our research and development programme in pulses has yet to appreciate and address this issue adequately.

☆ **Infrastructural Constraints**: Farmers also lack provision for safe storage of pulses. Warehousing facilities are either inadequate or inaccessible. Investments in market infrastructure, warehouses, market information systems both in public and private partnership through PPP models need to be encouraged in India. For example, in Madhya Pradesh, under Bundelkhand package, many water harvesting and grain storages structures were developed.

Shift from Technology Transfer to Agricultural Innovation Systems

Achieving self sufficiency in pulse production and nutritional security of people will require transforming the traditional top- down, technology-driven (TOT) approach to a more decentralized and farmer-led system. Historically, non-adoption of recommendations was attributed to farmers' ignorance, to be overcome through more and better extension, and then to farm level constraints, with the solution in easing the constraints. The salient feature of the new approach is the reversal of learning, where researcher and extension workers are learning from farmers.

In agricultural knowledge system consisting of four components set in a larger context (Figure 13.3). The components are technology generation, technology transfer (knowledge and input transfer), technology utilization, and agricultural policy (Swanson *et al.*, 1990). The organizations that constitute the components, as well as others in the system environment influence each other in complex ways. Such publicly funded systems are established by governments to improve the conditions of life and well-being of rural and urban populations and to increase agricultural productivity. The functions and linkages related to the flow and feedback of technology and information in the system define the components. In the public sector, agricultural research organizations (technology generation) and extension (technology transfer) are major actors, although private sector and NGOs are also involved.

Technology generation consists of planning, administration, and implementation of research activities that develop, assess, adapt, and test improved agricultural technology for farmers and other users. In the public sector, these tasks, as well as some dissemination work, are carried out by agricultural research organizations under NARS and technology dissemination thorough frontline extension system of KVKs.

Technology transfer further evaluates and adapts research outputs for users and then widely disseminates the knowledge and inputs to different target adopters - farmers of different categories, private companies, and so on so forth. Figure 13.3 shows two parts of the transfer component, namely, knowledge and

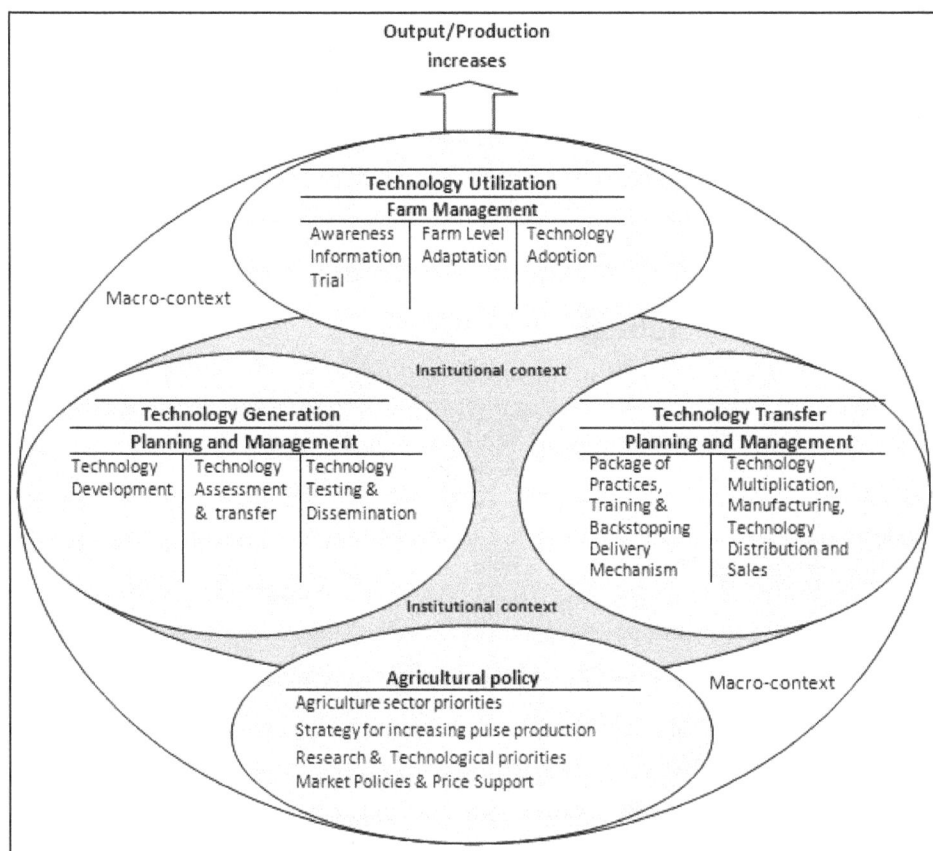

Figure 13.3: Agricultural Knowledge System Model.
(Adapted from Peterson, 1998).

inputs. Government extension and private organizations are also involved in the technology transfer, although farmer coverage is more limited and input transfer may play a bigger role.

The technology utilization component encompasses the users of the agricultural technology, mainly farmers. User awareness, adaptation, and adoption of improved technology from various sources affect farm-level productivity and profitability and, ultimately, economic growth at the national level. Interaction and feedback between users and research and transfer organizations improve cooperation and the relevance of technology.

The policy component relates to government development goals and strategies, market and price policies, and the levels of resource investments in the system. Various government bodies play a role in setting development policy. Technology development and transfer organizations are affected by the policy in fundamental ways. The key elements of the new paradigm are to put emphasis on people rather than 'things', to decentralize, empower the participants, to value and work on what matters to participants and to learn from the beneficiaries rather than to teach them. The purpose of an Agricultural Innovation System (AIS) is to strengthen the *capacity to innovate* and create novelty throughout the agricultural production and marketing system. Potentially all actors in the public and private sectors are involved in the creation, diffusion, adaptation and use of all types of knowledge relevant to agricultural production. Both AKIS and AIS approaches see the agricultural knowledge system as made up of actors who exchange knowledge. Farmers and many other actors interact back and forth to create and share agricultural knowledge.

Strategy for Self-Sufficiency in Pulses

There is an utmost need of a proactive strategy from researchers, planners, policy-makers, extension workers, market forces and farmers aiming not only at boosting the per unit productivity of land, but also at reduction in the production costs. Under complex rain fed areas, the farmer participatory research (FPR) needs to be developed and involve farmers more closely in on-farm research. Farmer-participatory testing will help refine the technologies, pinpoint and eliminate adoption constraints. The strategies which can have substantial in order to minimize risk and improved practice of pulse based cropping system and increasing Pulses production, the following points to be important:

  ☆ Incorporation of high yielding varieties and short duration pulse crops like black gram, to make the different production system profitable and improve soil health.

  ☆ Intensification of production system through popularizing summer green gram cultivation during under assured irrigated condition in potato fellow land to adopt crop rotation potato green gram in northern India (especially Uttar Pradesh, West Bengal and Bihar).

  ☆ Easy and timely availability of critical input like quality seeds, fertilizer and nutrients and insecticides and pesticides at nearby market. Creation of informal seed village system, where farmer to farmer seed production and distribution chain will ensure easy availability of quality seed.

  ☆ Production of quality seed of improved pulse varieties by private agencies needs to be encouraged to meet the demand of the farmers. Improved seed should be made available to the farmers at his doorstep at reasonable price.

  ☆ Research efforts to develop low temperature and drought tolerant high yielding varieties of pulses with pest resistance should be undertaken for Rajasthan, Haryana and Uttar Pradesh and Madhya Pradesh.

  ☆ It would be profitable to have encourages pulses processing industry instituted between groups of villagers so that proper milling is done soon after harvesting and storage is made of split pulses (*Dal*) rather than whole seed.

  ☆ Transfer of technology in relation to pulses should be strengthened in farmer participatory mode with active involvement of multidisciplinary team of scientists. Effective IPM module to control insect pest and diseases to avoid losses at the entire levels field and after post-harvest and storage.

☆ Delivery of improved technology, inputs, credits need to be stream lined through appropriate policy interventions. Benefit of Pradhan Mantri Fasal Beema Yojana need to be extended to pulses farmers.

Policy Decision and Pulses Production

Given the uncertainty of global supply of pulses and rising domestic demand in India, it would be prudent to plan future domestic pulse production in such a way that major share of demand is fulfilled by domestic production. There is overwhelming scientific evidence suggesting a vast gap between farmer's yield of pulses and front line demonstration plot yield. Further, a large chunk of rice fallow lands can be brought under pulses provided addition land and water resources for pulse production. Exploitation of promising intercropping systems in rainfed/ partial irrigated areas offers a vast opportunity for improving pulse production. Scientific management of "utera" cultivation of pulses in rice based cropping systems and utilization of the period between harvesting of timely planted wheat and planting of *kharif* crop for growing short duration pulses such as mungbean/ urdbean in north-western states are other avenues for augmenting production of pulses.

The government has taken several initiatives to motivate farmers to grow more pulses and to increase pulse production in the country. Increase in Minimum Support Price of pulses has also led to increase in the area coverage under pulses from 22.76 million hectares in 2004-05 to 25.23 million hectares in 2013-14. The productivity of pulses has significantly increased from 577 kg per hectare in 2004-05 to 789 kg per hectare in 2012-13. Pulses production strategies, along with a mix of policy and programmatic support have contributed significantly to the path breaking achievement of 19.27 million tons in 2012-13.

Initiation of new initiative of Integrated Development of 60,000 Pulse Villages Scheme in 2010-11 to enhance pulses production in selected watershed areas in major pulses growing states. During 2011-12, eleven pulses growing states constituting nearly 90 per cent of the pulses areas are provided funds for *in situ* moisture conservation- new farm ponds with polythene lining and/or dug wells, seed minikit's A3P units, and market linked extension support through Small Farmers Agri-Business Consortium.

Conclusion

There is an urgent need to address the current barriers to pulses cultivation by facilitating by demonstrating the benefits of pulses cultivation to farmers. Farmers perceive pulses as having a lower cost benefit ratio vis-à-vis cereals and other crops. This mindset needs to be changed by sensitizing farmers that pulses can be a profitable venture, with use of modern technology. This can be achieved by participatory technology development with active involvement of farmers in planning and implementation stage. Demonstration of new technologies and package of practices in form of cluster demonstrations with coverage of larger area. Efforts should be directed towards transfer the proven and demonstrated technologies at farmers' fields accompanied by efficient system of linking institutional credit with timely availability of quality inputs and marketing services that could enhance pulse production and productivity. Innovative use of ICT tools for pest and disease monitoring, information sharing, capacity building through training, field visits can help to provide the benefits of modern pulses cultivation practices to farmers.

Effective and continuous efforts are needed to increase the area under cultivation as well as the yield of pulses. To intensify/increase area and production of pulses crops, need based crop-specific and region-specific approaches should be adopted. A more formal strategic alliance involving multiple players in collaborative approach with or without equity participation so that research expertise and reach of various public institutes can meet the demand of farmers in terms of availability of technology, inputs, extension services, management and marketing. The minimum support price should be in coordination with farm harvest price and market price.

REFERENCES

Aivalli G. 2015. India needs a pulses revolution, Business line Retrieved from *http: //www.thehindubusinessline.com/opinion/india-needs-a-pulses-revolution/ article-7678140.ece* on May 6, 2016.

Anderson, J.R. 2007. Agricultural Advisory Services, Background paper for the World Development Report 2008, Agriculture and Rural Development Department, World Bank, Washington, D.C.

Anderson, Jock R., Gershon Feder and Sushima Ganguly. 2006. The Rise and Fall of Training and Visit extension: An Asian mini-drama with an African Epilogue? In A.W. Van den Ban and R.K. Samanta eds., Changing Roles of Agricultural Extension in Asian Nations, New Delhi: B.R. Publishing Company: 149-172.

Birner, R., Davis, K., Pender, J., Nkonya, E., Anandajayasekeram, P., Ekboir, J., Mbabu, A., Spielman, D., Horna, D. and Benin, S. 2006. From best practice to best fit: A framework for analyzing agricultural advisory services worldwide. Development Strategy and Governance Division, Discussion Paper No. 39, International Food Policy Research Institute (IFPRI), Washington, D.C.

Chatterjee, J. and Prabhakar, T. V. 2009. On to Action – Building Digital Eco-system for Knowledge Diffusion in Rural India. Retrieved from *http: //www.cse.iitk. ac.in/users/tvp/papers/vof_pasadena.pdf* on May 7, 2016.

Haque, M.A., Monayem Miah, M.A., Ali, A.M. and Luna, A.N. 2014. Adoption of Mungbean Technologies and Technical Efficiency of Mungbean (*Vigna Radiata*) Farmers in selected areas of Bangladesh. *Bangladesh J. Agril. Res.* 39(1): 113-125.

Hashemi, S.M.K. 2011. Agricultural Knowledge and Information System in the Context of Sustainable Agriculture : Sustainable Agricultural Knowledge and Information System Framework and Effective Factors. *Cercetări Agronomice în Moldova* Vol. XLIV, No. 4 (148): 99-110.

IIPR. 2013. Vision 2050, Indian Institute of Pulses Research's Vision document. Retrieved from *http: //www.iipr.res.in/pdf/vision_2050.pdf*

Kumar, S. 2009. A study on technological gap in adoption of the improved cultivation practices by the soybean growers. Thesis M.Sc. (Ag.), Department of Agricultural Extension Education, College of Agriculture, UAS, Dharwad.

Mandell, M.P. 1990. 'Network Management: Strategic Behaviour in the Public Sector' in Gage, R.W. and Mandell, M.P. (eds). 1990. *Strategies for Managing Intergovernmental Policies and Networks*. Praeger. N.Y.

Million, T. and Belay, K. 2004. Determinants of fertilizer use in Gununo area, Ethiopia. pp. 21-31. In Tesfaye Zegeye, Legesse Dadi and Dawit Alemu (eds). Proceedings of agricultural technology evaluation adoption and marketing. Workshop held to discuss results of 1998-2002, August 6-8, 2002.cf:

NCAP, 2014. India's Pulses Scenario, National Council of Applied Economic Research, New Delhi. Retrieved from *nfsm.gov.in/Meetings./7 per cent 20India's per cent 20pulses per cent 20Scenerio.docx on May 18, 2016*

NSAE. 2009. National Seminar on Agriculture Extension, Background Note. February27-28, 2009, New Delhi. Retrieved from *http: //www.syngentafoundation. org/db/1/657.pdf on May 6,2016.*

Patidar, N. 2011. A study on impact of front line demonstrations in adoption of production technology by pulses growers in Jabalpur district (M.P.). Thesis M.Sc. (Ag.), Department of Extension Education, JNKVV College of Agriculture, Jabalpur.

Peterson, W. 1998. The context of extension in agricultural and rural development, Improving agricultural extension. A reference manual, FAO, Rome.

Röling, N. 1988. *Extension Science*. Cambridge University Press, Cambridge. pp. 233.

Sharma, D. 2015. Why production of pulses is not increasing in India: An interview. Retrieved from *http//devinder-sharma.blogspot.in201510why-production-of-pulses-is-not.html.pdf on May 9, 2016.*

Singh, K.M. and Swanson, B.E. 2006. Developing a market-driven extension system in India. Annual Conference Proceedings of the Association for International Agricultural and Extension Education, 22, 627–637.

Singh, R. P. 2013. Status paper on pulses. Directorate of pulse development, Bhopal.

Swanson, B. E., Sands, C. M., and Peterson, W. E. 1990. Analyzing agricultural technology systems: Some methodological tools. In R. Echeverria (Ed.), *Methods for diagnosing research system constraints and assessing the impact of agricultural research: Vol. I. Diagnosing agricultural research system constraints*. The Hague: ISNAR.

Swanson, B.E. 2009. Changing Extension Paradigms within a Rapidly Changing Global Economy, in Proceeding of the 19th European Seminar on Extension Education. Assisi, Italy. Retrieved from *http: //www.agraria.unipg.it/ ESEE2009PERUGIA/files/Proceedings.pdf on May 9, 2016.*

Varma, R. 2011. A study on Technological gap of recommended chickpea production technology among tribal farmers of Mandla block of Mandla district M.P.). Thesis M.Sc. (Ag.), Department of Extension Education, JNKVV College of Agriculture, Jabalpur.

World Bank, 2000. Agricultural knowledge and information systems for rural development : strategic vision and guiding principles. Washington, DC: World Bank.

World Bank, 2007. World Development Report: Agriculture for development. Washington, D.C.: World Bank.

Appendices

Appendix I
Weighing Factor (W) for Use in Modified Penman Combination Method

Altitude (m)	Temperature (°C)									
	2	4	6	8	10	12	14	16	18	20
0	0.43	0.46	0.49	0.52	0.55	0.58	0.61	0.64	0.66	0.68
500	0.44	0.48	0.51	0.54	0.57	0.60	0.62	0.65	0.67	0.70
1000	0.46	0.49	0.52	0.55	0.58	0.61	0.64	0.66	0.69	0.71
2000	0.49	0.52	0.55	0.58	0.61	0.64	0.66	0.69	0.71	0.73
3000	0.52	0.55	0.58	0.61	0.64	0.66	0.69	0.71	0.73	0.75
4000	0.54	0.58	0.61	0.64	0.66	0.69	0.71	0.73	0.75	0.77

Altitude (m)	Temperature (oC)									
	22	24	26	28	30	32	34	36	38	40
0	0.71	0.73	0.75	0.77	0.78	0.80	0.82	0.83	0.84	0.85
500	0.72	0.74	0.76	0.78	0.79	0.81	0.82	0.84	0.85	0.86
1000	0.73	0.75	0.77	0.79	0.80	0.82	0.83	0.85	0.86	0.87
2000	0.75	0.77	0.79	0.81	0.82	0.84	0.85	0.86	0.87	0.88
3000	0.77	0.79	0.81	0.82	0.84	0.85	0.86	0.87	0.88	0.89
4000	0.79	0.81	0.81	0.84	0.85	0.86	0.87	0.88	0.90	0.90

Appendix II
Wind Function [f (u)] values for wind speed (km/day) at 2m Height

Wind Speed	0	10	20	30	40	50	60	70	80	90
<100	-	0.30	0.32	0.35	0.38	0.41	0.43	0.46	0.49	0.51
100	0.54	0.57	0.59	0.62	0.65	0.67	0.70	0.73	0.76	0.78
200	0.81	0.84	0.86	0.89	0.92	0.94	0.97	1.00	1.03	1.05
300	1.08	1.11	1.13	1.16	1.19	1.21	1.24	1.27	1.30	1.32
400	1.36	1.38	1.40	1.43	1.46	1.49	1.51	1.54	1.57	1.59
500	1.62	1.65	1.67	1.70	1.73	1.76	1.78	1.81	1.84	1.90
600	1.89	1.92	1.94	1.97	2.00	2.02	2.05	2.08	2.11	2.15
700	2.16	2.19	2.21	2.24	2.27	2.29	2.32	2.35	2.38	2.40
800	2.43	2.46	2.49	2.51	2.54	2.56	2.59	2.62	2.64	2.65
900	2.70	-	-	-	-	-	-	-	-	-

Appendix III
Adjustment Factor (C) to Account for Day and Night Weather Conditions (Modified Penman Combination Method)

U day, m/s	RS(mm/day⁻¹)											
	3	6	9	12	3	6	9	12	3	6	9	12
U day/U night = 4.0												
0	0.86	0.90	1.00	1.00	0.96	0.98	1.05	1.05	1.02	1.06	1.10	1.10
3	0.79	0.84	0.97	0.97	0.92	1.00	1.11	1.19	0.99	1.10	1.27	1.32
6	0.68	0.77	0.93	0.93	0.85	0.96	1.11	1.19	0.94	1.10	1.25	1.33
9	0.55	0.65	0.90	0.90	0.76	0.88	1.02	1.14	0.88	1.01	1.16	1.27
U day/U night = 3.0												
0	0.86	0.90	1.00	1.00	0.96	0.98	1.05	1.05	1.02	1.06	1.10	1.10
3	0.76	0.81	0.94	0.94	0.87	0.96	1.06	1.12	0.94	1.04	1.18	1.28
6	0.61	0.68	0.88	0.88	0.77	0.88	1.02	1.10	0.86	1.01	1.15	1.22
9	0.46	0.56	0.82	0.82	0.67	0.79	0.88	1.05	0.78	0.92	1.06	1.18
U day/U night = 2.0												
0	0.86	0.90	1.00	1.00	0.96	0.98	1.05	1.05	1.02	1.06	1.10	1.10
3	0.69	0.76	0.92	0.92	0.83	0.91	0.99	1.05	0.89	0.98	1.10	1.14
6	0.53	0.61	0.84	0.84	0.70	0.80	0.94	1.02	0.79	0.92	1.05	1.12
9	0.37	0.48	0.76	0.76	0.59	0.70	0.84	0.95	0.71	0.84	0.96	1.06
U day/U night = 1.0												
0	0.86	0.90	1.00	1.00	0.96	0.98	1.05	1.05	1.02	1.06	1.10	1.10
3	0.64	0.71	0.89	0.89	0.78	0.86	0.94	0.99	0.85	0.92	1.01	1.05
6	0.43	0.53	0.79	0.79	0.62	0.70	0.84	0.93	0.72	0.82	0.95	1.00
9	0.27	0.41	0.70	0.70	0.50	0.60	0.75	0.87	0.62	0.72	0.87	0.96

Index

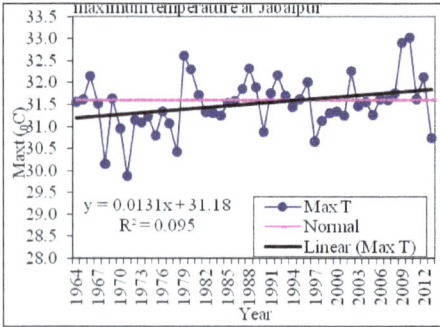

Figure 3.1: Annual Variability and Trend of Maximum Temperature at Jabalpur. p. (63)

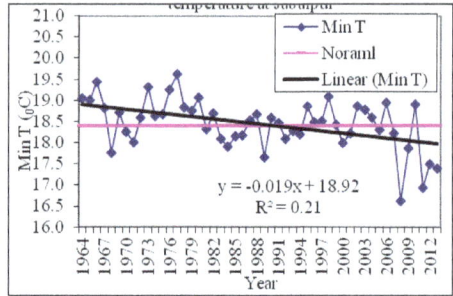

Figure 3.2: Annual Variability and Trend of Minimum Temperature at Jabalpur. p. (63)

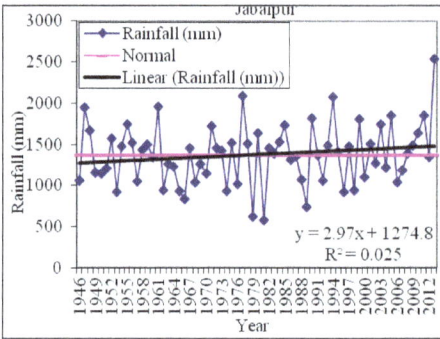

Figure 3.3: Annual Variability and Trend of Rainfall at Jabalpur. p. (63)

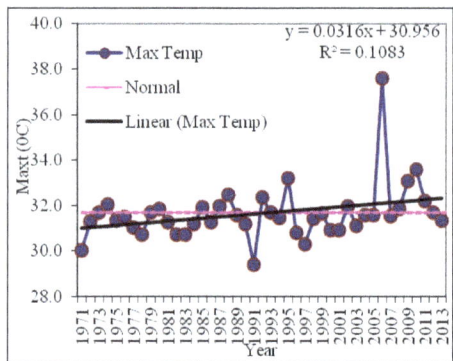

Figure 3.4: Annual Variability and Trend of Maximum Temperature at Rewa. p. (64)

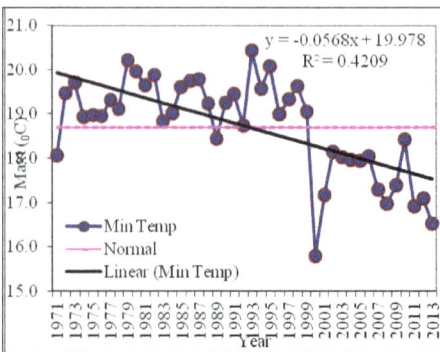

Figure 3.5: Annual Variability and Trend of Minimum Temperature at Rewa. p. (64)

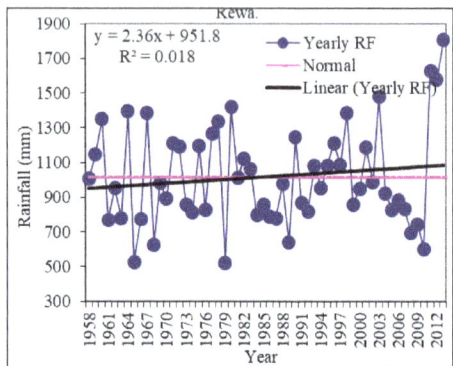

Figure 3.6: Annual Variability and Trend of Rainfall at Rewa. p. (64)

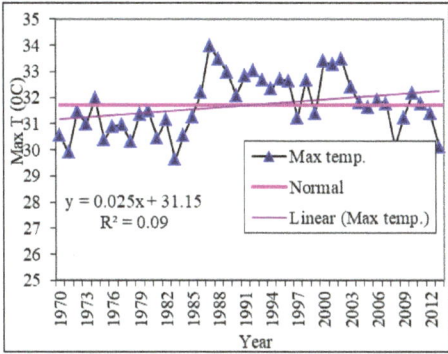

Figure 3.7: Annual Variability and Trend
of Maximum Temperature at Indore
p. (66)

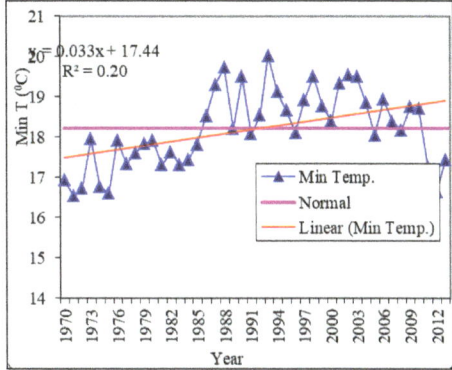

Figure 3.8: Annual Variability and Trend
of Minimum Temperature at Indore
p. (66)

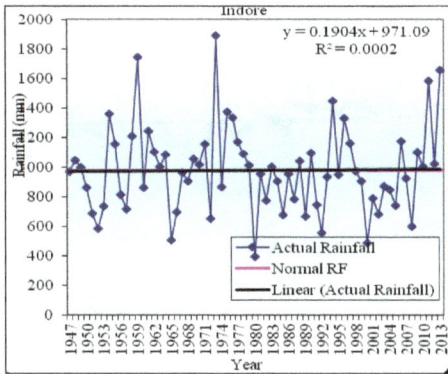

Figure 3.9: Annual Variability and Trend
of Rainfall at Indore. p. (66)

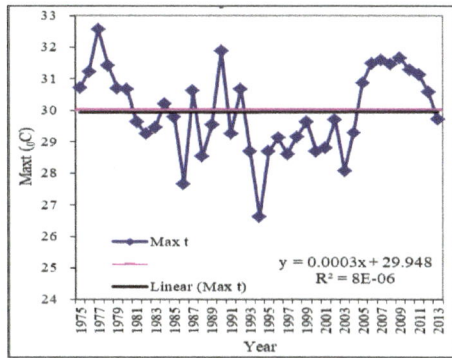

Figure 3.10: Annual Variability and Trend
of Maximum Temperature at Chindwara.
p. (67)

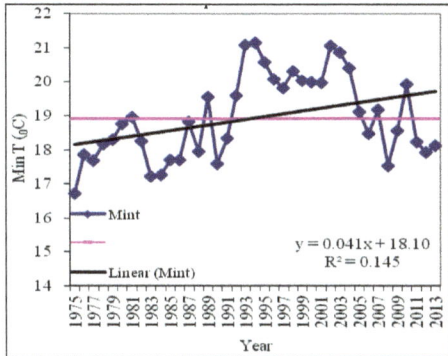

Figure 3.11: Annual Variability and
Trend of Minimum Temperature at
Chindwara. p. (67)

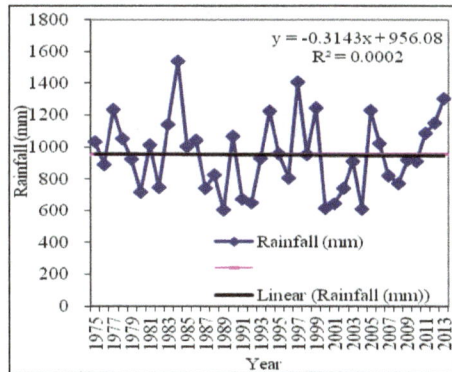

Figure 3.12: Annual Variability and Trend
of Rainfall at Chindwara. p. (67)

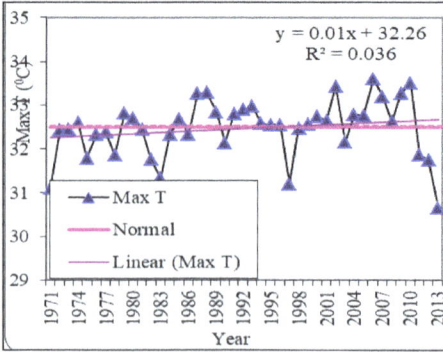

Figure 3.13: Annual Variability and Trend of Maximum Temperature at Tikamgarh. p. (68)

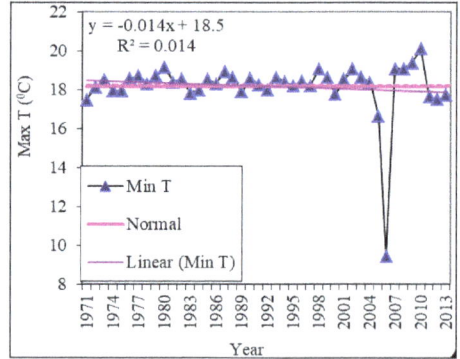

Figure 3.14: Annual Variability and Trend of Minimum Temperature at Tikamgarh. p. (68)

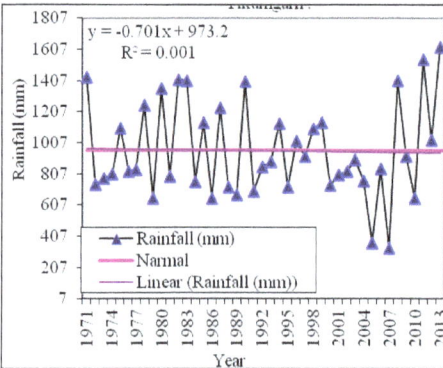

Figure 3.15: Annual Variability and Trend of Rainfall at Tikamgarh. p. (68)

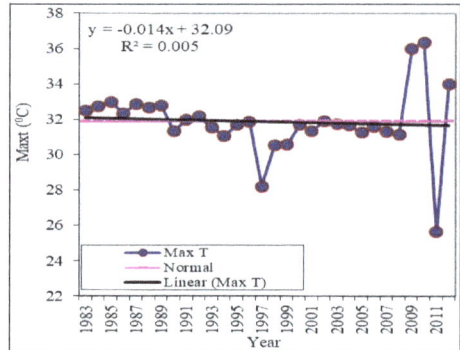

Figure 3.16: Annual Variability and Trend of Maximum Temperature at Hoshangabad. p. (69)

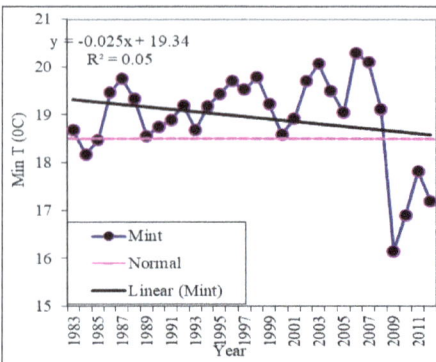

Figure 3.17: Annual Variability and Trend of Minimum Temperature at Hoshangabad. p. (69)

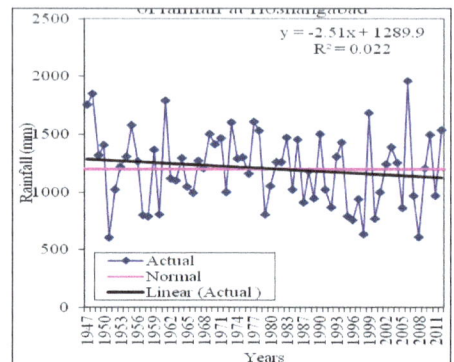

Figure 3.18: Annual Variability and Trend of Rainfall at Hoshangabad. p. (69)

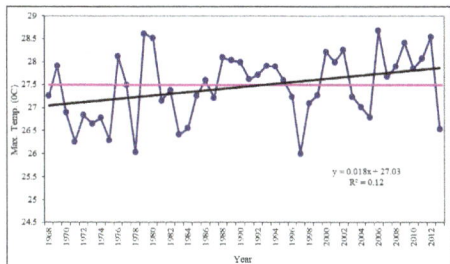

Figure 3.19: Maximum Temperature Trend in Rabi Season at Jabalpur. p. (70)

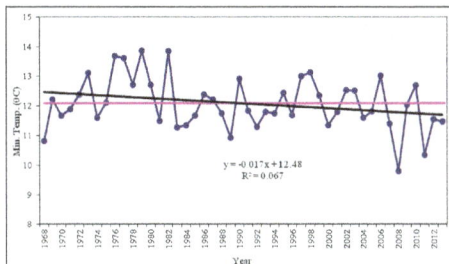

Figure 3.20: Minimum Temperature Trend in Rabi Season at Jabalpur. p. (70)

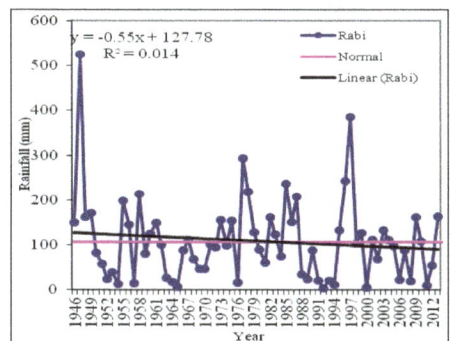

Figure 3.21: Rainfall Trend in Rabi Season at Jabalpur. p. (71)

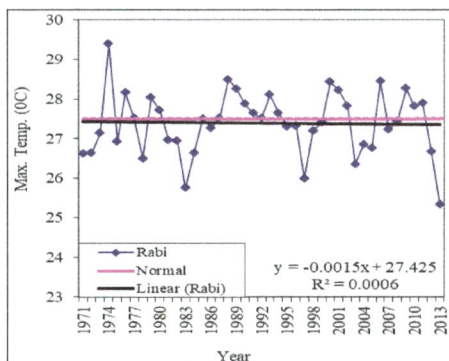

Figure 3.22: Maximum Temperature Trend in Rabi Season at Rewa. p. (71)

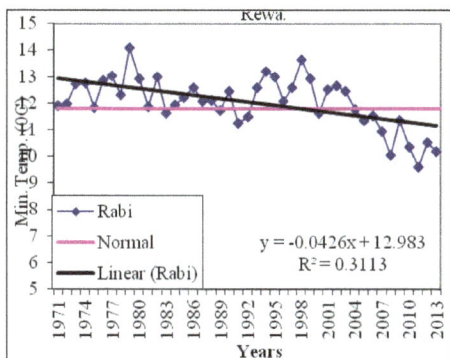

Figure 3.23: Minimum Temperature Trend in Rabi Season at Rewa. p. (71)

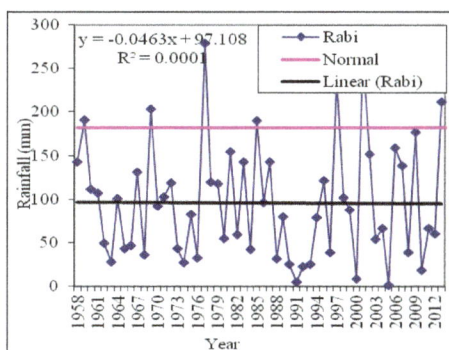

Figure 3.24: Rainfall Trend in Rabi Season at Rewa. p. (72)

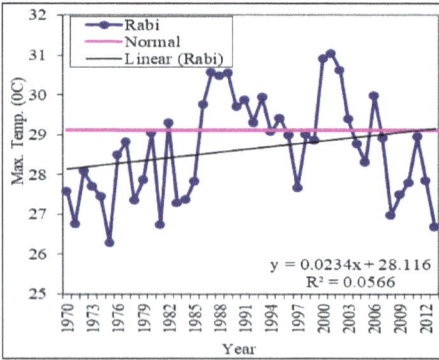

Figure 3.25: Maximum Temperature Trend in Rabi Season at Indore. p. (72)

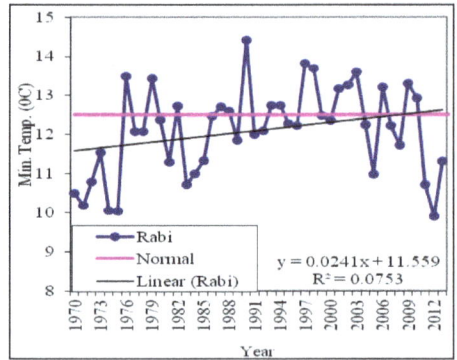

Figure 3.26: Minimum Temperature Trend in Rabi Season at Indore. p. (72)

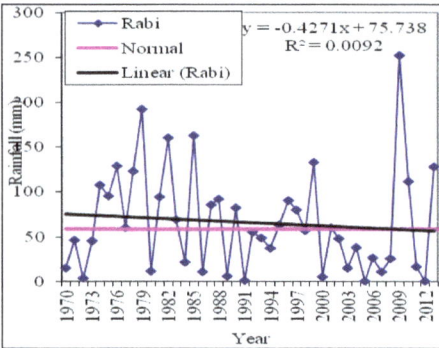

Figure 3.27: Rainfall Trend in Rabi Season at Indore. p. (73)

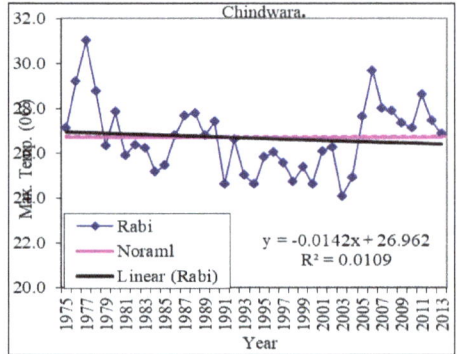

Figure 3.28: Maximum Temperature Trend in Rabi Season at Chindwara. p. (73)

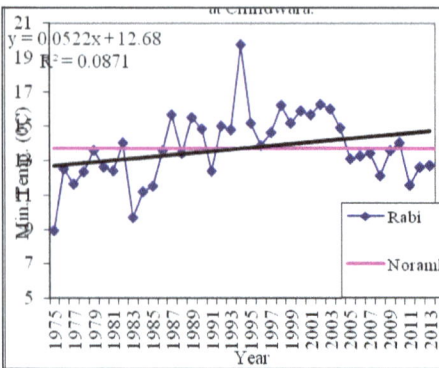

Figure 3.29: Minimum Temperature Trend in Rabi Season at Chindwara. p. (73)

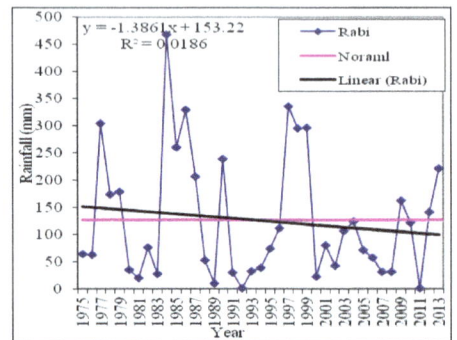

Figure 3.30: Rainfall Trend in Rabi p. (74)

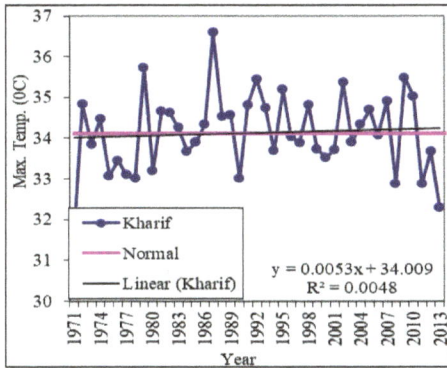

Figure 3.31: Maximum Temperature
Trend in Rabi Season at Tikamgarh.
p. (74)

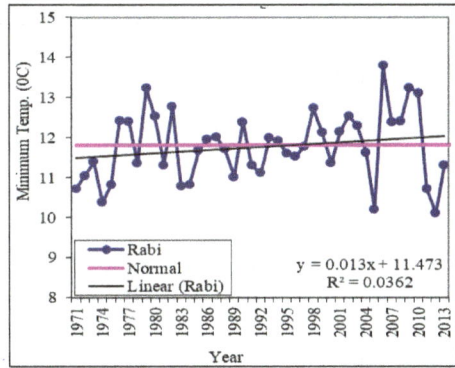

Figure 3.32: Minimum Temperature
Trend in Rabi Season at Tikamgarh.
p. (74)

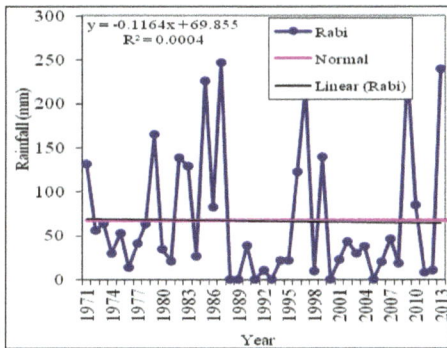

Figure 3.33: Rainfall Trend in Rabi
Season at Tikamgarh. p. (75)

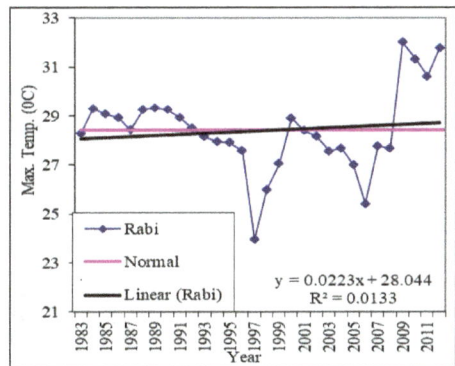

Figure 3.34: Maximum Temperature
Trend in Rabi Season at Hoshangabad.
p. (75)

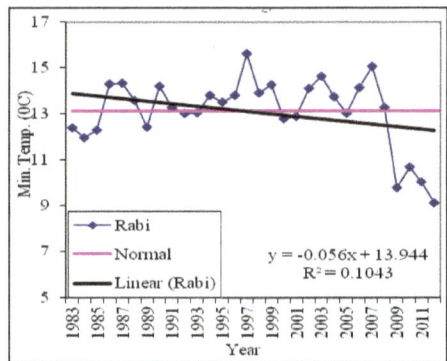

Figure 3.35: Minimum Temperature
Trend in Rabi Season at Hoshangabad.
p. (75)

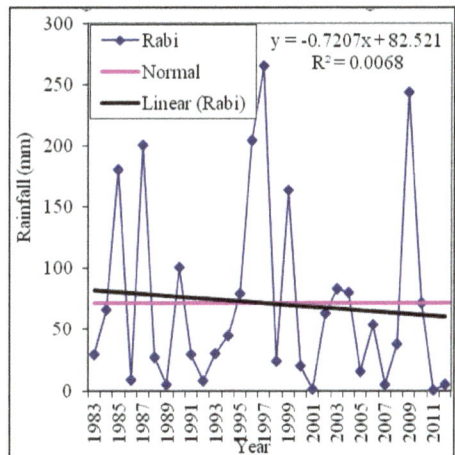

Figure 3.36: Rainfall Trend in Rabi
Season at Hoshangabad. p. (76)

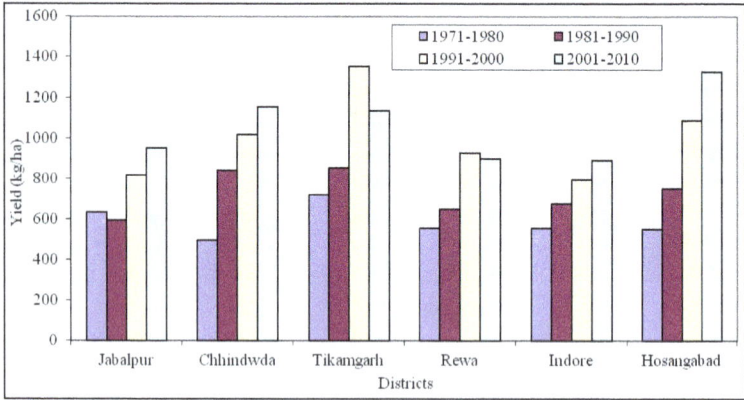

Figure 3.49: Decadal Yield Variability of Chickpea. p. (85)

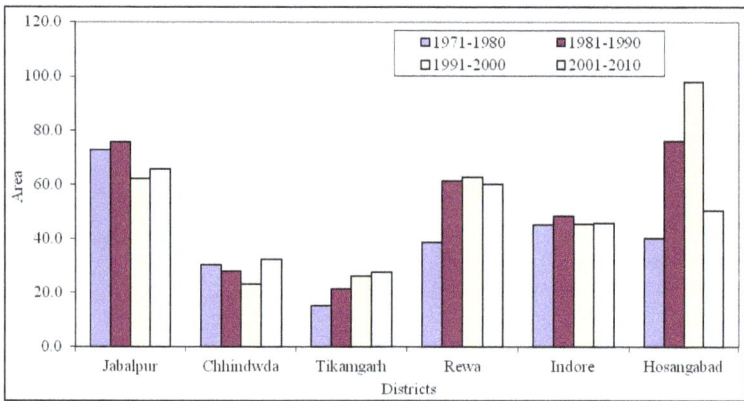

Figure 3.50: Decadal Acerage Variability of Chickpea. p. (85)

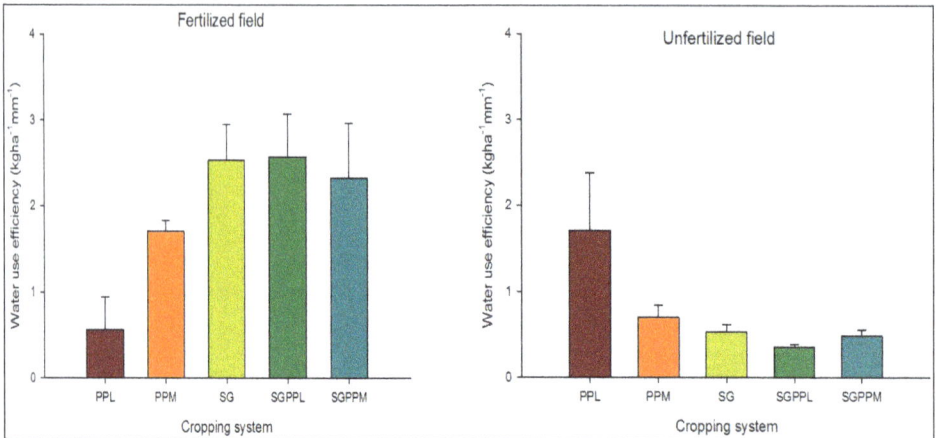

Figure 4.1: Water Use Efficiency of Pigeon Pea under different Cropping Systems. p. (100)

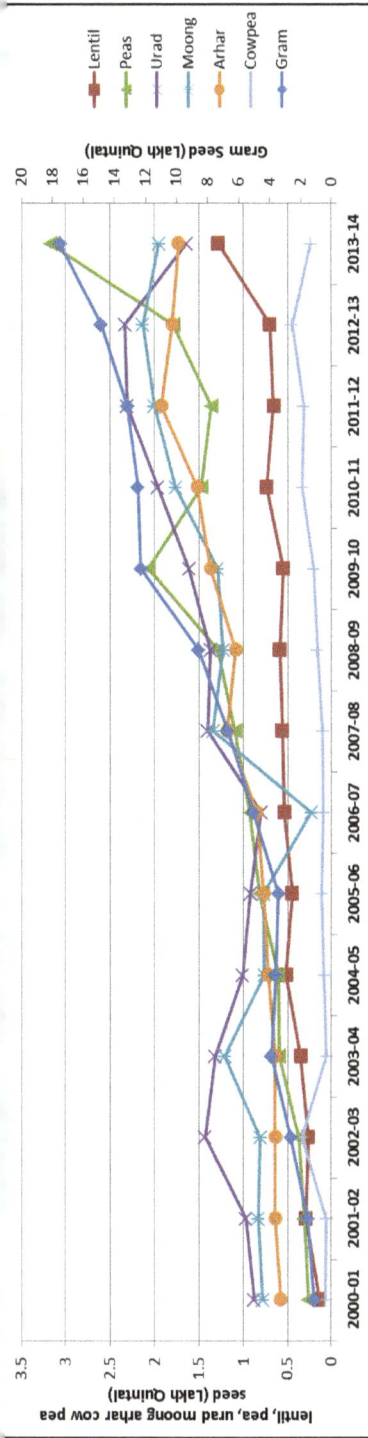

Figure 8.1: Quality Seed Availability Trend of Major Pulses in India. p. (171)

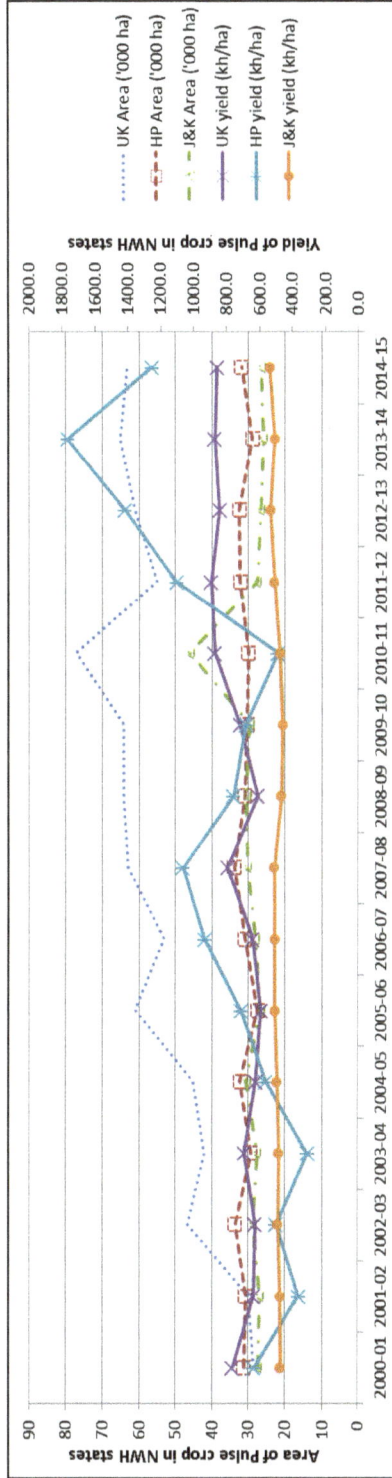

Figure 8.3: Area and Yield Trends of Total Pulse Crop in NWH States. p. (174)

www.ingramcontent.com/pod-product-compliance
Lightning Source LLC
Chambersburg PA
CBHW050511190326
41458CB00005B/1504